SpringerWienNewYork

CISM COURSES AND LECTURES

Series Editors:

The Rectors
Giulio Maier - Milan
Jean Salençon - Palaiseau
Wilhelm Schneider - Wien

The Secretary General
Bernhard Schrefler - Padua

Executive Editor
Carlo Tasso - Udine

The series presents lecture notes, monographs, edited works and proceedings in the field of Mechanics, Engineering, Computer Science and Applied Mathematics.
Purpose of the series is to make known in the international scientific and technical community results obtained in some of the activities organized by CISM, the International Centre for Mechanical Sciences.

INTERNATIONAL CENTRE FOR MECHANICAL SCIENCES

COURSES AND LECTURES - No. 470

PHENOMENOLOGICAL AND MATHEMATICAL MODELLING OF STRUCTURAL INSTABILITIES

EDITED BY

MARCELLO PIGNATARO
UNIVERSITÀ DI ROMA " LA SAPIENZA", ITALY

VICTOR GIONCU
TECHNICAL UNIVERSITY OF TIMIȘOARA, ROMANIA

SpringerWien NewYork

The publication of this volume was co-sponsored and co-financed by the UNESCO Venice Office - Regional Bureau for Science in Europe (ROSTE) and its content corresponds to a CISM Advanced Course supported by the same UNESCO Regional Bureau.

This volume contains 270 illustrations

In order to make this volume available as economically and as rapidly as possible the authors' typescripts have been reproduced in their original forms. This method unfortunately has its typographical limitations but it is hoped that they in no way distract the reader.

ISBN-10 3-211-25292-4 SpringerWienNewYork
ISBN-13 978-3-211-25292-5 SpringerWienNewYork

PREFACE

The study of structural instability plays a role of primary importance in the field of applied mechanics. Despite the remarkable progresses made in the recent past years, the structural instability remains one of the most challenging topics in applied mechanics. Many problems have been solved in the last decades but still many others remain to be solved satisfactorily. The increasing number of papers published in journals and conferences organized by ECCS, SSRC, IUTAM, and EUROMECH strongly indicates the interest of scientists and engineers in the subject. A careful examination of these publications shows that they tend to fall into one of the two categories.

The first is that of **practical design direction** in which methods for analyzing specific stability problems related to some specific structural typologies are developed. The research works are restricted to determining the critical load, considering that it is sufficient to know the limits of stability range. These studies are invaluable since their aim is to provide solutions to practical problems, to supply the designer with data useful for design and prepare norms, specifications and codes.

The second direction is that of **theoretical studies**, aiming at a mathematical modeling of the instability problems, for a better understanding of the phenomena. In these studies, special emphasis is placed on the behavior of structures after the loss of stability in the post-critical range. This approach is less familiar to designers as its results have not yet become part of current structural design practice.

A wide range of researches has been developed in both directions: by mathematicians and engineers specialized in applied mathematics and mechanics, by engineers who have been working in the field of space and naval construction, etc... Each of these two directions has scored many remarkable achievements, but some incompatibilities exist between them because in the first direction mainly structural designers are involved, while in the second one essentially academic researchers alien to practice are working.

The purpose of the course "**Phenomenological and Mathematical Modeling in Structural Instabilities**" is to present some recent progress in the filed of structural instability, with regard both to practical applications and to the transfer of theoretical results to practice, in order to fill the gap existing between the accumulated theoretical knowledge and practical applications. The course progressively covers topics such as phenomenological, mathematical and numeric modeling of instability analysis, static and dynamic instabilities, structural instability and catastrophe theory.

The first section "**Mathematical Modelling of Instability Phenomena**", elaborated by Marcello Pignataro and Giuseppe Ruta, begins with the theory of motion and stability of equilibrium. For the continuous systems, the bifurcation and post-buckling analysis is presented and the effect of initial imperfections is evaluated. The examples refer to post-buckling of frames and thin walled compression members.

Section two "**Phenomenological Modelling of Instability Phenomena**", elaborated by Victor Gioncu, presents the main directions of research works in the field of structural stability, new phenomenological models of evolving systems, instability types and a phenomenological methodology for instability design.

Section three "**Modelling Buckling Interaction**", elaborated by Eduardo de Miranda Batista, refers to the light steel structures where many modes of stability are possible: flexural, torsional-flexural, local and distortional buckling modes. The paper presents the effect of interaction between these instability modes from theoretical and experimental point of view.

Section four "**Computational Asymptotic Post-buckling Analysis of Slender Elastic Structures**", elaborated by Raffaele Casciaro, introduces the computational treatment of asymptotic strategy for post-buckling analysis of elastic structures, using finite element method.

Section five "**Mechanical Models for the Subclasses of Catastrophes**" elaborated by Zsolt Gaspar presents the behavior of some mechanical models aiming to show the relation between theory of catastrophes and structural instability.

The lectures are addressed to post-graduate students (PhD and postdoc), to researchers as well as to civil, mechanical, naval and aeronautical engineers involved in structural design.

The coordinators of CISM course wish to thank warmly all the colleagues for the excellence of the work performed. Special thanks are also due to CISM Rector, Prof. M.G. Velarde, to the Editor of the Series, Prof C. Tasso, and to the entire CISM staff in Udine.

Marcello Pignataro
Victor Gioncu

CONTENTS

Mathematical Modelling of Instability Phenomena

M.Pignataro and G.C.Ruta

University of Roma "La Sapienza", Roma, Italy

Abstract Liapunov theory is first presented and discrete mechanical systems are in particular analysed. Then buckling and postbuckling analysis of continuous mechanical system using the general theory formulated by Koiter are discussed in some detail following Budiansky presentation. Finally, the influence of multiple interactive buckling modes on postbuckling behaviour is analysed in some detail for frames, thin-walled members and panels.

1 Theory of Stability of Motion

1.1 Introduction

In the development of the theory of differential equations, it is possible to distinguish two quite different approaches. The first is characterised by the search for a solution in closed form or through a process of approximation. The second can be distinguished from the first by the fact that information on the solution is sought without actually solving the problem. This qualitative analysis was introduced by Poincaré around 1880 (Poincaré, 1885) and developed in the following decades, especially in Russia.

The central problem in qualitative analysis is to investigate the relationship between the solution and its *neighbourhood*. A solution is a curve or a trajectory C in a certain space. The question is whether any D trajectory, which at the time $t=0$ starts *near* C, tends to remain near C or moves away from it. In the first case, the trajectory C is said to be *stable*, in the second *unstable*. Liapunov is credited with creating a systematic qualitative analysis, which is generally called the *theory of stability*. In 1892 he published the first of a series of fundamental papers *"General Problem on the Stability of Motion"* (Liapunov, 1966), in which he treated the problem of stability in two different ways. His so-called *first method* presupposes explicit knowledge of the solution and is applied only to a limited but important number of cases; the *second method*, or *direct method*, is altogether general and does not require knowledge of the solution.

1.2 Differential Equations

From a historical point of view differential equations were introduced by Newton through the laws of mechanics which define the motion of a body subjected to a system of forces. Subsequent developments in physics have shown how a wide range of problems in completely different fields is governed by laws which are altogether analogous to those of mechanics. Thus, it is desirable, as a first step, to describe the types of equations on which we shall be working and their properties.

The ordinary differential equations which are the basis of the problems we are to study are essentially of two types (La Salle and Lefschetz, 1961; Pontriaguine, 1969). The first is represented by an equation of n-th order

$$y^{(n)} = f(y, \dot{y}, ..., y^{(n-1)}; t) \tag{1.1}$$

where t is a variable and generally, but not necessarily, represents time, and $\dot{y}, ..., y^{(k)}$ represent the first,..., k-th derivative of y with respect to t. The second type is a system of n equations of the first order

$$\dot{y}_i = Y_i(y_j; t) \quad (i, j = 1, 2, \ldots, n) \tag{1.2}$$

where, unless otherwise specified, the Latin indices are understood to vary from 1 to n.

The first type can be reduced to the second one if we introduce the new variables $y_1, y_2, ..., y_n$ defined by

$$y_i = y^{(i-1)} \tag{1.3}$$

In this case equation (1.1) is replaced by the system

$$\begin{aligned} \dot{y}_i &= y_{i+1} \quad (i, j = 1, 2, \ldots, n-1) \\ \dot{y}_n &= f(y_j; t) \end{aligned} \tag{1.4}$$

As an example the well known equation of van der Pol

$$\ddot{y} + k(y^2 - 1)\dot{y} + y = 0 \tag{1.5}$$

can be replaced by the system

$$\begin{aligned} \dot{y}_1 &= y_2 \\ \dot{y}_2 &= -k(y_1^2 - 1)y_2 - y_1 \end{aligned} \tag{1.6}$$

If we consider $y_1, y_2, ..., y_n$ as components of a vector $\mathbf{y}$, and $Y_1, Y_2, ..., Y_n$ as components of a vector $\mathbf{Y}$, the system (1.2), can be written in the compact form

$$\dot{\mathbf{y}} = \mathbf{Y}(\mathbf{y}; t) \tag{1.7}$$

In many problems the variable t does not appear explicitly in (1.7). In this case, the system becomes

$$\dot{\mathbf{y}} = \mathbf{Y}(\mathbf{y}) \tag{1.8}$$

A system of this type is called *autonomous*. For example, the system deduced from van der Pol's equation is autonomous. A system of the type (1.7) is *non-autonomous*.

Once the solution $y_1 = f_1(t), y_2 = f_2(t), \ldots, y_n = f_n(t)$ has been determined from (1.7) or (1.8) a curve, called the *integral curve*, in the space $E_{y,t}^{n+1}$ can be associated with it (Figure 1). The projection of this curve in the sub-space E_y^n of the $\mathbf{y}$ coordinates is defined as the *trajectory* or simple the *motion*, and the space E_y^n is the space of the *phases* (Figure 2). For the existence and uniqueness of the solution, the *Cauchy-Lipschitz* theorem holds.

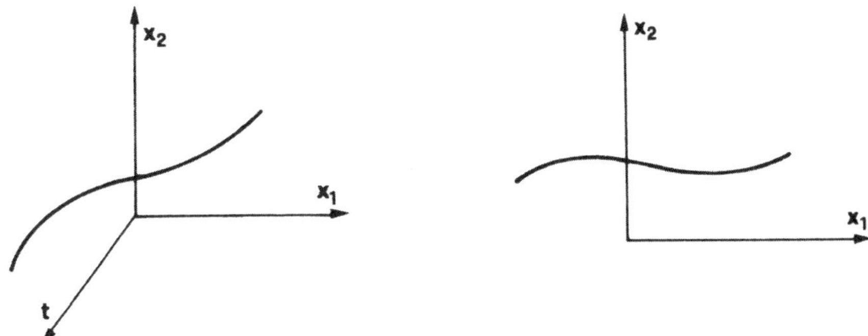

Figure 1. Integral curve

Figure 2. Trajectory

Theorem 1.1. *Let $E_{y,t}^{n+1}$ be the $n+1$ dimensions space y_i, t, and let Ω be a simply connected open region in such a space. Let the functions Y_i be continuous and admit partial derivatives $\partial Y_i / \partial y_h$ at each point of Ω. If $(\mathbf{y}_0, t_0)$ is a point in Ω there exists a unique solution of system (1.7) such that $\mathbf{y}(t_0) = \mathbf{y}_0$. Such a solution is a continuous solution of $(\mathbf{y}_0, t_0)$ as such a point varies in Ω.*

Let us consider a particular solution $\mathbf{v}(t)$ of system (1.7) and introduce new variables

$$\mathbf{x}(t) = \mathbf{y}(t) - \mathbf{v}(t) \tag{1.9}$$

By substituting (1.9) into (1.7) we obtain

$$\dot{\mathbf{x}} = \mathbf{X}(\mathbf{x}, t) \tag{1.10}$$

where

$$\mathbf{X}(\mathbf{x}; t) = \mathbf{Y}(\mathbf{x} + \mathbf{v}; t) - \mathbf{Y}(\mathbf{v}; t) \tag{1.11}$$

Since in (1.11) $\mathbf{X}(\mathbf{0}, t) = \mathbf{0}$, eqs. (1.10) admit as a solution $\mathbf{x}(t) = \mathbf{0}$, which is called *undisturbed motion* or *position of equilibrium* and furnish the differential equations of the *disturbed motion* $\mathbf{x}(t) \neq \mathbf{0}$. In the study of the stability of motion (eqs. (1.7), one can always refer to the study of stability of the undisturbed motion.

Let us now consider the motion defined by the autonomous system (1.8) and assume that for $\mathbf{y} = \mathbf{c}$, with $\mathbf{c}$ constant, $\mathbf{Y}(\mathbf{c}) = \mathbf{0}$. If we replace $\mathbf{y}$ by $\mathbf{c}$ in (1.8) we can see that the system is satisfied, and consequently $\mathbf{y} = \mathbf{c}$ is a solution to the system. From a physical point of view this means that if the system is initially in $\mathbf{c}$ then it remains in this position, and therefore $\mathbf{c}$ is a configuration of equilibrium. The point $\mathbf{c}$ is defined as the *critical point* or *equilibrium point*. By introducing the change of coordinates

$$\mathbf{x}(t) = \mathbf{y}(t) - \mathbf{c} \tag{1.12}$$

one has from (1.8)

$$\dot{\mathbf{x}} = \mathbf{X}(\mathbf{x}) \tag{1.13}$$

with

$$\mathbf{X}(\mathbf{x}) = \mathbf{Y}(\mathbf{x} + \mathbf{c}) \tag{1.14}$$

whence $\mathbf{X}(\mathbf{0}) = \mathbf{0}$, and therefore the equilibrium point of eqs. (1.13) coincides with the origin. From now on in studying stability of autonomous system, we can always refer to the origin as equilibrium point.

Let us now introduce the *norm* of vector $\mathbf{x}$ which is indicated with $\|\mathbf{x}\|$. The most common norm is the Euclidean vector length

$$\|\mathbf{x}\| = \left(\sum_{i=1}^{n} x_i^2 \right)^{1/2} \tag{1.15}$$

Two other types of norms which are often encountered are

$$\|\mathbf{x}\| = max\,|x_i| \tag{1.16}$$

$$\|\mathbf{x}\| = \sum_{i=1}^{n} |x_i| \tag{1.17}$$

The concepts of stability and of asymptotic stability stated below have been introduced by Liapunov in 1893 and therefore we speak of stability in the Liapunov sense, even if other definitions have been introduced later.

Definition 1.2. *The solution* $\mathbf{x} = \mathbf{0}$ *of the system* (1.10) *or* (1.13) *is said to be stable in the Liapunov sense if for each positive number* ε *it is possible to find a positive number* $\delta(\varepsilon)$ *such that if*

$$\|\mathbf{x}(t_0)\| < \delta \tag{1.18}$$

holds, then

$$\|\mathbf{x}(t)\| < \varepsilon\,, \qquad \forall t > t_0 \tag{1.19}$$

Definition 1.3. *The solution* $\mathbf{x} = \mathbf{0}$ *of the system* (1.10) *or* (1.13) *is said asymptotically stable in the sense of Liapunov, if for each solution* $\mathbf{x}(t)$ *with initial conditions*

$$\|\mathbf{x}(t_0)\| < \delta \tag{1.20}$$

we have

$$\lim_{t \to \infty} \|\mathbf{x}\| = 0 \tag{1.21}$$

Definition 1.4. *The solution* $\mathbf{x} = \mathbf{0}$ *of the system* (1.10) *or* (1.13) *is said unstable in the sense of Liapunov if for each number* ε *and for a positive number* δ *however fixed, there exists at least a point* $\mathbf{x}(t_0)$ *with*

$$\| \mathbf{x}(t_0) \| < \delta \tag{1.22}$$

such that

$$\| \mathbf{x}(t) \| > \varepsilon, \qquad \forall t > t_0 \tag{1.23}$$

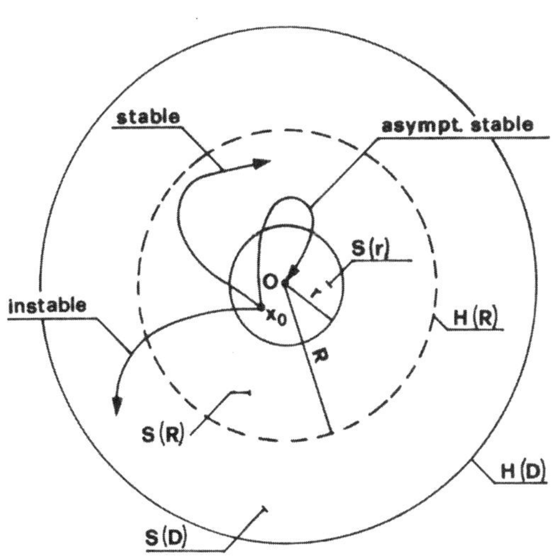

Figure 3. Types of equilibrium

Definition 1.5. *The domain of attraction of the solution of equilibrium* $\mathbf{x} = \mathbf{0}$ *of the system* (1.10) *or* (1.13) *is defined as the collection of the points* $\mathbf{x}(t_0)$, *such that motions starting from* $\mathbf{x}(t_0)$ *are asymptotically stable.*

Definitions $2, 3, 4$ may be visualized in (Figure 3), where $S(D)$ is an open spherical region in Ω in which the conditions requested by the theorem of existence and uniqueness of the solution are satisfied and $H(D)$ is its boundary. In addition $S(R)$ is the spherical region defined by $||\mathbf{x}|| < R$ and $||\mathbf{x}|| = R$ is the spherical surface $H(R)$. Then we have

Definition 1.6. *A motion is stable if for every* $R < D$ *there exists* $r < R$ *such that a trajectory* $g(t)$ *with its origin at a point* $\mathbf{x}_0 \in S(r)$ *remains in the spherical region* $S(R)$ *when* t *increases; that is to say, a trajectory with origin in* $S(r)$ *never reaches the boundary* $H(R)$ *of* $S(R)$.

Definition 1.7. *A motion is asymptotically stable if it is stable and, besides, each trajectory* $g(t)$ *with origin in* $S(r)$ *tends to the origin for* $t \to \infty$.

Definition 1.8. *A motion is unstable if for a fixed $R < D$ and for any r, however small, there always exists a point $\mathbf{x}_0$ in $S(r)$, such that a $g(t)$ trajectory which originates in $\mathbf{x}_0$ reaches the boundary $H(R)$.*

The stability of motion (1.10) and of equilibrium (1.13) may depend on a certain number of parameters, besides the above mentioned disturbances. This is the case, for instance, of a rigid bar loaded by a vertical force N and connected to the ground by a hinge to which it is applied a linear elastic spring initially unloaded (Figure 4). Eq. (1.7) in this case becomes

$$\dot{\mathbf{y}} = \mathbf{Y}(\mathbf{y}; \gamma_k; t) \qquad (k = 1, 2, \ldots, m) \tag{1.24}$$

γ_k being a parameter. This problem is more complicated and can be treated through a perturbation analysis. By denoting with $\mathbf{v}(\gamma_k; t)$ a particular solution of (1.24), introducing the change of coordinates

$$\mathbf{x}(\gamma_k; t) = \mathbf{y}(\gamma_k; t) - \mathbf{v}(\gamma_k; t) \tag{1.25}$$

and replacing (1.25) into (1.24) the following equations are obtained

$$\dot{\mathbf{x}} = \mathbf{X}(\mathbf{x}; \gamma_k; t) \tag{1.26}$$

having posed

$$\mathbf{X}(\mathbf{x}; \gamma_k; t) = \mathbf{Y}(\mathbf{x} + \mathbf{v}; \gamma_k; t) - \mathbf{Y}(\mathbf{v}; \gamma_k; t) \tag{1.27}$$

Since $\mathbf{X}(\mathbf{0}; \gamma_k; t) = \mathbf{0}$, eqs. (1.26) admit as a solution $\mathbf{x}(\gamma_k; t) = \mathbf{0}$ which is the undisturbed motion, and permit to determine the disturbed motion $\mathbf{x}(\gamma_k; t) \neq \mathbf{0}$.

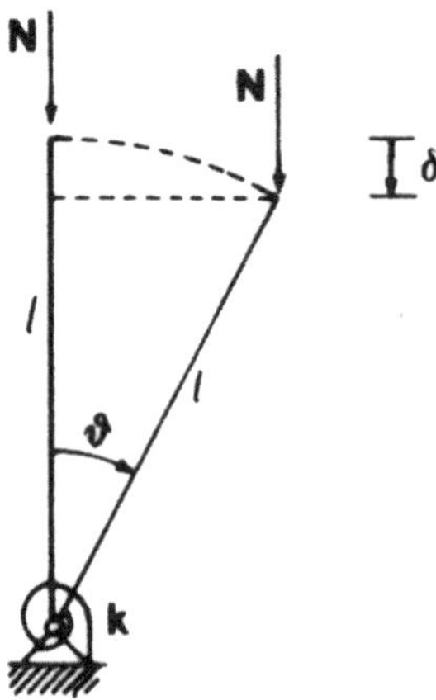

Figure 4. Stability depending on parameters

1.3 General Theorems on Stability

We shall give in this section few basic concepts regarding the *second method* or *direct method* of Liapunov on the qualitative response of a motion (eqs. (1.10) and (1.13)), without solving the relevant differential equations. We emphasize here that motion is any situation of a body defined by a set of state variables (chemical, electrical, mechanical processes). In particular, we shall refer first to motions defined by eqs. (1.13) and, following Liapunov, we introduce a scalar function $V(\mathbf{x})$ which is said to be *positive definite* in an open simply connected region Ω around the origin if the following conditions hold

(a) $V(\mathbf{x})$ together with its first partial derivatives is continuous in $\Omega\mathfrak{g}$;
(b) $V(\mathbf{0}) = 0$;
(c) $V(\mathbf{x})$ has an isolated minimum at the origin.

If, in addition, dV/dt is non-positive in Ω along the trajectories of motion of system (1.13), that is

$$\dot{V} = V_{,i}\,\dot{x}_i = V_{,i}\,X_i(x_1, x_2, \ldots, x_n) \leqslant 0 \tag{1.28}$$

the function $V(\mathbf{x})$ is called a *Liapunov function*. It is assumed that $X_i(x_j)$ and when necessary $\dot{X}_i(x_j)$ are continuous. It follows that $\dot{V}$ is a continuous function in Ω. In equation (1.28), a subscript preceded by a comma indicates differentiation with respect to the corresponding variable. Let us examine the quadratic form

$$V(\mathbf{x}) = a_{ij}x_i x_j \qquad (a_{ij} = a_{ji}\,, a_{ij} \in \mathbb{R}) \tag{1.29}$$

The necessary and sufficient condition for $V(\mathbf{x})$ to be positive definite is that the successive principal minors of the symmetrical matrix of the coefficients $[a_{ij}]$ have positive determinant (Sylvester).

Generally, the function $V(\mathbf{x})$ can be represented as a series of powers in $\mathbf{x}$ in the neighbourhood of the origin

$$V(\mathbf{x}) = V_p(\mathbf{x}) + V_{p+1}(\mathbf{x}) + \ldots \tag{1.30}$$

where $V_p(\mathbf{x})$ is a homogeneous polynomial in $\mathbf{x}$ of degree p. A necessary condition for $V(\mathbf{x})$ to be positive definite is that the lowest degree p of the series of powers (1.30) is an even number. Such a condition, however, is not sufficient. In fact, for $p = 2$ the function

$$V_p(\mathbf{x}) = x_2^2 - x_1^2 \tag{1.31}$$

is positive definite for $x_1 = 0$ and negative definite for $x_2 = 0$. If p is an odd number then $V(\mathbf{x})$ can never be a Liapunov function. Let us now pass to the enunciation of some basic theorems.

Theorem 1.9. *Stability (Liapunov).* *If in a certain neighbourhood Ω of the origin there exists a Liapunov function $V(\mathbf{x})$, then the origin is stable.*

Theorem 1.10. *Asymptotic stability (Liapunov).* *If there exists in Ω a Liapunov function $V(\mathbf{x})$ such that $\dot{V} < 0$ then the stability is asymptotic.*

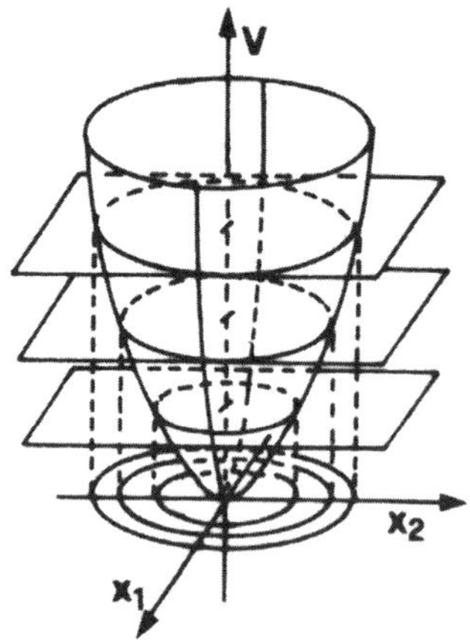

Figure 5. Liapunov function

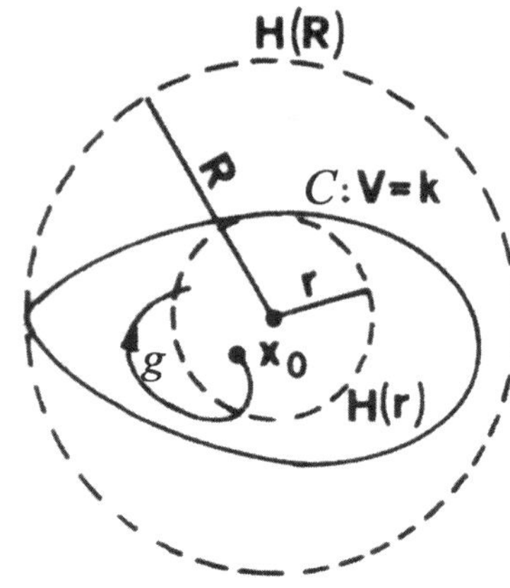

Figure 6. Plane of the phases

Let us first demonstrate the first theorem using the geometrical interpretation of a positive definite function $V(\mathbf{x})$, represented in (Figure 5) for $\mathbf{x} = \{x_1,\ x_2\}$. (Figure 6) the curve $V(\mathbf{x}) = k$ is represented by the solid line whilst the spheres $H(R)$ and $H(r)$ are indicated by dashes.

Given then $R < D$ (Figure 3) and $H(R)$, we can find a constant k such that the curve C defined by $V(\mathbf{x}) = k$ is contained in $H(R)$ and an $r > 0$ such that $H(r)$ is contained in C. Let us now consider a trajectory $g(\mathbf{x})$ with initial point $\mathbf{x}_0$ belonging to $S(r)$, the interior of $H(r)$. In $\mathbf{x}_0$ it is $V(\mathbf{x}_0) < k$. Furthermore, as $V(\mathbf{x})$ does not increase along the trajectories, $g(\mathbf{x})$ never reaches C and so will never reach $H(R)$. Therefore each trajectory with origin in $S(r)$ must remain in $S(R)$ and this implies stability.

The demonstration of the second theorem follows from the previous demonstration, since $\dot{V} < 0$ implies that the trajectory $g(\mathbf{x})$ which starts at $\mathbf{x}_0 \in S(r)$ tends to the origin as $t \to \infty$, and this implies asymptotic stability.

Theorem 1.11. *Instability (Liapunov).* *Let $V(\mathbf{x})$ with $V(\mathbf{0} = 0)$ have continuous first partial derivatives in Ω. Let $\dot{V}$ be positive definite and let $V(\mathbf{x})$ be able to assume positive values arbitrarily near the origin. Then the origin is unstable.*

The demonstration is omitted here. However, it is easy to guess that the condition $\dot{V} > 0$ implies that the trajectory $g(\mathbf{x})$ which starts from $\mathbf{x}_0 \in S(r)$ where $V(\mathbf{x}) > 0$ reaches C and therefore $H(R)$, and so we have instability.

Theorem 1.12. *Instability (Chetayev).* *Let Ω be a neighbourhood of the origin. Let $V(\mathbf{x})$ be a given function and Ω_1 a region in Ω with the following properties*

(a) *$V(\boldsymbol{x})$ and $\dot{V}$ are positive in Ω_1;*
(b) *$V(\boldsymbol{x})$ has continuous partial derivatives in Ω_1;*
(c) *at the boundary points of Ω_1 inside Ω $V(\mathbf{x}) = 0$;*

$$\text{i)} \quad V(x,y) => 0, \qquad \text{for } \mid x \mid > \mid y \mid$$
$$\text{ii)} \quad \dot{V} = 4x^3(y^3 + x^5) - 4y^3(x^3 + y^5) = 4(x^8 - y^8) > 0 \qquad \text{for } \mid x \mid > \mid y \mid \tag{1.35}$$

In a neighbourhood of the origin and for $\mid x \mid > \mid y \mid$ we have $V > 0, \dot{V} > 0$; thus the equilibrium point $x = y = 0$ is unstable.

Liapunov theorems presented for autonomous systems (1.13) can be extended to non autonomous system (1.10). To this end we introduce a positive definite function $W(\mathbf{x})$. $V(\mathbf{x};t)$ is then *positive definite* if

$$\text{(a)} \qquad V(\mathbf{0};t) = 0 \qquad \text{for } t \geqslant 0$$
$$\text{(b)} \qquad V(\mathbf{x};t) \geqslant W(\mathbf{x}) \qquad \text{for } t \geqslant 0 \text{ and } |\mathbf{x}| < r \tag{1.36}$$

where r is a sufficiently small quantity (Figure 3). The function $V(\mathbf{x};t)$ is *negative definite* if, under condition (a)

$$V(\mathbf{x};t) \leqslant -W(\mathbf{x}) \qquad \text{for } t \geqslant 0 \text{ and } |\mathbf{x}| < r \tag{1.37}$$

For instance the function

$$V = t(x_1^2 + x_2^2) - 2x_1 x_2 \cos t \tag{1.38}$$

is positive definite for $t > 2$. In fact, by choosing $W = x_1^2 + x_2^2$ one has

$$V - W = (t - 1)(x_1^2 + x_2^2) - 2x_1 x_2 \cos t > 0 \tag{1.39}$$

In a different example, the function $V = e^{-t}(x_1^2 + \ldots + x_n^2)$ is not positive definite in that $V \to 0$ when $t \to \infty$. In this case it is not possible to find any positive definite function W such that $V > W$. If, in addition to conditions (1.36), the function $V(\mathbf{x},t)$ satisfies the inequality $\dot{V} = \frac{\partial V}{\partial t} + V_{,i} X_i \leqslant 0$ along the trajectory of motion, we say that $V(\mathbf{x},t)$ is a *Liapunov function*.

The function $V(\mathbf{x};t)$ is said to be *decrescent* (or *uniformly small*) if it satisfies the condition

$$|V(\mathbf{x};t)| \leqslant W(\mathbf{x}) \qquad \text{for } t \geqslant 0 \text{ and } |\mathbf{x}| < r \tag{1.40}$$

where $W(\mathbf{x})$ is a positive definite function. For instance the function $V(\mathbf{x};t) = (\sin t)(x_1 + \ldots + x_n)$ is decrescent while the function $V(\mathbf{x};t) = \sin[t(x_1 + \ldots + x_n)]$ is not decrescent. In the following, theorems on stability for *non autonomous system* are presented without demonstration.

Theorem 1.15. *Stability (Liapunov).* *The equilibrium is stable if there exists a positive definite function $V(\mathbf{x};t)$ such that its total derivative $\dot{V}$ along the trajectory of motion (1.10) is not positive.*

(d) *the origin is a point belonging to the boundary of Ω_1.*

Under these conditions the origin is unstable.

It is not difficult to see that any trajectory $g(\mathbf{x})$ starting from a point situated in must leave Ω since it cannot cross the boundary of Ω_1 inside Ω. As the origin is situa on the boundary of Ω_1, we can choose some points inside Ω_1 arbitrarily close to the or from which trajectories $g(\mathbf{x})$ which start must leave Ω, and this implies instability.

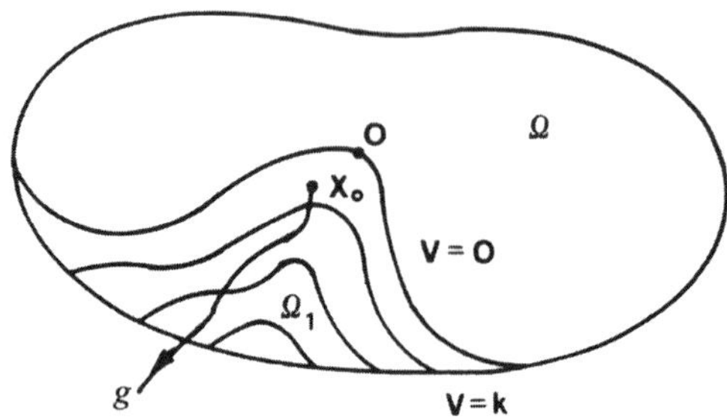

Figure 7. Representation of Chetayev's theorem on the plane of the phases

Example 1.13. Let us analyse the stability of the trivial solution to the system

$$\dot{x} = -y - x^3$$
$$\dot{y} = x - y^3$$

(1.3

The function $V(x,y) = x^2 + y^2$ satisfies the conditions of Liapunov's theorem asymptotic stability. In fact

$$\text{i) } V(x,y) = \geq 0, \qquad V(0,0) = 0$$
$$\text{ii) } \dot{V} = 2x(-y - x^3) + 2y(x - y^3) = -2(x^4 + y^4) \leq 0$$

(1.

At a point which is arbitrarily near the origin we have $\dot{V} < 0$, and so the origin asymptotically stable.

Example 1.14. Analyse the stability of the equilibrium point $x = y = 0$ of the syst of equations

$$\dot{x} = y^3 + x^5$$
$$\dot{y} = x^3 + y^5$$

(1

The function $V(x,y) = x^4 - y^4$ satisfies the conditions of Chetayev's theorem

Theorem 1.16. *Asymptotic stability (Liapunov).* *The equilibrium is asymptotically stable if a positive definite and decrescent function $V(\mathbf{x};t)$ exists such that its total derivative along the trajectory of motion (1.10) is negative.*

Theorem 1.17. *Instability (Liapunov).* *The equilibrium is unstable if a decrescent function $V(\mathbf{x};t)$ exists having the same sign of $\dot{V}$ along the trajectory of motion (1.10).*

It is important to remember that the existence of a Liapunov function is a sufficient but not necessary condition for stability.

1.4 Analysis of the Stability of Equilibrium by Linear Approximation

Let us consider the autonomous system (1.13) $\dot{\mathbf{x}} = \mathbf{X}(\mathbf{x})$ with $\mathbf{X}(0) = \mathbf{0}$. If the functions X_i are derivable in a neighbourhood of the origin of coordinates, then the second member of system (1.13) can be substituted by a series expansion

$$\dot{x}_i = a_{ij}x_j + R_i(x_1,\ldots,x_n) \tag{1.41}$$

where $a_{ij} = (\partial X_i/\partial x_j)_{x=0}$ and $||\mathbf{R}||$ is small with respect to $||\mathbf{x}||$, that is to say

$$\lim_{\mathbf{x}\to 0}\frac{||\mathbf{R}||}{||\mathbf{x}||} = \mathbf{0} \quad \Leftrightarrow ||\mathbf{R}|| = o(||\mathbf{x}||) \tag{1.42}$$

Instead of investigating the stability of the equilibrium point $\mathbf{x} = \mathbf{0}$ of system (1.41), the stability of the same point of the linear system

$$\dot{x}_i = a_{ij}x_j \tag{1.43}$$

is analysed. System (1.43) is called a *system of equations of linear approximation* with respect to system (1.41). The conditions of stability of this system were examined by Liapunov and successively generalised by Malkin, Chetayev and others.

The analysis of stability of the system of equations of linear approximation is a much simpler problem than the study of the original system. In this regard there are two useful theorems of great practical importance.

Let us suppose that the characteristic roots λ_i of the matrix of coefficients $[a_{ij}]$ are real and distinct, and let us apply to system (1.4.1) the linear transformation of coordinates $\mathbf{y} = \mathbf{P}\mathbf{x}$ with $\mathbf{P}$ non-singular and time independent. As $(d/dt)\mathbf{P}\mathbf{x} = P\dot{x}$, by making use of (1.41) we can write

$$\dot{\mathbf{y}} = \mathbf{P}\dot{\mathbf{x}} = \mathbf{PAP}^{-1}\mathbf{y} + \mathbf{PR} \tag{1.44}$$

We now choose the matrix $\mathbf{P}$ in such a way that $\mathbf{PAP}^{-1} = \mathrm{diag}(\lambda_1, \lambda_2,\ldots,\lambda_n)$ and assume $\mathbf{PR} = \mathbf{R}^*$. Then, the system (1.44) is written as

$$\dot{\mathbf{y}} = diag(\lambda_1, \lambda_2, ..., \lambda_n)\mathbf{y} + \mathbf{R}^* \tag{1.45}$$

where it can easily be shown that

$$||\mathbf{R}^*(\mathbf{y})|| = o(||\mathbf{y}||) \tag{1.46}$$

The transformation of system (1.41) into system (1.45) is useful for demonstrating the following theorems.

Theorem 1.18. *A sufficient condition for the origin of the non-linear system (1.45) to be asymptotically stable is that the characteristic roots are all negative. If there is a single positive characteristic root, then the origin is unstable.*

Two cases can be distinguished in the demonstration.

(a) *The λ_h roots are all negative*

The following Liapunov function is assumed

$$V = y_1^2 + y_2^2 + \ldots + y_n^2 \tag{1.47}$$

from which

$$\dot{V} = 2(\lambda_1 y_1^2 + \lambda_2 y_2^2 + \ldots + \lambda_n y_n^2) + r(\mathbf{y}) \tag{1.48}$$

where r is small with respect to the terms in parenthesis. In a sufficiently small region Ω around the origin V and $-\dot{V}$ are positive definite functions, and so the origin is asymptotically stable.

(b) *Some of the λ_h roots, for example λ_1, $\lambda_2,\ldots,\lambda_p$ ($p < n$) are positive and the rest negative.*

This time we take

$$V = y_1^2 + \ldots + y_p^2 - y_{p+1}^2 - \ldots - y_n^2 \tag{1.49}$$

from which

$$\dot{V} = 2(\lambda_1 y_1^2 + \ldots + \lambda_p y_p^2 - \lambda_{p+1} y_{p+1}^2 - \ldots - \lambda_n y_n^2) + r_1(\mathbf{y}) \tag{1.50}$$

where, as before, the r_1 term is small with respect to those in parenthesis. At some points which are arbitrarily near to the origin (those for which $y_{p+1} = \ldots = y_n = 0$) V is positive. As for $\dot{V}$, since $\lambda_1, \lambda_2,\ldots, \lambda_p > 0$, it is positive definite in that Ω_1 region in Ω where V is positive definite and therefore, according to the Chetayev theorem, the origin is unstable.

Let us now suppose that some of the λ_h are complex. For example, let $\lambda_1,\ldots,\lambda_p$ be real and $\lambda_{p+1}, \bar{\lambda}_{p+1}, \ldots, \lambda_{p+m}, \bar{\lambda}_{p+m}$ be complex with $p+2m = n$. If $\lambda_1,\ldots,\lambda_p$ are negative and $\lambda_{p+h}, \bar{\lambda}_{p+h}$ have negative real part, then we can choose the following Liapunov function

$$V = y_1^2 + \ldots + y_p^2 + y_{p+1}\bar{y}_{p+1} + \ldots + y_{p+m}\bar{y}_{p+m} \tag{1.51}$$

and everything proceeds as in case (a), with the origin asymptotically stable. If, on the other hand, some of the $\lambda_1,\ldots,\lambda_p$ are positive or some of the λ_{p+h} have a positive real part, then we proceed exactly as in case (b) and we find that the origin is unstable. We can therefore enunciate the following theorem.

Theorem 1.19. *A sufficient condition for the origin of the non-linear system (1.45) to be asymptotically stable is that the characteristic roots all have negative real parts. If there is a characteristic root with positive real part, then the origin is unstable.*

If a certain number of characteristic roots vanish or have a purely imaginary value, results from the analysis of the linear approximation system cannot be extended to the nonlinear system, as the nonlinear terms R_i influence the stability of the system.

Example 1.20. Analyse the stability of the equilibrium point $x = y = 0$ of the system

$$\dot{x} = 2x + 8\sin y$$
$$\dot{y} = 2 - e^x - 3y - \cos y \tag{1.52}$$

By expanding $\sin y, \cos y$ and e^x in a Taylor series around the origin we can write the system in the form

$$\dot{x} = 2x + 8y + R_1$$
$$\dot{y} = -x - 3y + R_2 \tag{1.53}$$

where $R_1 = -4y^3/3 + \ldots$ and $R_2 = (y^2 - x^2)/2 + \ldots$. Since the limitations (1.42) are satisfied we can analyse the stability of equilibrium point of the linear system

$$\dot{x} = 2x + 8y$$
$$\dot{y} = -x - 3y \tag{1.54}$$

The roots of the characteristic equation $\lambda^2 + \lambda + 2 = 0$ are $\lambda_{1,2} = -1/2 \pm i\sqrt{7/4}$; therefore the equilibrium point $x = y = 0$ of system (1.54) and (1.52) is asymptotically stable.

Example 1.21. Let us consider the system

$$\dot{x} = y - xf(x,y)$$
$$\dot{y} = -x - yf(x,y) \tag{1.55}$$

and suppose that the nonlinear terms $xf(x,y)$ and $yf(x,y)$ satisfy condition (1.42) and that $f(0,0) = 0$. The characteristic roots of the linear system are $\lambda_{1,2} = \pm i$ and therefore the analysis of the stability of equilibrium point $x = y = 0$ of system (1.55) depends on nonlinear terms. In fact, let us choose as Liapunov function $V = (x^2 + y^2)/2$, from which

$$\dot{V} = -(x^2 + y^2)f(x,y) \tag{1.56}$$

Three case can occur

$f \geq 0$ in an arbitrary vicinity of the origin, the origin is stable;

$f < 0$ in an arbitrary vicinity of the origin, the origin is unstable;

f is positive definite within a certain vicinity of the origin, the origin is asymptotically stable.

Note that the system of equations studied in Example(1.13) is of the same type as system (1.55). In fact, as the characteristic roots of the linearized system are $\lambda_{1,2} = \pm i$, the stability of the equilibrium point has been decided by non linear terms.

1.5 Criterion of Negative Real Parts of all the Roots of a Polynomial

In the previous section the problem of the stability of the trivial solution to a wide class of systems of differential equations was reduced to the analysis of the sign of the real parts of the roots of the characteristic equation.

If the characteristic equation is a polynominal of high degree, then its solution is very difficult, and therefore the methods which allow us to determine whether the roots do or do not have negative real parts are of great importance. With regard to this, we have the following

Theorem 1.22. (Hurwitz). *The necessary and sufficient condition for the real parts of all the roots of the polynomial*

$$p(z) = z^n + a_1 z^{n-1} + \ldots + a_{n-1} z + a_n \tag{1.57}$$

with positive real coefficients to be negative is that each principal minor of the Hurwitz matrix

$$\begin{pmatrix} a_1 & 1 & 0 & 0 & \ldots & 0 \\ a_3 & a_2 & a_1 & 1 & \ldots & 0 \\ a_5 & a_4 & a_3 & a_2 & \ldots & 0 \\ \ldots & \ldots & \ldots & \ldots & \ldots & \ldots \\ 0 & 0 & 0 & 0 & \ldots & a_n \end{pmatrix} \tag{1.58}$$

is positive.

Example 1.23. Let us consider the polynominal

$$p(z) = z^4 + a_1 z^3 + a_2 z^2 + a_3 z + a_4 \tag{1.59}$$

The Hurwitz conditions require

$$a_1 > 0, \quad a_1 a_2 - a_3 > 0, \quad (a_1 a_2 - a_3) a_3 - a_4 a_1^2 > 0, \quad a_4 > 0 \tag{1.60}$$

to be satisfied.

2 Equilibrium of Mechanical System

2.1 Stability of Equilibrium of Discrete Mechanical Systems. Lagrange and Hamilton Equations of Motion

Lagrange has demonstrated (Gantmacher, 1970) that the differential equations of motion of a system with n degrees of freedom can be written immediately if we know the *kinetic potential* or *Lagrange function* defined by

$$L = K - \Phi \tag{2.1}$$

where K is the kinetic energy and Φ is the potential energy of the forces acting on the system.

Let $q_1, q_2, \ldots, q_n$ be generalized coordinates with which it is possible to define the configuration of the discrete system, and suppose that the q_i $(i = 1, 2, \ldots, n)$ are chosen in such a way that in the position of equilibrium we have $q_i = 0$. Indicating the position vector by $\mathbf{r} = \mathbf{r}(\mathbf{q}; t)$, the kinetic energy of a system of N particles is expressed by the relation

$$K = \frac{1}{2} \sum_{i=1}^{N} m_i (\dot{r}_i)^2 = \frac{1}{2} \sum_{i=1}^{N} m_i \left(\frac{\partial r_i}{\partial q_j} \dot{q}_j + \frac{\partial r_i}{\partial t} \right)^2 \tag{2.2}$$

which can be written in the form

$$K = \frac{1}{2} a_{ij} \dot{q}_i \dot{q}_j + a_i \dot{q}_i + a_0 = K_2 + K_1 + K_0 \tag{2.3}$$

for *reonomous* systems or $K = K_2$ for *scleronomous systems*. In eq. (2.3) the summation convention with respect to repeated indices has been adopted.

The coefficients a_{ij}, a_i, a_0 are function of $\mathbf{q}$ and t in the first case, while a_{ij} are function of $\mathbf{q}$, only, in the second case. We shall always refer, in the future, to scleronomous systems. The external forces Q_i are supposed to be *conservative*, i.e. derivable from a potential $\Phi = \Phi(\mathbf{q})$

$$Q_i = -\Phi_{,i} \tag{2.4}$$

The Lagrange equations of motions are then

$$\frac{d}{dt} \frac{\partial L}{\partial \dot{q}_i} - \frac{\partial L}{\partial q_i} = 0 \tag{2.5}$$

which are of the type

$$\ddot{\mathbf{q}} = \mathbf{f}(\mathbf{q}; \dot{\mathbf{q}}) \tag{2.6}$$

which, according to notations of sect. 1.2, can always be reduced to the form

$$\dot{\mathbf{x}} = \mathbf{S}(\mathbf{x}) \tag{2.7}$$

The kinetic potential L from which the equations (2.5) have been deduced depends on the variables $\mathbf{q}, \dot{\mathbf{q}}$ which are called *Lagrange variables*. Hamilton proposed to assume as basic variables the quantities $\mathbf{q}$ and $\mathbf{p}$, where $\mathbf{p}$ is the generalized linear momentum defined by

$$p_i = \frac{\partial L}{\partial \dot{q}_i} = a_{ij}(q)\dot{q}_j \tag{2.8}$$

The quantities $\mathbf{q}$, $\mathbf{p}$ are called *Hamilton variables*. By simple steps it is possible to express the kinetic energy as a function of $\mathbf{q}$ and $\mathbf{p}$, arriving at the expression

$$K = \frac{1}{2}l_{ij}(\mathbf{q})p_i p_j \tag{2.9}$$

The potential energy $\Phi(\mathbf{q})$ in terms of the new variables remains unchanged.

By introducing the *Hamiltonian*

$$H(\mathbf{p};\mathbf{q}) = \Phi(\mathbf{q}) + K(\mathbf{p};\mathbf{q}) + C \tag{2.10}$$

with C arbitrary constant, it is easily demonstrated that the following equations hold

$$\dot{q}_i = \frac{\partial H}{\partial p_i}$$
$$\dot{p}_i = -\frac{\partial H}{\partial q_i} \tag{2.11}$$

Equations (2.11) constitute the Hamilton equations of motion and it is possible to use them as an alternative to (2.5) in order to study the stability of equilibrium. These equations are of the type (2.7).

2.2 Stability of Equilibrium According to Liapunov

Let us consider a system in the state of equilibrium $\mathbf{q} = \mathbf{0}$, $\dot{\mathbf{q}} = \mathbf{0}$ and suppose that we apply at the instant $t = 0$ a perturbation characterised by

$$\mathbf{q}(0) = \mathbf{q}_0, \quad \dot{\mathbf{q}}(0) = \dot{\mathbf{q}}_0 \tag{2.12}$$

We now introduce a norm ρ which measures the *distance* between the state of equilibrium and the current state and endow ρ with the following properties

$$\rho(\mathbf{q};\dot{\mathbf{q}}) > 0 \qquad\qquad \text{for } \mathbf{q} \neq \mathbf{0}, \dot{\mathbf{q}} \neq \mathbf{0}$$
$$\rho(\mathbf{q_1} + \mathbf{q_2}; \dot{\mathbf{q}}_1 + \dot{\mathbf{q}}_2) \leqslant \rho(\mathbf{q_1}; \dot{\mathbf{q}}_1) + \rho(\mathbf{q_2}; \dot{\mathbf{q}}_2) \quad \text{(triangle inequality)} \tag{2.13}$$
$$\rho(\alpha\mathbf{q}; \alpha\dot{\mathbf{q}}) = |\alpha|\,\rho(\mathbf{q}; \dot{\mathbf{q}}) \qquad\qquad (\alpha \text{ real})$$

Definition 2.1. Liapunov. *The configuration of equilibrium* $\mathbf{q} = \mathbf{0}$, $\dot{\mathbf{q}} = \mathbf{0}$ *is stable if, for every positive number* ε, *there exists a second positive number* $\delta(\varepsilon)$ *with the property*

$$\rho\,[\mathbf{q}(t); \dot{\mathbf{q}}(t)] \leqslant \varepsilon \tag{2.14}$$

for any $t > 0$ and for any motion with initial conditions which satisfy

$$\rho_0 = \rho\left(\mathbf{q}_0; \dot{\mathbf{q}}_0\right) \leqslant \delta(\varepsilon) \tag{2.15}$$

Expressions of ρ which are suitable for the solution to mechanical problems are, for example

$$\rho = \sqrt{q_i q_i + \dot{q}_i \dot{q}_i} \tag{2.16}$$

$$\rho = \max |q_i| + \max |\dot{q}_i| \tag{2.17}$$

2.3　Lagrange-Dirichlet Theorem

In this section we demonstrate the Lagrange-Dirichlet theorem by following the presentation furnished in La Salle and Lefschetz (La Salle and Lefschetz, 1961). We notice from (2.10) that along the motion

$$\dot{H} = \frac{\partial H}{\partial q_i}\dot{q}_i + \frac{\partial H}{\partial p_i}\dot{p}_i = 0 \tag{2.18}$$

having made use of (2.11). Relation (2.18) shows that during motion the sum of the kinetic energy and of the potential energy remains constant. The theorem is therefore enunciated as follows

Theorem 2.2. (Lagrange-Dirichlet). *If the potential energy* $\Phi(\mathbf{q})$ *of a conservative system is positive definite in the neighbourhood* $\Omega : \|\mathbf{q}\| < D$ *of an equilibrium configuration, then the configuration of equilibrium is stable.*

Let us assume that the origin $\mathbf{q} = \mathbf{0}$ is a configuration of equilibrium and consider the motion arising from the perturbation (2.12) impressed at the instant $t = 0$. We choose the constant C, which appears in (2.10), so that $H(\mathbf{0}; \mathbf{0}) = 0$. Of the two terms $\Phi(\mathbf{q})$ and $K(\mathbf{p}; \mathbf{q})$ forming the Hamiltonian $H(\mathbf{p}; \mathbf{q})$, the kinetic energy is always positive definite. If *the potential energy has an isolated minimum by correspondence with the configuration of equilibrium* $\mathbf{q} = \mathbf{0}$, then $H(\mathbf{p}; \mathbf{q})$ is also positive definite, and as $\dot{H} = 0$ in conservative systems, the function H is a Liapunov function. Therefore in accordance with the Liapunov theorem in section 1.3, the position of equilibrium is stable.

The Lagrange-Dirichlet theorem on the stability of equilibrium does not give any information on the behaviour of a mechanical system when the potential energy corresponding to a configuration of equilibrium does not exhibit a minimum. There are two theorems regarding this, accredited to Liapunov and Chetayev, respectively, which are enunciated here without proof.

Theorem 2.3. *Theorem on instability (Liapunov).* *If the potential energy* $\Phi(\mathbf{q})$ *of a conservative system has an isolated maximum corresponding to a configuration of equilibrium, then the configuration of equilibrium is unstable.*

Theorem 2.4. *Theorem on instability (Chetayev).* *If the potential energy* $\Phi(\mathbf{q})$ *of a conservative system is a homogeneous function of the coordinates* $\mathbf{q}$ *and if, corresponding to a configuration of equilibrium,* $\Phi(\mathbf{q})$ *does not have a minimum, then the configuration of equilibrium is unstable.*

Example 2.5. Let us suppose that the potential energy of a system is of the type $\Phi(\mathbf{q}) = Aq_1q_2\ldots q_n$, with A positive real constant, and that $\mathbf{q} = \mathbf{0}$ is a configuration of equilibrium. The aim is to examine the type of equilibrium of such a configuration. According to the Chetayev theorem, we can assert that the configuration of equilibrium $\mathbf{q} = \mathbf{0}$ of the system is unstable.

3 Stability of Equilibrium of Mechanical Autonomous Systems and Postcritical Behaviour

3.1 Discrete Systems

We have furnished in sect. 1.3 the necessary and sometimes sufficient conditions for the Liapunov function to be positive definite and therefore for the equilibrium of a general system to be stable. The same arguments hold true for a mechanical system when, according to Lagrange theorem, the total potential energy $\Phi(\mathbf{q})$ is employed instead of $V(\mathbf{x})$. More in general, by assuming that $\Phi(\mathbf{q})$ is a continuous regular function we write the series expansion

$$
\begin{aligned}
\Phi(\mathbf{q}) &= \Phi_2(\mathbf{q}) + \Phi_3(\mathbf{q}) + \Phi_4(\mathbf{q}) + \ldots \\
&= \frac{1}{2}\left(\frac{\partial^2\Phi}{\partial q_i\partial q_j}\right)_{\mathbf{q}=\mathbf{0}} q_iq_j + \frac{1}{6}\left(\frac{\partial^3\Phi}{\partial q_i\partial q_j\partial q_h}\right)_{\mathbf{q}=\mathbf{0}} q_iq_jq_h \\
&\quad + \frac{1}{24}\left(\frac{\partial^4\Phi}{\partial q_i\partial q_j\partial q_h\partial q_k}\right)_{\mathbf{q}=\mathbf{0}} q_iq_jq_hq_k + \ldots \qquad (i,j,h,k=1,\ldots,n)
\end{aligned}
\tag{3.1}
$$

where derivatives are evaluated at the origin and the first order derivative term vanishes because of the equilibrium at that point. In alternative form, eq. (3.1) is written as

$$
\Phi(\mathbf{q}) = C_{ij}q_iq_j + C_{ijh}q_iq_jq_h + C_{ijhk}q_iq_jq_hq_k + \ldots (i,j,h,k=1,\ldots,n)
\tag{3.2}
$$

The matrix $[C_{ij}]$ is called *stiffness matrix* in the configuration of equilibrium. If the quadratic form $C_{ij}q_iq_j$ is positive definite, negative definite or indefinite, then it prevails on higher order terms and consequently the equilibrium is stable in the first case or unstable in the second and third case. If it is positive semidefinite (positive definite in all directions except in one direction $\bar{\mathbf{q}}$ where $\Phi_2(\bar{\mathbf{q}}) = 0$), then higher order terms must be analysed. The total potential energy is then positive definite if

$$
\begin{aligned}
\Phi_3(\bar{\mathbf{q}}) &= 0 \tag{3.3} \\
\Phi_4(\bar{\mathbf{q}}) &> 0 \tag{3.4}
\end{aligned}
$$

where (3.3) is a *necessary and sufficient condition* and (3.4) is a *necessary condition only*.

For (3.4) to be also sufficient, $\Phi_4(\bar{\mathbf{q}})$ must be *"sufficiently larger"* than zero. As an example, let us consider the function

$$\Phi(\mathbf{q}) = q_2^2 + q_1^2 q_2 + c q_1^4 \qquad (3.5)$$

of a two degrees of freedom system where c is a constant. Along the direction $q_2 = 0$ the quadratic and cubic terms vanish and besides $\Phi(\mathbf{q}) > 0$ if $c > 0$. This condition is only necessary but not sufficient. To show this, we observe that (3.5) can be rewritten in the form

$$\Phi(\mathbf{q}) = (q_2 + \frac{1}{2}q_1^2)^2 + (c - \frac{1}{4})q_1^4 \qquad (3.6)$$

Along the curve

$$q_2 = -\frac{1}{2}q_1^2 \qquad (3.7)$$

the energy is positive definite if $c > 1/4$ and negative definite if $c < 1/4$. Therefore the sufficient condition for stability is $c > 1/4$. The results of this analysis are represented in (Figure 8).

The variational equation

$$\delta\Phi_2(\mathbf{q})\delta(\mathbf{q}) = 0 \qquad (3.8)$$

is an eigenvalue problem which furnishes the *bifurcation points* along the *fundamental path* and one or more coincident or nearly coincident *buckling modes*. In addition, initial imperfections may by present in the structures as geometric imperfections, loads eccentricity and so on.

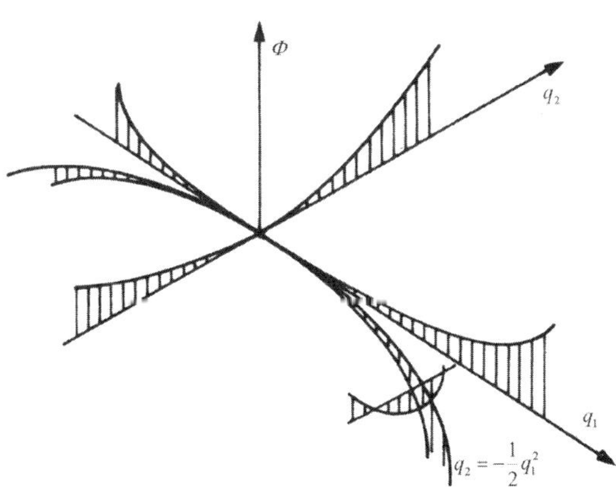

Figure 8. Representation of the function (3.6) for $c < 1/4$

The entire analysis regarding the solution to eq. (3.8), the evaluation of all equilibrium paths for perfect structures, the effect of the interaction between several simultaneous buckling modes in the presence or without initial imperfections, will be analysed in detail in dealing with continuous systems. Actually, the theory of buckling and postbuckling behaviour for discrete and continuous systems follows parallel directions and is based on perturbation theory.

4 Equilibrium of Mechanical Continuous Systems

4.1 Introduction

In this chapter we present a resumé of Koiter general theory of elastic stability (Koiter, 1945) in the form reformulated by Budiansky (Budiansky, 1974), focusing in particular our attention on interactive buckling.

The subject of interactive buckling has received a great deal of attention in the last decades, after Koiter and Skaloud (Koiter and Skaloud, 1963) have pointed out the danger of naive optimization without due regard to imperfection sensitivity. Van der Neut (Van der Neut, Springer Verlag, Berlin 1969) formulated a simple mechanical model to investigate the behaviour of a thin-walled column. Graves-Smith (Graves-Smith, 1967) investigated the full range behaviour of a locally buckled box column including the interaction of the overall mode as well as plasticity effects.

After these pioneering works there has been a spread of studies on this subject. With regard to framed structures we mention the works by Pignataro and Rizzi (Pignataro and Rizzi, 1983; Rizzi and Pignataro, 1982) who investigated symmetric and asymmetric structures. Interaction between two and three overall buckling modes in thin walled members was studied by Grimaldi and Pignataro (Grimaldi and Pignataro, 1979). Stiffened panels have been analysed by Tveergaard (Tvergaard, 1973) and successively by Koiter and Pignataro (Koiter and Pignataro, 1976). Axially stiffened cylindrical shells have been investigated by Byskov and Hutchinson (Byskov and Hutchinson, 1977). All these works furnish an analitycal solution to the problem.

More recently, in order to override mathematical difficulties, a semianalytical approach has been utilised by many researchers who have employed the finite strip method to study local-overall interaction in plated structures such as thin-walled members. Among these authors we mention Hancock (Hancock, 1981), Bradford and Hancock (Bradford and Hancock, 1984), Sridharan et al. (Benito and Sridharan, 1984-1985; Sridharan, 1983; Sridharan and Ali, 1985; Sridharan and Benito, 1984), Pignataro et al. (Pignataro and Luongo, 1987; Pignataro, Luongo et al., 1985). The problem of the interaction of infinitely many buckling modes has been finally studied by Byskov (Byskov, 1986) and Luongo and Pignataro (Luongo and Pignataro, 1988) who have confirmed the occurrence of localization phenomena previously pointed out by Tvergaard and Needleman (Tvergaard and Needleman, 1980) and Potier-Ferry (Potier-Ferry, 1984) after the experimental results obtained by Moxham (Moxham, 1971).

A few of the previously listed works make use of the direct equilibrium method while most of them utilize the Koiter / Budiansky perturbation theory to calculate the post-buckling equilibrium paths. For asymmetric structures, the analysis is usually carried out up to third order terms in order to evaluate the slope of the bifurcated paths. This is in general sufficient to describe the postcritical behaviour of the systems. If the system is symmetric, then the analysis is more involved since the evaluation of the curvature of the bifurcated paths is necessary. There are however a few cases in which the slope of the bifurcated paths, even if different from zero, is so small that the evaluation of the curvature is necessary.

4.2 Bifurcation and Postbuckling Analysis

Let $\Phi[\mathbf{u};\lambda]$ be the total potential energy of a hyperelastic body subjected to conservative loads, where $\mathbf{u}$ is the displacement field measured from the stress free configuration and λ a parameter governing the external force field. The equilibrium condition is obtained by requiring the functional $\Phi[\mathbf{u};\lambda]$ to be stationary with respect to all kinematically admissible displacement fields, that is

$$\Phi'[\mathbf{u};\lambda]\delta\mathbf{u} = 0 \qquad \forall\, \delta\mathbf{u} \tag{4.1}$$

where a prime denotes Fréchet differentiation with respect to $\mathbf{u}$. Eq. (4.1) furnishes all possible equilibrium paths $\mathbf{u} = \mathbf{u}(\lambda)$. In stability theory it is usually assumed that an equilibrium path $\mathbf{u}_0 = \mathbf{u}_0(\lambda)$ is known (*fundamental path*). Then, a second bifurcated equilibrium path is detected by writing

$$\mathbf{u}(\lambda) = \mathbf{u}_0(\lambda) + \mathbf{v}(\lambda) \tag{4.2}$$

$\mathbf{v}(\lambda)$ being an additional displacement measured from the fundamental configuration (Figure 9). By replacing eq. (4.2) into eq. (4.1) and performing the series expansion with respect to $\mathbf{v}$, we have

$$\Phi_0''\mathbf{v}\delta\mathbf{u} + \frac{1}{2}\Phi_0'''\mathbf{v}^2\delta\mathbf{u} + \ldots = 0 \tag{4.3}$$

where $\Phi_0'' \equiv \Phi''[\mathbf{u}_0(\lambda);\lambda], \ldots$ and use has been made of eq. (4.1). From eq. (4.3), by expanding each term with respect to λ starting from the bifurcation value $\lambda = \lambda_c$ we obtain

$$[\Phi_c'' + (\lambda - \lambda_c)\,\dot{\Phi}''_c + \frac{1}{2}(\lambda - \lambda_c)^2\ddot{\Phi}_c'' + \ldots]\mathbf{v}\delta\mathbf{u} + \frac{1}{2}[\Phi_c''' + (\lambda - \lambda_c)\dot{\Phi}_c''' + \ldots]\mathbf{v}^2\delta\mathbf{u}$$
$$+ \frac{1}{6}[\Phi_c^{IV} + \ldots]\mathbf{v}^3\delta\mathbf{u} = 0 \tag{4.4}$$

In eq. (4.4) $\dot{\Phi}''_c \equiv (d/d\lambda)\Phi_0''|_{\lambda=\lambda_c}, \ldots$ etc. It is now convenient to express the dependence of $\mathbf{v}$ on λ through a parameter ξ

$$\lambda = \lambda(\xi) \qquad \mathbf{v} = \mathbf{v}(\xi) \tag{4.5}$$

Then under the assumption of regularity and keeping in mind that we are looking for an asymptotic solution to our problem, we write eqs. (4.5) as series expansions from $\xi = 0$

$$\lambda = \lambda_c + \lambda_1\xi + \frac{1}{2}\lambda_2\xi^2 + \ldots$$
$$\mathbf{v} = \mathbf{v}_1\xi + \frac{1}{2}\mathbf{v}_2\xi^2 + \frac{1}{6}\mathbf{v}_3\xi^3 + \ldots \tag{4.6}$$

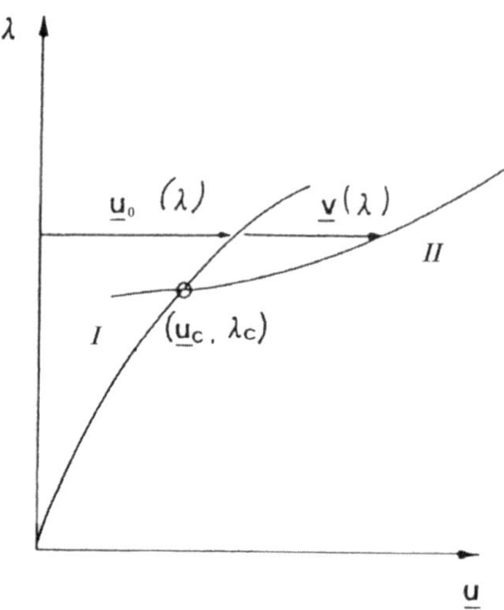

Figure 9. Equilibrium paths in a structure

where $\lambda(0) = \lambda_c$ and $\mathbf{v}(0) = \mathbf{0}$. Besides $\lambda_n = d^n\lambda/d\xi^n|_{\lambda=\lambda_c}$ and $v_n = d^n v/d\xi^n|_{\lambda=\lambda_c}$. By replacing (4.6) into (4.4) and collecting terms with equal power of ξ we have

$$
\xi\,\{\,\Phi_c''\mathbf{v}_1\,\}\,\delta\mathbf{u} + \frac{1}{2}\xi^2\,\{\,\Phi_c''\mathbf{v}_2 + 2\lambda_1\dot\Phi_c''\mathbf{v}_1 + \Phi_c'''\mathbf{v}_1^2\,\}\,\delta u + \frac{1}{6}\xi^3\,\{\,\Phi_c''\mathbf{v}_3 + 3\lambda_1\dot\Phi_c''\mathbf{v}_2
$$
$$
+ 3\lambda_2\dot\Phi_c''\mathbf{v}_1 + 3\lambda_1^2\ddot\Phi_c''\mathbf{v}_1 + 3\Phi_c'''\mathbf{v}_1\mathbf{v}_2 + 3\lambda_1\dot\Phi_c'''\mathbf{v}_1^2 + \Phi_c^{IV}\mathbf{v}_1^3\,\}\,\delta\mathbf{u} + \ldots = 0 \tag{4.7}
$$

whence the first, second and third order perturbation equations are obtained by equating separately to zero terms with equal power of ξ

$$
\Phi_c''\mathbf{v}_1\delta\mathbf{u} = 0
$$
$$
\Phi_c''\mathbf{v}_2\delta\mathbf{u} = -\{\,2\lambda_1\dot\Phi_c''\mathbf{v}_1 + \Phi_c'''\mathbf{v}_1^2\,\}\,\delta\mathbf{u}
$$
$$
\Phi_c''\mathbf{v}_3\delta\mathbf{u} = -3\{\,\lambda_1\dot\Phi_c''\mathbf{v}_2 + \lambda_2\dot\Phi_c''\mathbf{v}_1 + \lambda_1^2\ddot\Phi_c''\mathbf{v}_1 \tag{4.8}
$$
$$
+ \Phi_c'''\mathbf{v}_1\mathbf{v}_2 + \lambda_1\dot\Phi_c'''\mathbf{v}_1^2 + \frac{1}{3}\Phi_c^{IV}\mathbf{v}_1^3\,\}\,\delta\mathbf{u}
$$

By denoting with $(\hat{\ })$ partial differentiation with respect to λ eqs. (4.8) change into

$$\Phi_c'' \mathbf{v}_1 \delta\mathbf{u} - 0$$

$$\Phi_c'' \mathbf{v}_2 \delta\mathbf{u} = -\left\{ \Phi_c''' \mathbf{v}_1^2 + 2\lambda_1 \Phi_c''' \hat{\mathbf{u}}_c \mathbf{v}_1 + 2\lambda_1 \hat{\Phi}_c'' \mathbf{v}_1 \right\} \delta\mathbf{u}$$

$$\begin{aligned}
\Phi_c'' \mathbf{v}_3 \delta\mathbf{u} = -3 \Big\{ &\Phi_c''' \mathbf{v}_1 \mathbf{v}_2 + \frac{1}{3}\Phi_c^{IV} \mathbf{v}_1^3 \\
&+ \lambda_1 \left[\Phi_c''' \hat{\mathbf{u}}_c \mathbf{v}_2 + \hat{\Phi}_c'' \mathbf{v}_2 + \Phi_c^{IV} \hat{\mathbf{u}}_c \mathbf{v}_1^2 + \hat{\Phi}_c''' \mathbf{v}_1^2 \right] \\
&+ \lambda_1^2 \left[\Phi_c^{IV} \hat{\mathbf{u}}_c^2 \mathbf{v}_1 + 2\hat{\Phi}_c''' \hat{\mathbf{u}}_c \mathbf{v}_1 + \Phi_c''' \hat{\hat{\mathbf{u}}}_c \mathbf{v}_1 + \hat{\hat{\Phi}}_c'' \mathbf{v}_1 \right] \\
&+ \lambda_2 \left[\Phi_c''' \hat{\mathbf{u}}_c \mathbf{v}_1 + \hat{\Phi}_c'' \mathbf{v}_1 \right] \Big\} \delta\mathbf{u}
\end{aligned} \tag{4.9}$$

If Φ is bilinear in $\mathbf{u}$ and λ and besides $\mathbf{u}$ is a linear function of λ along the fundamental path as in most practical problems, it results $\hat{\Phi}_c'' = \hat{\Phi}_c''' = \hat{\hat{\Phi}}_c'' = \hat{\hat{u}}_c = 0$ and eqs. (4.9) simplify into

$$\Phi_c'' \mathbf{v}_1 \delta\mathbf{u} = 0$$

$$\Phi_c'' \mathbf{v}_2 \delta\mathbf{u} = -\left\{ \Phi_c''' \mathbf{v}_1^2 + 2\lambda_1 \Phi_c''' \hat{\mathbf{u}}_c \mathbf{v}_1 \right\} \delta\mathbf{u}$$

$$\begin{aligned}
\Phi_c'' \mathbf{v}_3 \delta\mathbf{u} = -3 \Big\{ &\Phi_c''' \mathbf{v}_1 \mathbf{v}_2 + \frac{1}{3}\Phi_c^{IV} \mathbf{v}_1^3 + \lambda_1 \left[\Phi_c''' \hat{\mathbf{u}}_c \mathbf{v}_2 + \Phi_c^{IV} \hat{\mathbf{u}}_c \mathbf{v}_1^2 \right] \\
&\lambda_1^2 \Phi_c^{IV} \hat{\mathbf{u}}_c^2 \mathbf{v}_1 + \lambda_2 \Phi_c''' \hat{\mathbf{u}}_c \mathbf{v}_1 \Big\} \delta\mathbf{u}
\end{aligned} \tag{4.10}$$

Eq. $(4.8)_1$ or $(4.9)_1$ is an eigenvalue problem whose solution furnishes the critical load λ_c and the buckling mode $\mathbf{v}_1$. Suppose that the solution to this problem has several linearly independent eigenmodes $\mathbf{v}_{1i}$ $(i = 1, \ldots, m)$ all associated with the lowest eigenvalue λ_c. Then the most general solution to eq. $(4.8)_1$ can be expressed as a linear combination

$$\mathbf{v}_1 = \nu_i \mathbf{v}_{1i} \qquad (i = 1, 2, \ldots, m) \tag{4.11}$$

where repeated indices denote summation from 1 to m and ν_i are arbitrary parameters. Without loss of generality these modes can be orthonormalised according to

$$\Pi_2'' \mathbf{v}_{1i} \mathbf{v}_{1j} = \delta_{ij} \qquad (i, j = 1, 2, \ldots, m) \tag{4.12}$$

where δ_{ij} is the Kronecker delta and Π_2 collects all quadratic terms of the series expansion around the stress-free configuration of the elastic energy. By requesting $\Pi_2'' \mathbf{v}_1^2 = 1$, the condition

$$\nu_i \nu_i = 1 \tag{4.13}$$

follows.

To evaluate the second coefficient $\mathbf{v}_2$ of the series expansion $(4.6)_2$ we use the differential equation

$$\Phi_c'' \mathbf{v}_2 \delta\mathbf{u} = -\left\{ 2\lambda_1 \dot{\Phi}_c'' \nu_i \mathbf{v}_{1i} + \Phi_c''' (\nu_i \mathbf{v}_{1i})^2 \right\} \delta\mathbf{u} \tag{4.14}$$

obtained from $(4.8)_2$ by replacing $\mathbf{v}_1$ with $\nu_i\mathbf{v}_{1i}$. By assuming $\delta\mathbf{u} = \mathbf{v}_{11}, \mathbf{v}_{12}, \ldots, \mathbf{v}_{1m}$ successively, eq. (4.14) in conjunction with eq. $(4.8)_1$ yields a set of m equations of the type (Fredholm orthogonality conditions)

$$A_{ijk}\nu_i\nu_j + \lambda_1 B_{ik}\nu_i = 0 \qquad (i,j,k = 1,\ldots,m) \tag{4.15}$$

where

$$A_{ijk} = \Phi_c'''\mathbf{v}_{1i}\mathbf{v}_{1j}\mathbf{v}_{1k}, \qquad B_{ik} = 2\dot{\Phi}_c''\mathbf{v}_{1i}\mathbf{v}_{1k} \tag{4.16}$$

Eqs. (4.15) together with eq. (4.13) permit the evaluation of the $m+1$ coefficients ν_i and λ_1. Since eqs. (4.15) are nonlinear it is shown that, unless all the A_{ijk}'s vanish, there are at the most $2^m - 1$ essentially different real solutions and at least one real solution each one describing a bifurcated path (Van der Waerden, 1950). In the following we shall distinguish the case in which all the A_{ijk} vanish ($\lambda_1=0$, symmetric postbuckling behaviour) from the case in which one at least of these coefficients is different from zero ($\lambda_1 \neq 0$, asymmetric behaviour).

4.3 Asymmetric Postcritical Behaviour

The general integral of the differential equation (4.14) corresponding to any of the r solutions of eqs. (4.15) can be written as

$$\mathbf{v}_2 = \beta_i\mathbf{v}_{1i} + \mathbf{v}_{2p} \tag{4.17}$$

where $\mathbf{v}_{2p}$ denotes a particular integral and $\beta_i\mathbf{v}_{1i}$ is the general solution of the homogeneous equation with β_i arbitrary constants. If the orthogonality condition

$$\Pi_2''\mathbf{v}_1\mathbf{v}_2 = 0 \tag{4.18}$$

is imposed, then by using (4.11), (4.12), (4.17) the following condition on the coefficients β_i is obtained

$$\nu_i(\Pi_2''\mathbf{v}_{1i}\mathbf{v}_{2p} + \beta_i) = 0 \tag{4.19}$$

In order to evaluate the constants β_i and the second load rate coefficient λ_2 corresponding to each of the r bifurcated paths we make use of the equation

$$\begin{aligned}
\Phi_c''\mathbf{v}_3\delta\mathbf{u} = -3\Big\{ &\lambda_1\dot{\Phi}_c''\left(\mathbf{v}_{2p} + \beta_i\mathbf{v}_{1i}\right) + \lambda_2\dot{\Phi}_c''\nu_i\mathbf{v}_{1i} + \lambda_1^2\ddot{\Phi}_c''\nu_i\mathbf{v}_{1i} \\
&+ \Phi_c'''(\nu_i\mathbf{v}_{1i}\mathbf{v}_{2p} + \nu_i\beta_j\mathbf{v}_{1i}\mathbf{v}_{1j}) + \lambda_1\dot{\Phi}_c'''(\nu_i\mathbf{v}_{1i})^2 \\
&+ \frac{1}{3}\Phi_c^{IV}(\nu_i\mathbf{v}_{1i})^3 \Big\}\delta\mathbf{u}
\end{aligned} \tag{4.20}$$

obtained by replacing eqs. (4.11), (4.17) in $(4.8)_3$. Note that there are r equations (4.20) each one corresponding to a particular solution ν_i, λ_1 of eqs. (4.15). By successively identifying $\delta\mathbf{u}$ with $\mathbf{v}_{11},\ldots, \mathbf{v}_{1m}$ and by imposing the orthogonality condition on the right hand member of each of these equations we obtain

$$(2A_{ijk}\nu_j + \lambda_1 B_{ik})\beta_i + \lambda_2 B_{ik}\nu_i = \begin{aligned}&2\breve{A}_{ijkl}\nu_i\nu_j\nu_l\\ &- \lambda_1(2B_{ijk}\nu_i\nu_j + B_{pk})\\ &- \lambda_1^2 C_{ik}\nu_i - 2A_{pik}\nu_i\\ &(i,\,j,\,k,\,l = 1,\ldots,m)\end{aligned} \tag{4.21}$$

where A_{ijk} and B_{ik} are given by (4.16) and

$$\begin{aligned}\breve{A}_{ijkl} &= \frac{1}{3}\Phi_c^{IV}\mathbf{v}_{1i}\mathbf{v}_{1j}\mathbf{v}_{1k}\mathbf{v}_{1l}\\ B_{ijk} &= \dot{\Phi}_c'''\mathbf{v}_{1i}\mathbf{v}_{1j}\mathbf{v}_{1k}\\ C_{ik} &= 2\ddot{\Phi}_c''\mathbf{v}_{1i}\mathbf{v}_{1k}\end{aligned} \tag{4.22}$$

Besides

$$\begin{aligned}A_{pik} &= \Phi_c'''\mathbf{v}_{2p}\mathbf{v}_{1i}\mathbf{v}_{1k}\\ B_{pk} &= 2\dot{\Phi}_c''\mathbf{v}_{2p}\mathbf{v}_{1k}\end{aligned} \tag{4.23}$$

Eqs. (4.19), (4.21) are r linear nonhomogeneous systems each one containing $m+1$ equations (m orthogonality conditions plus a constraint equation) and $m+1$ unknowns β_i, λ_2. For each bifurcated path the coefficients $\mathbf{v}_2$ and λ_2 of the series expansions (4.6) can thus be evaluated.

4.4 Symmetric Postcritical Behaviour

Symmetric postcritical behaviour arises in the particular case in which all coefficients A_{ijk}'s previously examined vanish. In this situation from eqs. (4.15) the solution $\lambda_1 = 0$ is obtained and the coefficients ν_i's, which are undetermined at this level, must be evaluated from the third order perturbation equation. Eq. (4.14) admits now the particular solution

$$\mathbf{v}_{2p} = \nu_i\nu_j\mathbf{v}_{2ij} \tag{4.24}$$

where the $\mathbf{v}_{2ij}$'s satisfy the equations

$$\Phi_c''\mathbf{v}_{2ij}\delta\mathbf{u} = -\Phi_c'''\mathbf{v}_{1i}\mathbf{v}_{1j}\delta\mathbf{u} \tag{4.25}$$

Also, due to the arbitrariness of the ν_i's, the orthogonality condition (4.19) furnishes all β_i's as function of ν_i's

$$\beta_i = p_{ijk}\nu_j\nu_k \tag{4.26}$$

where

$$p_{ijk} = \Pi_2''\mathbf{v}_{1i}\mathbf{v}_{2jk} \tag{4.27}$$

By replacing (4.24) into eq. (4.20) and accounting for $\lambda_1 = 0$, we have the Fredholm orthogonality conditions

$$2A_{ijkl}\nu_i\nu_j\nu_l + \lambda_2 B_{ik}\nu_i = 0 \qquad (i,\, j,\, k,\, l = 1, \ldots, m) \tag{4.28}$$

where the B_{ik}'s are given by $(4.16)_2$ and

$$A_{ijkl} = \Phi_c''' \mathbf{v}_{1i}\mathbf{v}_{1l}\mathbf{v}_{2jk} + \Phi_c^{IV}\mathbf{v}_{1i}\mathbf{v}_{1j}\mathbf{v}_{1k}\mathbf{v}_{1l} \tag{4.29}$$

Note that the solution to eqs. (4.28) is unaffected by the β_i's because of the assumption $A_{ijk} = 0$.

Eqs. (4.28) together with condition (4.13) are a set of $m + 1$ equations in the $m + 1$ unknowns ν_i's and λ_2 and therefore, according to Bézout (Van der Waerden, 1950), they furnish at the most $3^m{-}1$ essentially different real solutions and at least one real solution. By correspondence with each set ν_i, the corresponding β_i's are determined from eq. (4.26) and consequently $\mathbf{v}_2$ from eqs. (4.17), (4.24) and (4.25).

If the second hand member in eq. (4.25) vanishes for any $\delta\mathbf{u}$, then $\mathbf{v}_{2ij} = \mathbf{0}$. Consequently $\beta_i = 0$ from eqs. (4.26), (4.27), $\mathbf{v}_{2p} = \mathbf{0}$ from eq. (4.24) and $\mathbf{v}_2 = \mathbf{0}$ follows from eq. (4.17).

4.5 Single Buckling Mode

If a single buckling mode $\mathbf{v}_1$ occurs for $\lambda = \lambda_c$, explicit expressions for the first and second load rate λ_1 and λ_2 are obtained by solving the Fredholm orthogonality conditions relative to eqs. $(4.8)_2$ and $(4.8)_3$, respectively. It is found that

$$\lambda_1 = -\frac{1}{2}\frac{\Phi_c'''\mathbf{v}_1^3}{\dot\Phi_c''\mathbf{v}_1^2} \tag{4.30}$$

$$\lambda_2 = -\frac{\frac{1}{3}\Phi_c^{IV}\mathbf{v}_1^4 + \Phi_c'''\mathbf{v}_1^2\mathbf{v}_2 + \lambda_1(\dot\Phi_c''\mathbf{v}_1\mathbf{v}_2 + \dot\Phi_c'''\mathbf{v}_1^3 + \lambda_1\ddot\Phi_c''\mathbf{v}_1^2)}{\dot\Phi_c''\mathbf{v}_1^2} \tag{4.31}$$

which for $\lambda_1 = 0$ reduces to

$$\lambda_2 = -\frac{\frac{1}{3}\Phi_c^{IV}\mathbf{v}_1^4 + \Phi_c'''\mathbf{v}_1^2\mathbf{v}_2}{\dot\Phi_c''\mathbf{v}_1^2} \tag{4.32}$$

5 Initial Imperfections

5.1 General Theory

If the structure under analysis is not perfect, in that it contains a displacement $\bar{\mathbf{u}}$ before the application of load, its potential energy functional $\Phi[\mathbf{u};\lambda]$ is modified as follows

$$\bar\Phi = \Phi\left[\mathbf{u};\lambda\right] + \Psi\left[\mathbf{u},\bar{\mathbf{u}};\lambda\right] \tag{5.1}$$

where $\Psi[\mathbf{u},\mathbf{0};\lambda] = 0$ for any $\mathbf{u}$. The equilibrium eq. (4.1) reads then

$$\Phi'[\mathbf{u};\lambda]\,\delta\mathbf{u} + \Psi'[\mathbf{u},\bar{\mathbf{u}};\lambda]\,\delta\mathbf{u} = 0 \tag{5.2}$$

Under the assumption that $\bar{\mathbf{u}}$ is small, let us take the series expansion of (5.2) in terms of $\bar{\mathbf{u}}$ from $\bar{\mathbf{u}} = \mathbf{0}$ by retaining only linear terms

$$\Phi'[\mathbf{u};\lambda]\,\delta\mathbf{u} + \tilde{\Psi}'[\mathbf{u},\mathbf{0};\lambda]\,\bar{\mathbf{u}}\delta\mathbf{u} = 0 \tag{5.3}$$

In eq. (5.3) the symbol $(\sim)$ denotes differentiation with respect to $\bar{\mathbf{u}}$. Besides use has been made of the property that Ψ vanishes for $\bar{\mathbf{u}} = \mathbf{0}$.

Let

$$\mathbf{u}(\lambda) = \mathbf{u}_0(\lambda) + \bar{\mathbf{v}}(\lambda) \tag{5.4}$$

be an equilibrium path where $\mathbf{u}_0(\lambda)$ is the fundamental path of the perfect structure and $\bar{\mathbf{v}}(\lambda)$ an additional displacement measured from it. By replacing (5.4) into (5.3) and expanding in terms of $\bar{\mathbf{v}}$ we have

$$\left\{ \Phi_0''\,\bar{\mathbf{v}} + \frac{1}{2}\Phi_0'''\,\bar{\mathbf{v}}^2 + \frac{1}{6}\Phi_0^{IV}\,\bar{\mathbf{v}}^3 + \ldots + \tilde{\Psi}_0'\bar{\mathbf{u}} + \tilde{\Psi}_0''\,\bar{\mathbf{u}}\,\bar{\mathbf{v}} + \ldots \right\}\delta\mathbf{u} = 0 \tag{5.5}$$

where $\tilde{\Psi}_0' = \tilde{\Psi}'[\mathbf{u}_0,\mathbf{0};\lambda],\ldots$ Further expansion of eq. (5.5) in terms of λ about the critical load $\lambda = \lambda_c$ gives

$$[\Phi_c'' + (\lambda-\lambda_c)\dot{\Phi}_c'' + \frac{1}{2}(\lambda-\lambda_c)^2\ddot{\Phi}_c'' + \ldots]\,\bar{\mathbf{v}}\delta\mathbf{u}$$

$$+ \frac{1}{2}[\Phi_c''' + (\lambda-\lambda_c)\dot{\Phi}_c''' + \ldots]\,\bar{\mathbf{v}}^2\delta\mathbf{u} + \frac{1}{6}[\Phi_c^{IV} + \ldots]\,\bar{\mathbf{v}}^3\delta\mathbf{u} + \ldots \tag{5.6}$$

$$+ \left[\tilde{\Psi}_c' + (\lambda-\lambda_c)\dot{\tilde{\Psi}}_c' + \ldots\right]\bar{\mathbf{u}}\delta\mathbf{u} + \left[\tilde{\Psi}_c'' + \ldots\right]\bar{\mathbf{u}}\,\bar{\mathbf{v}}\delta\mathbf{u} + \ldots = 0$$

being $\dot{\tilde{\Psi}}_c' = (d/d\lambda)\,\tilde{\Psi}_0'|_{\lambda=\lambda_c},\ldots$

It is now convenient to choose an initial imperfection in the form

$$\bar{\mathbf{u}} = \bar{\xi}\mathbf{u}^* = \alpha\xi^\gamma\mathbf{u}^* \tag{5.7}$$

where $\bar{\xi}$ is the imperfection amplitude and $\mathbf{u}$ gives the shape of the imperfection which is normalised according to $\Pi_2''\mathbf{u}^{*2} = 1$. Besides α is a scalar parameter and the exponent $\gamma > 0$ will be chosen to suit our convenience. Under suitable regularity condition we write

$$\lambda = \lambda_c + \bar{\lambda}_1\xi + \frac{1}{2}\bar{\lambda}_2\xi^2 + \ldots$$

$$\bar{\mathbf{v}} = \bar{\mathbf{v}}_1\xi + \frac{1}{2}\bar{\mathbf{v}}_2\xi^2 + \ldots \tag{5.8}$$

Then replacing eqs. (5.8) into (5.6) and collecting terms with equal powers of ξ one gets

$$\xi\left\{\Phi_c''\bar{\mathbf{v}}_1\right\}\delta\mathbf{u} + \frac{1}{2}\xi^2\left\{\Phi_c''\bar{\mathbf{v}}_2 + 2\bar{\lambda}_1\dot{\Phi}_c''\bar{\mathbf{v}}_1 + \Phi_c'''\bar{\mathbf{v}}_1^2\right\}\delta\mathbf{u}$$

$$+\frac{1}{6}\xi^3\left\{\Phi_c''\bar{\mathbf{v}}_3 + 3\bar{\lambda}_1\dot{\Phi}_c''\bar{\mathbf{v}}_2 + 3\bar{\lambda}_2\dot{\Phi}_c''\bar{\mathbf{v}}_1 + 3\bar{\lambda}_1^2\ddot{\Phi}_c''\bar{\mathbf{v}}_1 + 3\Phi_c'''\bar{\mathbf{v}}_1\bar{\mathbf{v}}_2\right.$$

$$\left.+ 3\bar{\lambda}_1\dot{\Phi}_c'''\bar{\mathbf{v}}_1^2 + \Phi_c^{IV}\bar{\mathbf{v}}_1^3\right\}\delta\mathbf{u} + \ldots + \alpha\xi^\gamma\tilde{\Psi}_c'\mathbf{u}^*\delta\mathbf{u}$$

$$+ \alpha\xi^{\gamma+1}(\bar{\lambda}_1\dot{\tilde{\Psi}}_c'\mathbf{u}^*\delta\mathbf{u} + \tilde{\Psi}_c''\mathbf{u}^*\bar{\mathbf{v}}_1\delta\mathbf{u}) + \ldots = 0 \tag{5.9}$$

5.2 Asymmetric Postcritical Behaviour

Let us consider first the case of asymmetric postcritical behaviour. Assume $\gamma = 2$ in eq. (5.9) and write the first, the second and the third order perturbation equations by equating to zero terms with ξ, ξ^2 and ξ^3 as a factor, respectively. We have

$$\Phi_c''\bar{\mathbf{v}}_1\delta\mathbf{u} = 0$$

$$\Phi_c''\bar{\mathbf{v}}_2\delta\mathbf{u} = -\left\{2\,\bar{\lambda}_1\dot{\Phi}_c''\bar{\mathbf{v}}_1 + \Phi_c'''\bar{\mathbf{v}}_1^2 + 2\alpha\tilde{\Psi}_c'\mathbf{u}^*\right\}\delta\mathbf{u}$$

$$\Phi_c''\bar{\mathbf{v}}_3\delta\mathbf{u} = -3\left\{\bar{\lambda}_1\dot{\Phi}_c''\bar{\mathbf{v}}_2 + \bar{\lambda}_2\dot{\Phi}_c''\bar{\mathbf{v}}_1 + \bar{\lambda}_1^2\ddot{\Phi}_c''\bar{\mathbf{v}}_1\right.$$

$$+ \Phi_c'''\bar{\mathbf{v}}_1\bar{\mathbf{v}}_2 + \bar{\lambda}_1\dot{\Phi}_c'''\bar{\mathbf{v}}_1^2 + \frac{1}{3}\Phi_c^{IV}\bar{\mathbf{v}}_1^3 \tag{5.10}$$

$$\left.+ 2\alpha\bar{\lambda}_1\dot{\tilde{\Psi}}_c'\mathbf{u}^* + 2\alpha\tilde{\Psi}_c''\mathbf{u}^*\bar{\mathbf{v}}_1\right\}\delta\mathbf{u}$$

Eq. $(5.10)_1$ is the same eigenvalue problem as eq. $(4.8)_1$ and therefore furnishes the same eigenvalue λ_c and eigenvectors. In case of multiple buckling modes the general solution is expressed by

$$\bar{\mathbf{v}}_1 = \bar{\nu}_i\mathbf{v}_{1i} \qquad (i = 1, 2, \ldots, m) \tag{5.11}$$

under the condition, analogous to eq. (4.13),

$$\bar{\nu}_i\bar{\nu}_i = 1 \tag{5.12}$$

We assume that the initial imperfection shape vector is a linear combination of the critical modes

$$\mathbf{u}^* = \eta_i\mathbf{v}_{1i} \tag{5.13}$$

where the coefficient η_i are subjected to the condition $\eta_i\eta_i = 1$.

By substituting (5.11) and (5.13) in $(5.10)_2$ we have

$$\Phi_c''\bar{\mathbf{v}}_2\delta\mathbf{u} = -\left\{2\,\bar{\lambda}_1\dot{\Phi}_c''\bar{\nu}_i\mathbf{v}_{1i} + \Phi_c'''(\bar{\nu}_i\mathbf{v}_{1i})^2 + 2\alpha\tilde{\Psi}_c'\eta_i\mathbf{v}_{1i}\right\}\delta\mathbf{u} \tag{5.14}$$

and successively, by identifying $\delta\mathbf{u}$ with $\mathbf{v}_{11},\ldots,\mathbf{v}_{1m}$ and imposing Fredholm orthogonality conditions we obtain the m equations

$$A_{ijk}\bar{\nu}_i\bar{\nu}_j + \bar{\lambda}_1 B_{ik}\bar{\nu}_i + \alpha D_{ik}\eta_i = 0 \qquad (i, j, k = 1, \ldots, m) \tag{5.15}$$

where A_{ijk} and B_{ik} are given by (4.16) and

$$D_{ik} = 2\tilde{\Psi}'_c \mathbf{v}_{1i}\mathbf{v}_{1k} \tag{5.16}$$

Eqs. (5.15), (5.12) are $m+1$ non linear nonhomogenous equations in the unknowns $\bar{\nu}_i, \bar{\lambda}_1$. Their solution depends on the coefficients η_i and on the amplitude parameter α. The general solution to the differential equation (5.14) is

$$\bar{\mathbf{v}}_2 = \bar{\beta}_i \mathbf{v}_{1i} + \bar{\mathbf{v}}_{2p} \tag{5.17}$$

where $\bar{\mathbf{v}}_{2p}$ is a particular integral and the $\bar{\beta}_i$'s are arbitrary constants. By requiring $\Pi''_2 \bar{\mathbf{v}}_1 \bar{\mathbf{v}}_2 = 0$ the following condition is obtained

$$\bar{\nu}_i (\Pi''_2 \bar{\mathbf{v}}_{1i}\bar{\mathbf{v}}_{2p} + \bar{\beta}_i) = 0 \tag{5.18}$$

The Fredholm orthogonality condition relative to the third order perturbation equation $(5.10)_3$ leads to

$$\begin{aligned}
(2A_{ijk}\bar{\nu}_j + \bar{\lambda}_1 B_{ik})\bar{\beta}_i + \bar{\lambda}_2 B_{ik}\bar{\nu}_i = & -2\breve{A}_{ijkl}\bar{\nu}_i\bar{\nu}_j\bar{\nu}_l - \bar{\lambda}_1(2B_{ijk}\bar{\nu}_i\bar{\nu}_j + \bar{B}_{pk}) \\
& - \bar{\lambda}_1^2 C_{ik}\bar{\nu}_i - 2\bar{A}_{pik}\bar{\nu}_i - 2\alpha\bar{\lambda}_1 E_{ik}\eta_i \\
& - 2\alpha D_{ijk}\eta_i\bar{\nu}_j = 0 \quad (i,j,k,l = 1,\dots,m)
\end{aligned} \tag{5.19}$$

where

$$\begin{aligned}
\bar{A}_{pik} &= \Phi'''_c\, \bar{\mathbf{v}}_{2p}\mathbf{v}_{1i}\mathbf{v}_{1k} \\
\bar{B}_{pk} &= 2\Phi'''_c\, \bar{\mathbf{v}}_{2p}\mathbf{v}_{1k} \\
E_{ik} &= 2\dot{\tilde{\Psi}}'_c \mathbf{v}_{1i}\mathbf{v}_{1k} \\
D_{ijk} &= 2\tilde{\Psi}''_c \mathbf{v}_{1i}\mathbf{v}_{1j}\mathbf{v}_{1k}
\end{aligned} \tag{5.20}$$

the other coefficients being defined by eqs. (4.16), (4.22). Eqs. (5.18), (5.19) are a linear system of $m + 1$ equations in the $m + 1$ unknowns $\bar{\beta}_i, \bar{\lambda}_2$, which depend on η_i and α. After evaluating $\bar{\nu}_i, \bar{\beta}_i, \bar{\lambda}_1, \bar{\lambda}_2$, for given initial imperfections parameters $\bar{\xi}_i, \eta_i$, equilibrium paths (5.8) can be determined by using equations (5.11) and (5.17) and remembering that $\alpha = \bar{\xi}/\xi^2$. In most practical problems it is sufficient to determine only the first order terms $\bar{\lambda}_1, \bar{\mathbf{v}}_1$ of the asymptotic expansions (5.8). In this case the solution to eqs. (5.19) and the evaluation of the particular integral $\bar{\mathbf{v}}_{2p}$ are not required. Thus, to solve eqs. (5.15) it is convenient to multiply all terms by ξ^2 and remember that $\lambda - \lambda_c = \bar{\lambda}_1\xi$, $\alpha\xi^2 = \bar{\xi}$. The following equations are obtained

$$(\lambda - \lambda_c)B_{ik}a_i + A_{ijk}a_i a_j + \bar{\xi}D_{ik}\eta_i = 0 \quad (i,j,k = 1,\dots,m) \tag{5.21}$$

where $a_i = \bar{\nu}_i\xi$ are the amplitudes of the buckling modes. The m nonlinear equations (5.21) can be solved perhaps numerically to determine the equilibrium paths.

Some experimental (Roorda, 1965) and theoretical results (Koiter, 1967) are available regarding the post-buckling analysis of a simple two-bar frame hinged at the ends.

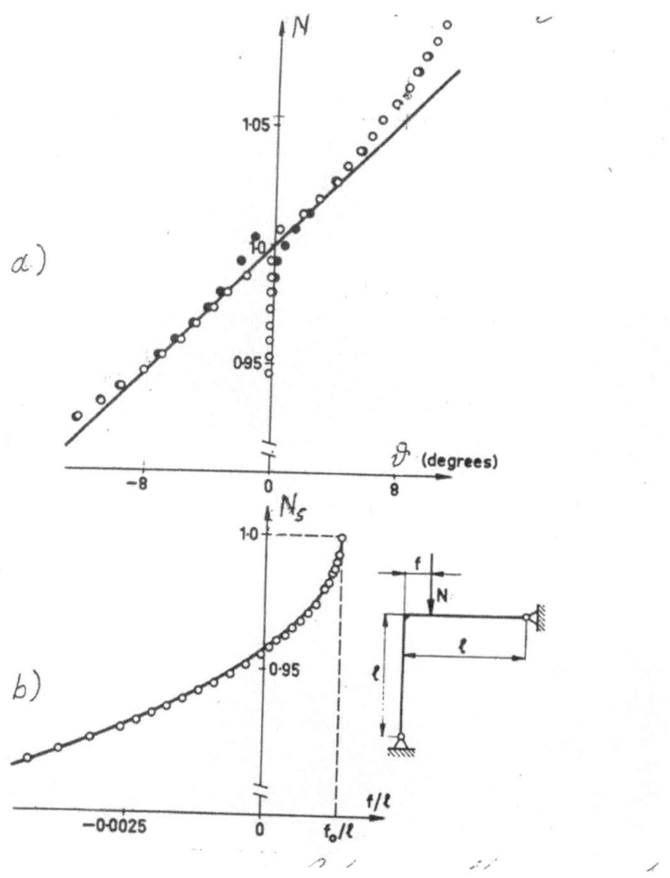

Figure 10. Comparison between theory and Roorda's test; a) post-buckling curves, b) effect of load eccentricities

These results are summarized in the following diagrams where small circles come from experimental results and the heavy lines from the general theory carried out up to the evaluation of $\lambda_1, \mathbf{v}_1$ (Figure 10a)). Figure 10b) shows the dependence of the snapping loads N_s on initial imperfections. One can see that the agreement between experiments and theory is very good.

5.3 Symmetric Postcritical Behaviour

In this case the easiest procedure is to take $\gamma = 3$ in eq. (5.7) and repeat all previous steps. The term of the initial imperfection in the second perturbation equation $(5.10)_2$ will now be shifted to the third order equation $(5.10)_3$. Second order equation is identical to that of the perfect structure, thus giving $\bar{\lambda}_1 = \lambda_1 = 0$ and

$$\bar{\mathbf{v}}_2 = \bar{\beta}_i \mathbf{v}_{1i} + \bar{\nu}_i \bar{\nu}_j \mathbf{v}_{2ij} \tag{5.22}$$

where, as in eq. (4.26),

$$\bar{\beta}_i = p_{ijk}\bar{\nu}_j\bar{\nu}_k; \qquad p_{ijk} = -\Pi_2''\mathbf{v}_{1i}\mathbf{v}_{2jk} \tag{5.23}$$

From third order equation

$$\Phi_c''\bar{\mathbf{v}}_3\delta\mathbf{u} = -3\left\{\bar{\lambda}_2\,\dot{\Phi}_c''\bar{\mathbf{v}}_1 + \Phi_c'''\bar{\mathbf{v}}_1\bar{\mathbf{v}}_2 + \frac{1}{3}\Phi_c^{IV}\,\bar{\mathbf{v}}_1^3 + 2\alpha\,\tilde{\Psi}_c'\mathbf{u}^*\right\}\delta\mathbf{u} \tag{5.24}$$

by taking successively $\delta\mathbf{u} = \mathbf{v}_{11},\dots,\mathbf{v}_{1m}$ and imposing Fredholm orthogonality conditions the following equations are obtained

$$2A_{ijkl}\bar{\nu}_i\bar{\nu}_j\bar{\nu}_l + \bar{\lambda}_2 B_{ik}\bar{\nu}_i + 2\alpha D_{ik}\eta_i = 0 \qquad (i,\,j,\,k,\,l = 1,\dots,m) \tag{5.25}$$

where the coefficients are defined by (4.29), (4.16)$_2$, (5.16). Eqs. (5.25) together with (5.12) are a nonlinear nonhomogeneous system of $m+1$ equations in the $m+1$ unknowns $\bar{\nu}_i$, $\bar{\lambda}_2$, depending on α and η_i. As in the previous case the series expansions (5.8) furnish the equilibrium paths as function of $\bar{\xi}$ and η_i. Eqs. (5.25) can be alternatively utilized by multiplying all terms by ξ^3. One gets

$$(\lambda - \lambda_c)B_{ik}a_i + A_{ijkl}a_ia_ja_l + \bar{\xi}D_{ik}\eta_i = 0 \qquad (i,\,j,\,k,\,l = 1,\dots,m) \tag{5.26}$$

if use is made of the relations $2(\lambda - \lambda_c) = \bar{\lambda}_2\xi^2$, $\alpha\xi^3 = \bar{\xi}$, $a_i = \bar{\nu}_i\xi$.

5.4 Post buckling analysis of o.d.f. systems

The analysis of some simple models may be useful to highlight the post-buckling behaviour of structures. Three o.d.f. systems will be therefore investigated to some extent in detail.

a) Rigid rod elastically restrained at the hinge

The system analysed is illustrated in the Figure 11. The total potential energy in the generic configuration is

$$\Phi(\theta) = \frac{1}{2}k\theta^2 - Nl(1 - \cos\theta) \tag{5.27}$$

Due to the simplicity of the problem, we can evaluate directly all equilibrium paths be requesting $\Phi(\theta)$ to be stationary, without resorting to perturbation analysis. The two equilibrium paths are obtained

$$\theta = 0, \qquad \forall N$$
$$N = \frac{k}{l}\frac{\theta}{\sin\theta} \tag{5.28}$$

where (5.28)$_1$ is the fundamental path, (5.28)$_2$ the bifurcated path and the critical load is given by $N_c = k/l$. An approximate solution can be obtained by replacing the energy $\Phi(\theta)$ with a series expansion up to the fourth order terms starting from $\theta = 0$

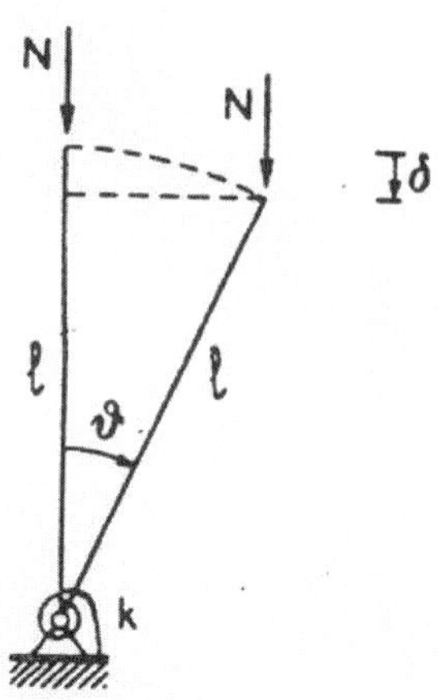

Figure 11. Rigid rod elastically restrained at the hinge

$$\Phi(\theta) = \frac{1}{2}(k - Nl)\theta^2 + \frac{Nl}{24}\theta^4 \tag{5.29}$$

The counterparts of eqs (5.28) are now

$$\theta = 0, \qquad \forall N$$
$$N = \frac{k}{l}\left(1 + \frac{\theta^2}{6}\right) \tag{5.30}$$

where $(5.30)_2$ can be obtained from $(5.28)_2$ by expanding in series $\sin\theta$. By Applying the Lagrange-Dirichlet theorem, we may now investigated stability along the equilibrium paths (5.30). Since $\Phi(\theta)$ is a one-variable function, it will be positive definite around an equilibrium configuration if $d^2\Phi/d\theta^2 > 0$. It is found that $d^2\Phi/d\theta^2 > 0$ along the fundamental path for $0 \leq N \leq N_c$ (stable equilibrium) and $d^2\Phi/d\theta^2 < 0$ for $N > N_c$ (unstable equilibrium). Along the bifurcated path it is always $d^2\Phi/d\theta^2 > 0$ and hence the equilibrium is unstable.

When a geometric imperfection is present as a small initial slope θ_0 of the road, the total potential energy (5.27) is modified into

$$\Phi(\theta, \theta_0) = \frac{1}{2}k(\theta^2 - 2\theta\theta_0) - Nl(1 - \cos\theta) \tag{5.31}$$

from which we obtain the equilibrium relation

$$N = \frac{k}{l}\frac{\theta}{\sin\theta} - \frac{k}{l}\frac{\theta_0}{\sin\theta} \tag{5.32}$$

which differs from $(5.28)_2$ because of the presence of the additional term accounting for initial imperfections. Eq (5.32) has been plotted in Figure 12 where four families of curves are observed, two corresponding to $\theta_0 > 0$ and two to $\theta_0 < 0$ in addition to the equilibrium path of the perfect structure corresponding to $\theta_0 = 0$. It can be shown that all equilibrium configuration on one side of the curve S obtained by plotting the function

$$\frac{d^2\Phi}{d\theta^2} = k - Nl\cos\theta = 0 \tag{5.33}$$

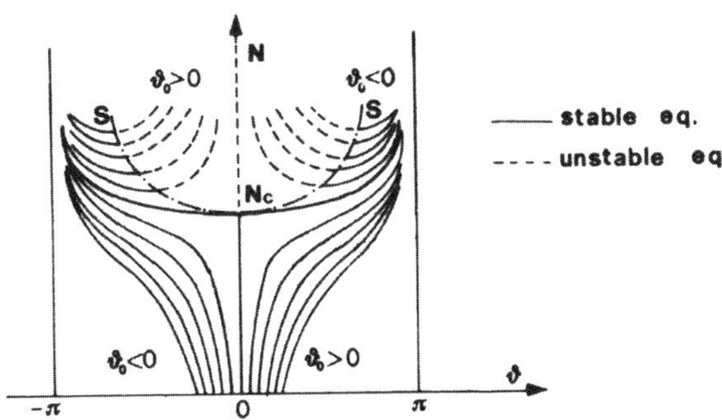

Figure 12. Equilibrium paths for the perfect and imperfect structures

are unstable and all those on the other side are stable. An elementary analysis which accounts for the second order terms only in the energy expansions (5.29) and (5.31) provides the equilibrium paths shown in Figure 13.

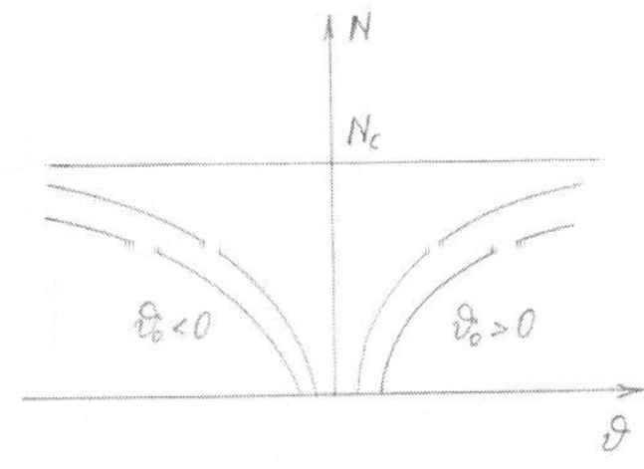

Figure 13. Equilibrium paths from elementary analysis

It is apparent from the figure the qualitative change of behaviour with respect to the actual structure in that here the critical load is never attained. All those structures which exhibit bifurcated equilibrium paths of the type examined are called *structures with stable symmetrical post-critical behaviour.*

b) Rigid rod elastically restrained at the free end

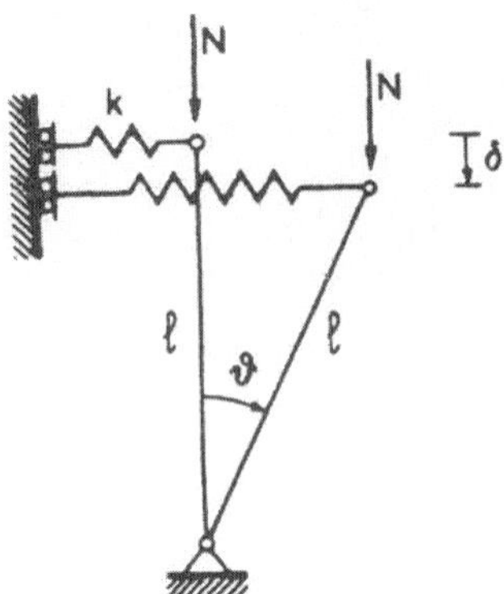

Figure 14. Rigid rod elastically restrained at the free end

We analyse here the system illustrated in Figure 14. The total potential energy relative to the generic configuration is given by

$$\Phi(\theta) = \frac{1}{2}k(l\sin\theta)^2 - Nl(1 - \cos\theta) \tag{5.34}$$

from which two equilibrium paths are derived by imposing to $\Phi(\theta)$ to be stationary

$$\theta = 0, \qquad \forall N$$
$$N = kl\cos\theta \tag{5.35}$$

the first one representing the fundamental path and the second one the bifurcated path. Here the critical load is $N_c = kl$. By evaluating $d^2\Phi/d\theta^2$ along the equilibrium path it is found that the equilibrium is stable along $(5.35)_1$ for $0 \le N < N_c$ and unstable for $N \ge N_c$. Along the bifurcated path the equilibrium is always unstable.

As in the previous case , an approximate solution can be obtained by replacing the energy expression (5.34) with a series expansions up to fourth order terms starting from $\theta = 0$. We have

$$\Phi(\theta) = \frac{1}{2}\left(kl^2 - Nl\right)\theta^2 - \frac{1}{6}\left(kl^2 + \frac{Nl}{4}\right)\theta^4 \tag{5.36}$$

from which the two equilibrium paths follow

$$\theta = 0, \quad \forall N$$
$$N = kl\left(1 - \frac{\theta^2}{2}\right) \tag{5.37}$$

As in the previous case, eq. $(5.37)_2$ can be obtained from a series expansions of $(5.35)_2$. Stability along the paths (5.37) is the same as for paths (5.35).

In the presence of small geometric initial imperfections such as a slope θ_0 of the rod, the total potential energy (5.36) is replaced by

$$\Phi(\theta, \theta_0) = \frac{1}{2}kl^2 \left(\theta^2 - \frac{\theta^4}{3} - 2\theta\theta_0\right) - Nl(\frac{\theta^2}{2} - \frac{\theta^4}{24}) \tag{5.38}$$

from which we get the equation of equilibrium

$$N = kl(1 - \frac{\theta^2}{2} - \frac{\theta_0}{\theta}) \tag{5.39}$$

which differs from $(5.37)_2$ because of the initial imperfections additional term. The function (5.39) has been represented in Figure 15 where, as in the case previously discussed, we can see four families of curves, two corresponding to $\theta_0 > 0$ and two to $\theta_0 < 0$ in addition to the equilibrium path relative to the perfect structure.

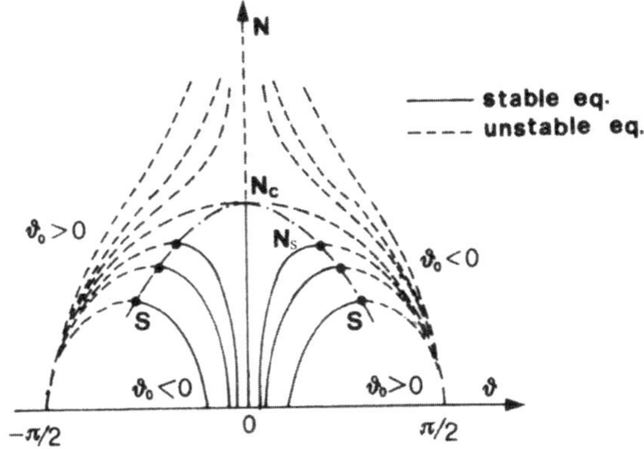

Figure 15. Equilibrium paths for the perfect and imperfect structures

In contrast with the previous model, the equilibrium curves here reach a maximum N_s, called *collapse load*, or *snapping load*, or *limit load* and then decrease. The maxima are located along the curve S defined by

$$\frac{d^2\Phi}{d\theta^2} = kl(1 - 2\theta^2) - N\left(1 - \frac{\theta^2}{2}\right) = 0 \tag{5.40}$$

It can be shown that all equilibrium configuration on one side of S are stable and all those on the other side are unstable. The dependence of the snapping load on θ_0 is furnished by

$$N_s = kl\left(1 - \frac{3}{2}\theta_0^{2/3}\right) \tag{5.41}$$

and is shown in Figure 16.

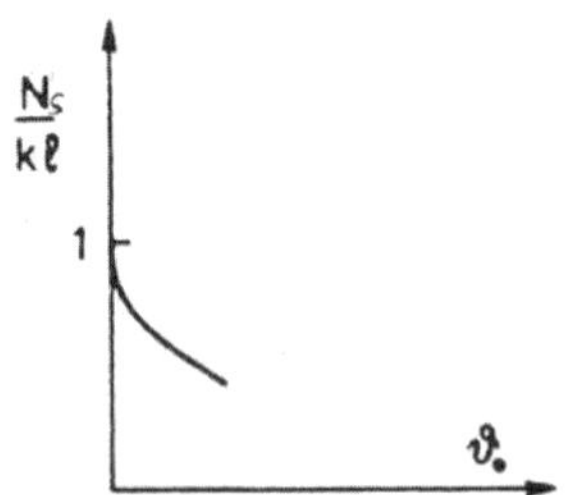

Figure 16. Dependence of the snapping load on the initial imperfections.

Here one can see that a small initial imperfection may cause a remarkable erosion of the critical load due to the fact that the curve is tangent to the vertical axis. All structures which present post-buckling equilibrium paths of the type analysed are called *structures with unstable symmetrical post-critical behaviour.*

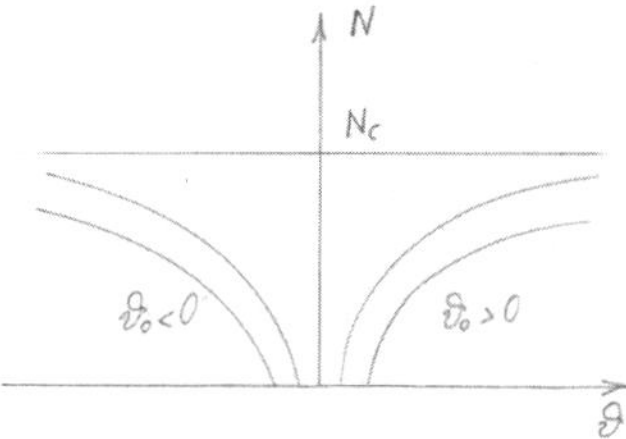

Figure 17. Equilibrium paths from elementary analysis

If an elementary analysis is performed by taking into account only the second order terms in the energy expansion (5.36) and (5.38), we obtain the equilibrium paths shown in Figure 17 which are qualitatively the same as those shown in Figure 13. It is evident that this type of analysis completely obscures the actual behaviour of the structure with the risk, not present in the previous model, that this naif approach may cause damages or disasters due to the fact that the presence of the collapse load is ignore

c) Rigid rod asymmetric elastic restraint at the free end

The system here investigated is illustrated in Figure 18. The total potential energy relative to the generic configuration is

$$\Phi(\theta) = \frac{1}{2}kl^2\left(\sqrt{1 + \frac{1}{\tan\varphi}\left(\frac{1}{\tan\varphi} + 2\sin\theta\right)} - \frac{1}{\sin\varphi}\right) - Nl(1 - \cos\theta) \qquad (5.42)$$

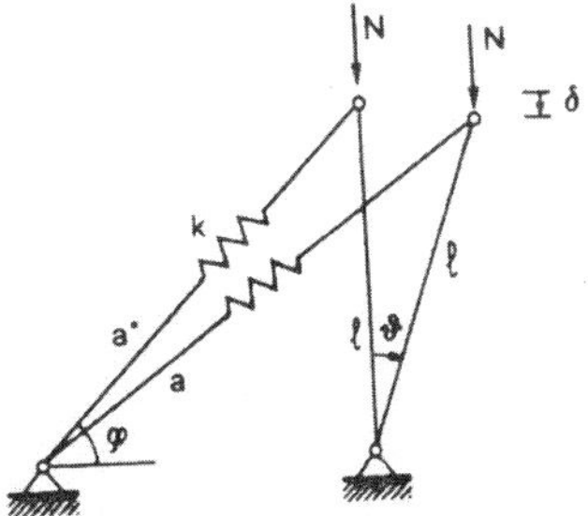

Figure 18. Rigid rod with asymmetric elastic restraint at the free end.

Assuming $\varphi = \pi/4$ and taking the series expansions of eq. (5.42) around $\theta = 0$ we have

$$\Phi(\theta) = \frac{1}{4}kl^2\left(\theta^2 - \frac{\theta^3}{2} - \frac{\theta^4}{48}\right) - Nl\left(\frac{\theta^2}{2} - \frac{\theta^4}{24}\right) \tag{5.43}$$

From the stationarity condition applied to (5.43) the fundamental path and the bifurcated path are obtained

$$\theta = 0, \qquad \forall N$$
$$N = \frac{kl}{2}\left(1 - \frac{3}{4}\theta + \frac{\theta^2}{8}\right) \tag{5.44}$$

In contrast with the two previous models,here we have a third order term appearing in the potential energy. Whenever this is the case, the fourth order term is neglected with respect to it and consequently, the square term disappears in $(5.44)_2$ which becomes linear. For $\theta = 0$, eq. $(5.44)_2$ furnishes $N_c = \frac{kl}{2}$. By evaluating now $d^2\Phi/d\theta^2$ along the paths (5.44) it is found that the equilibrium along the fundamental path is stable for $0 \le N < N_c$ and unstable for $N \ge N_c$. Along the bifurcated path the equilibrium is unstable for $\theta > 0$ and stable for $\theta < 0$ (see Figure 19). In the presence of small geometric initial imperfection characterized by a deviation θ_0 of the road, the total potential energy modifies into

$$\Phi(\theta; \theta_0) = \frac{kl^2}{2}\left(\frac{\theta^2}{2} - \frac{\theta^3}{4} - \theta\theta_0\right) - \frac{Nl}{2}\theta^2 \tag{5.45}$$

from which the equilibrium equation is derived

$$N = \frac{kl}{2}\left(1 - \frac{3}{4}\theta - \frac{\theta_0}{\theta}\right) \tag{5.46}$$

Eq. (5.46) has been plotted in Figure 19. Four families of curves are observed, two corresponding to $\theta_0 > 0$ and two to $\theta_0 < 0$ and the straight line relative to the perfect

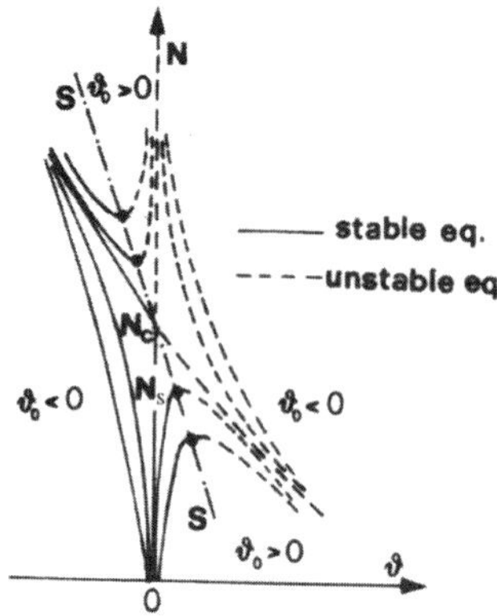

Figure 19. Equilibrium paths for the perfect and imperfect structure

structure. The curves corresponding to $\theta_0 > 0$ reach a maximum N_s and then decrease. The maxima are along the straight line S

$$\frac{d^2\Phi}{d\theta^2} = \frac{kl}{2}\left(1 - \frac{3}{2}\theta\right) - N = 0 \tag{5.47}$$

It can been seen that the equilibrium is stable to the left of S and unstable on the right. The dependence of the snapping load on θ_0 is given by

$$N_s = \frac{kl}{2}(1 - \sqrt{3\theta_0}) \tag{5.48}$$

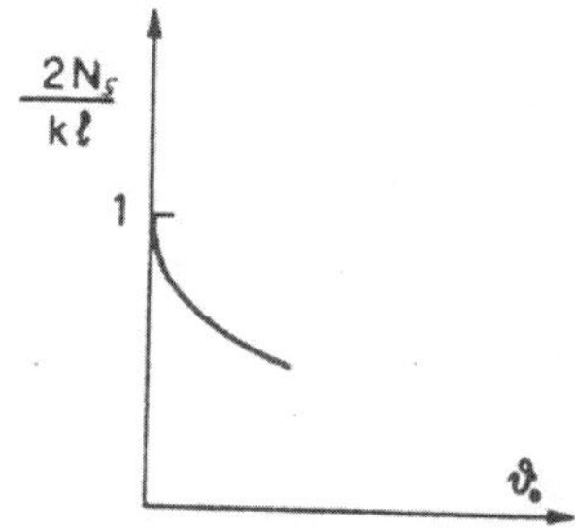

Figure 20. Dependence of the snapping load on the initial imperfection

and is represented in the Figure 20. As in the previous model, the curve is tangent to the vertical axis but drops down more rapidly in comparison with it. Consequently this model is more sensitive to initial imperfections. All structures which exhibit this type of behaviour in the post-buckling range are called *structure with asymmetrical post-critical behaviour*.

An ingenuous interpretation of the equilibrium paths shown in Figure 19 might bring to the firm belief that one could manufacture a structure with *ad hoc* initial imperfections corresponding to the ascending, stable equilibrium path. Unfortunately, actual structures are much more complex and initial imperfections are unknown since they come from industrial manufacturing processes. It is therefore advisable for the designer to assume that the structure will turn into the unstable equilibrium path with a possible collapse.

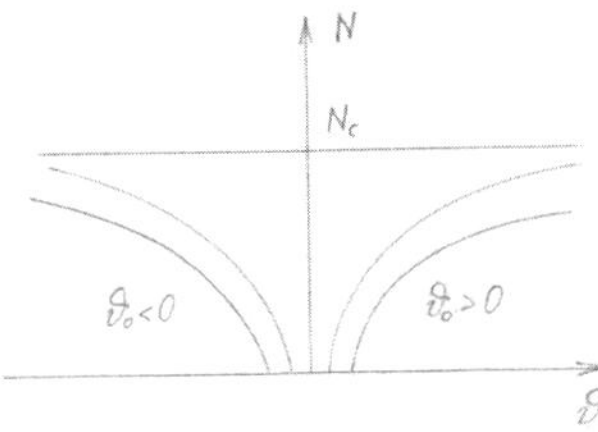

Figure 21. Equilibrium paths from elementary analysis

By retaining only quadratic terms in the potential energy (5.43) we obtain the equilibrium paths shown in the Figure 21 where, as in the previous model, the collapse load has disappeared. From comparison of Figures 13,17,21 it is apparent that if elementary analysis is performed, the three models behave exactly in the same way without any distinction between structures insensitive (safe) or sensitive (unsafe) to initial imperfection. This should caution designers against the utilization of elementary theories which hide completely the actual behaviour of structures.

5.5 Influence of Initial Imperfections on the Collapse Load - Special Cases of Initial Imperfections

The main influence of initial imperfections on post-buckling behaviour of the structure consists in an erosion of the critical load. What becomes really significant in these cases is the *collapse load* which is sometimes much below the critical load, especially for structures with asymmetric postbuckling behaviour. The effect is more detrimental when many buckling modes occur simultaneously or nearly simultaneously like, for instance, in cylinders under axial or radial compression, spheres under hydrostatic pressure and so on.

From an analytical point of view, the solution to the nonlinear algebric problems (5.15) or (5.25) is in general very difficult and requires use of numerical methods. There is however a particular class of structures and initial imperfections for which a solution in closed form can be easily found.

We start from the basic and quite common assumption that the fundamental path and the eigenvalue problem $(4.8)_1$ are linear in λ. We find it convenient to write it in the form

$$\Pi_{2c}'' \mathbf{v}_{1i}\delta\mathbf{u} + \lambda_c \dot{\Phi}_c'' \mathbf{v}_{1i}\delta\mathbf{u} = 0 \tag{5.49}$$

By posing $\delta\mathbf{u} = \mathbf{v}_{1j}$, using the orthonormalization condition (4.12) and replacing into eq. $(4.16)_2$ the following result is obtained

$$B_{ij} = -\frac{2}{\lambda_c}\delta_{ij} \tag{5.50}$$

A second useful result is derived by observing that $\tilde{\Psi}_c' \mathbf{v}_{1i}\mathbf{v}_{1j} = -\Pi_{2c}''\mathbf{v}_{1i}\mathbf{v}_{1j}$. Then from eq. (5.16) on account of (4.12) one has

$$D_{ij} = -2\delta_{ij} \tag{5.51}$$

From results (5.50) and (5.51) nonlinear problems (5.15) and (5.25) are written as

$$A_{ijk}\bar{\nu}_i\bar{\nu}_j - 2\frac{\bar{\lambda}_1}{\lambda_c}\bar{\nu}_k - 2\alpha\eta_k = 0 \qquad (i,j,k = 1,\ldots,m) \tag{5.52}$$

$$A_{ijkl}\bar{\nu}_i\bar{\nu}_j\bar{\nu}_l - \frac{\bar{\lambda}_2}{\lambda_c}\bar{\nu}_k - 2\alpha\eta_k = 0 \qquad (i,j,k,l = 1,\ldots,m) \tag{5.53}$$

respectively. By specializing eqs. (4.15) and (4.28), the corresponding equations relative to the perfect structure are obtained

$$A_{ijk}\nu_i\nu_j - 2\frac{\lambda_1}{\lambda_c}\nu_k = 0 \qquad (i,j,k = 1,\ldots,m) \tag{5.54}$$

$$A_{ijkl}\nu_i\nu_j\nu_l - \frac{\lambda_2}{\lambda_c}\nu_k = 0 \qquad (i,j,k,l = 1,\ldots,m) \tag{5.55}$$

Let us now consider the case of initial imperfections having the same shape as the buckling mode associated with one of the r bifurcated paths of the perfect structure. Then eq. (5.13) is replaced by

$$\mathbf{u}^* = \nu_i\mathbf{v}_{1i} \tag{5.56}$$

where the ν_i's are a solution to eqs. (5.54) or (5.55). It is of particular interest to examine the set of ν_i's corresponding to the path of steepest descent or largest negative curvature since the associated imperfections are likely to be the most dangerous for the structure. By virtue of eq. (5.56) we replace η_k with ν_k in eqs. (5.52), (5.53) and look for a trial solution $\bar{\nu}_i = \nu_i$. Then on account of eqs. (5.54), (5.55) we have

$$2\nu_k\left(\frac{\lambda_1}{\lambda_c} - \frac{\bar{\lambda}_1}{\lambda_c} - \alpha\right) = 0 \qquad (k = 1,\ldots,m) \tag{5.57}$$

$$2\nu_k\left(\frac{\lambda_2}{2\lambda_c} - \frac{\bar{\lambda}_2}{2\lambda_c} - \alpha\right) = 0 \qquad (k = 1,\ldots,m) \tag{5.58}$$

These equations are satisfied for

$$\bar{\lambda}_1 = \lambda_1 - \alpha\lambda_c \tag{5.59}$$

$$\bar{\lambda}_2 = \lambda_2 - 2\alpha\lambda_c \tag{5.60}$$

and therefore $\bar{\nu}_i = \nu_i$ is a solution of the nonlinear equations (5.52) and (5.53). In the following the two cases $\lambda_1 \neq 0$ and $\lambda_1 = 0$ are discussed separately.

a) *Asymmetric postbuckling behaviour* $(\lambda_1 \neq 0)$

Evaluation of the coefficient $\bar{\lambda}_1$ is usually sufficient to describe the equilibrium path of asymmetric structures and to furnish an adequate estimate of the snapping load.

By replacing eq. (5.59) into (5.8)$_1$ and accounting for $\alpha = \bar{\xi}/\xi^2$ one gets

$$\lambda = \lambda_c + \lambda_1\xi - \lambda_c\frac{\bar{\xi}}{\xi} \tag{5.61}$$

The maximum of eq. (5.61) is attained by correspondence with the value of ξ

$$\xi_s = \sqrt{-\bar{\xi}\frac{\lambda_c}{\lambda_1}} \tag{5.62}$$

to which corresponds the load parameter λ_s

$$\frac{\lambda_s}{\lambda_c} = 1 - 2\sqrt{-\bar{\xi}\frac{\lambda_1}{\lambda_c}} \tag{5.63}$$

Eqs. (5.62) and (5.63) are valid for $\bar{\xi}\lambda_1 < 0$. If $\bar{\xi}\lambda_1 > 0$ eq. (5.61) does not exhibit a maximum. Within the same order of approximation of eq. (5.61) the displacement field is furnished by

$$\mathbf{v} = \nu_i\mathbf{v}_{1i}\xi \tag{5.64}$$

The foregoing approximation can be improved by evaluating $\bar{\lambda}_2$ and $\bar{\mathbf{v}}_2$ as for the general case by means of eqs. (5.17), (5.18), (5.19) once $\bar{\lambda}_1$ has been determined from (5.59) and accounting for $\bar{\nu}_i = \nu_i$, but in general this is not necessary.

b) *Symmetric postbuckling behaviour* $(\lambda_1 = 0)$

By following the same procedure as in the previous case and remembering that now $\bar{\xi} = \alpha\xi^3$ one gets

$$\lambda = \lambda_c + \frac{1}{2}\lambda_2\xi^2 - \lambda_c\frac{\bar{\xi}}{\xi} \tag{5.65}$$

For $\lambda_2 < 0$ the snapping load is

$$\frac{\lambda_s}{\lambda_c} = 1 - \frac{3}{2}(\bar{\xi}^2\frac{\lambda_2}{\lambda_c})^{\frac{1}{3}} \tag{5.66}$$

which occurs at

$$\xi_s = (-\bar{\xi}\frac{\lambda_c}{\lambda_2})^{\frac{1}{3}} \tag{5.67}$$

The displacement field $(5.8)_2$ is furnished by

$$\mathbf{v} = \nu_i \mathbf{v}_{1i}\xi + \frac{1}{2}\nu_i\nu_j(\mathbf{v}_{2ij} + p_{kij}\mathbf{v}_{1k})\xi^2 \qquad (i,j,k = 1,\dots,m) \tag{5.68}$$

where use has been made of eqs. (5.11), (5.22), (5.23) and of the result $\bar{\nu}_i = \nu_i$.

Eqs. (5.63), (5.66) which explicitly describe the structure sensitivity to initial imperfections have been derived for a particular imperfection shape under the assumption that B_{ij}, D_{ij} are diagonal matrices. Whenever this condition is not verified it is in general $\bar{\nu}_i \neq \nu_i$ and the solution to the nonlinear problem is much more involved. However, if initial imperfections are still of the type (5.56), it is possible to find an approximate solution of the nonlinear eqs. (5.15), (5.25) in a rather simple way according to the following procedure.

Let $\bar{\nu}_i = \nu_i$. Then by multiplying the k-th of eqs. (5.15) or (5.25) by ν_k, by adding and solving with respect to $\bar{\lambda}_1$ or $\bar{\lambda}_2$ one gets

$$\bar{\lambda}_1 = -\frac{A_{ijk}\nu_i\nu_j\nu_k + \alpha D_{ik}\nu_i\nu_k}{B_{ik}\nu_i\nu_k} \qquad (i,j,k = 1,\dots,m) \tag{5.69}$$

$$\bar{\lambda}_2 = -\frac{2A_{ijkl}\nu_i\nu_j\nu_k\nu_l + 2\alpha D_{ik}\nu_i\nu_k}{B_{ik}\nu_i\nu_k} \qquad (i,j,k,l = 1,\dots,m) \tag{5.70}$$

respectively, from which

$$\bar{\lambda}_1 = \lambda_1 - \rho\alpha\lambda_c \tag{5.71}$$

$$\bar{\lambda}_2 = \lambda_2 - 2\rho\alpha\lambda_c \tag{5.72}$$

being λ_1 and λ_2 the load coefficients of the perfect structure and

$$\rho = \frac{1}{\lambda_c}\frac{D_{ik}\nu_i\nu_k}{B_{ik}\nu_i\nu_k} \tag{5.73}$$

Note that eqs. (5.71), (5.72) are analogous to eqs. (5.59), (5.60). It should be observed, however, that in this case $\bar{\nu}_i \cong \nu_i$ and that the values of $\bar{\lambda}_1$, $\bar{\lambda}_2$ furnished by (5.71), (5.72) satisfy the nonlinear eqs. (5.15) or (5.25) only in the average. The meaning of the previous approximations can be better understood if eqs. (5.69), (5.70) are obtained through an alternative procedure.

Let us consider again the perturbation eqs. (5.14) and (5.24) and instead of proceeding as in subsect. 5.2 and 5.3 we impose a unique orthogonality condition by taking $\delta\mathbf{u} = \nu_i\mathbf{v}_{1i}$ where coefficients ν_i select the shape of the critical mode of the perfect structure corresponding to the equilibrium path of steepest descent along with the associated initial imperfection. The orthogonality condition furnishes again eqs. (5.69) and (5.70) from the two perturbation equations, provided that $\bar{\nu}_i = \nu_i$. The technique employed may therefore be viewed as a discretization procedure which reduces the continuous body to a one degree of freedom system. Note that the possible solution to the higher order perturbation equation requires the same discretization technique. In fact if Fredholm conditions are not satisfied, particular integrals cannot in general be determined.

If the series expansion of λ is limited to coefficients (5.71), (5.72) one obtains through the same procedure previously outlined

$$\frac{\lambda_s}{\lambda_c} = 1 - 2\sqrt{-\bar{\xi}\rho\frac{\lambda_1}{\lambda_c}} \tag{5.74}$$

valid for $\lambda_1 \neq 0$ and

$$\frac{\lambda_s}{\lambda_c} = 1 - \frac{3}{2}(\bar{\xi}^2\rho^2\frac{\lambda_2}{\lambda_c})^{\frac{1}{3}} \tag{5.75}$$

valid for $\lambda_1 = 0$ and $\lambda_2 \neq 0$.

6 The Effect of Multiple Buckling Modes on the Postbuckling Behaviour of Plane Elastic Frames

6.1 Introduction

The postbuckling behaviour of plane frames exhibiting a single buckling mode at the critical stress has been analyzed in many papers utilizing the general theory of elastic stability developed by Koiter (Koiter,1945). Numerical solutions have been furnished in general, even if analytical solutions are sometimes available for some simple frames. It happens quite often in practice, however, that a "naive" approach to the optimum design of structures leads engineers to select the geometric parameter of the frame in such a way that two or more buckling modes occur simultaneously under the same critical stress.

It was pointed out by Koiter and Skaloud (Koiter and Skaloud, 1963) that the nonlinear interaction of simultaneous buckling modes at the same value of the critical stress is mostly responsible for imperfection sensitivity of structures even if single buckling modes show a stable postbuckling behaviour. Several authors have contributed to the developments of research in this area in recent years (Bonito, 1983; Byskov and Hutchinson, 1977; Casciaro, Di Carlo et al., 1976; Koiter and Pignataro, 1976; Koiter and Skaloud, 1963; Kotelko and Kolakowski, 2000; Krolak, Kolakowski et al., 2001; Menken,Groot et al., 1991; Pignataro and Rizzi 1983; Rizzi and Pignataro 1982; Thompson, Tulk et al., 1976; Tvergaard, 1973; Van der Neut, 1976) after the first detailed investigation of Van der Neut on nonlinear interaction between overall buckling of a built-up column and local buckling of the plate flanges (Van der Neut, 1969). The results of Van der Neut simplified model have been confirmed by Koiter and Kuiken approximate analysis (Koiter and Kuiken, 1971) based on the general nonlinear theory (Koiter, 1945).

6.2 Postbuckling Analysis of a Symmetric one-bay, two-Storey Frame

In this section we discuss the buckling and a postbuckling analysis of a symmetric one-bay, two-storey frame Figure 22 keeping in mind that qualitative results here achieved can be extended to predict the postbuckling behaviour of multi-bay, multi-storey frames as are usually encountered in civil and industrial realizations.

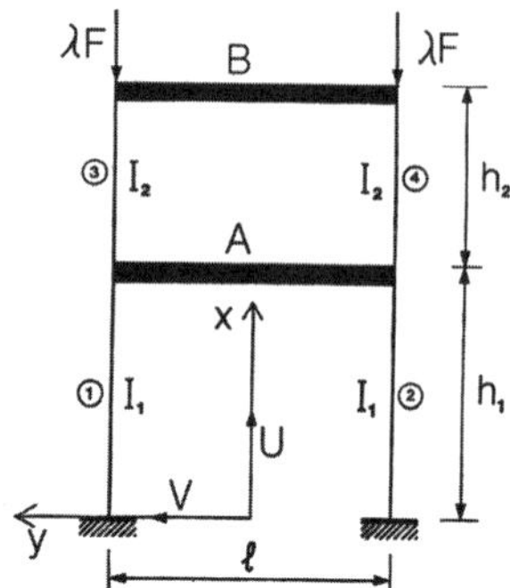

Figure 22. Symmetric one-bay, two-storey frame

The bulk of theoretical, as well as experimental research on postbuckling of beams and frames, is based on the assumption of inextensible, shear undeformable beam model. This model will be adopted here and analytical solutions to our frame will be derived.

We define first the axial and shear deformation measures. Starting from three-dimensional theory and assuming the cross-section to be rigid, we obtain (Pignataro, Di Carlo, et al., 1982)

$$\varepsilon_{11} = u' + \frac{1}{2}(u'^2 + v'^2) - y\varphi'[(1 + u')cos\varphi + v'sin\varphi] + \frac{1}{2}y^2\varphi'^2$$
$$2\varepsilon_{12} = v'cos\varphi - (1 + u')sin\varphi \tag{6.1}$$

where symbols are illustrated in Figure 23 and a prime denotes differentiation with respect to x. In deriving eqs. (6.1) out-of-plane displacements have been excluded.

On the other hand, by following the direct approach treatment by Antman (Antman, 1977) we get

$$\varepsilon = (1 + u')cos\varphi + v'sin\varphi - 1$$
$$\gamma = v'cos\varphi - (1 + u')sin\varphi \tag{6.2}$$

By enforcing now the Bernouilli hypothesis ($\varepsilon_{12} = 0$, $\gamma = 0$) and the axis inextensibility constraint ($\varepsilon_{11} = 0$, $\varepsilon = 0$), both (6.1) and (6.2) lead to

$$u' = cos\varphi - 1$$
$$v' = sin\varphi \tag{6.3}$$

Eqs. (6.3) show that kinematics can be simply described in terms of the cross-section rotation φ. Obviously, their nonlinear nature implies nonlinear boundary conditions on u and v which can be accounted for by means of Lagrangean multipliers (Di Carlo, Pignataro et al., 1981).

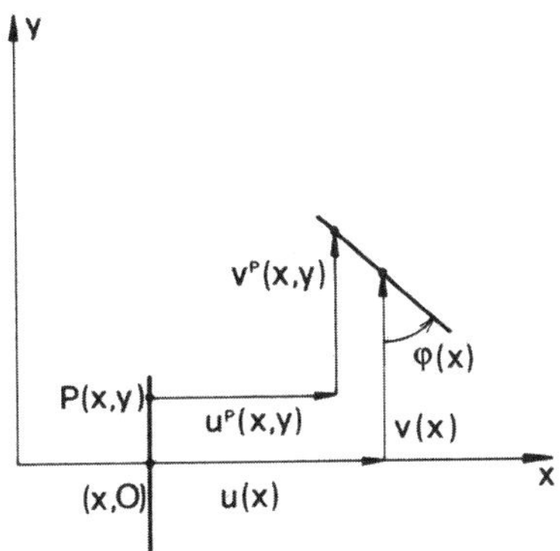

Figure 23. Development from the three-dimensional theory

a) Single buckling mode

By assuming local reference frames for the columns, solution to the eigenvalue problem $(4.8)_1$ furnishes

$$\varphi_{1c} = \frac{\pi}{2} \sin \pi \frac{x}{h_1}, \qquad \varphi_{3c} = 0$$
$$\varphi_{2c} = \frac{\pi}{2} \sin \pi \frac{x}{h_1}, \qquad \varphi_{4c} = 0 \tag{6.4}$$

$$\lambda_c F = \frac{\pi^2 E I_1}{h_1^2} \tag{6.5}$$

or

$$\varphi_{1c} = 0, \qquad \varphi_{3c} = \frac{\pi}{2} \sin \pi \frac{x}{h_2}$$
$$\varphi_{2c} = 0, \qquad \varphi_{4c} = \frac{\pi}{2} \sin \pi \frac{x}{h_2} \tag{6.6}$$

h/l	0,1	0.2	0.5	1.0	2.0	3.076	5.0	10.0
$\frac{1}{2}\frac{\lambda_2}{\lambda_c}$	0.3081	0.3071	0.3003	0.2758	0.1780	0.0000	-0.5065	-2.951

Table 1. Post-buckling equilibrium path curvature in terms of h/l

$$\lambda_c F = \frac{\pi^2 E I_2}{h_2^2} \tag{6.7}$$

according to whether the critical load (6.5) is smaller or larger than the critical load (6.7). In the previous equations, φ_i ($i = 1, 2, 3, 4$) denote rotation of columns as labelled in Figure 22, I_1, I_2 are the moments of inertia and h_1, h_2 the heights of columns

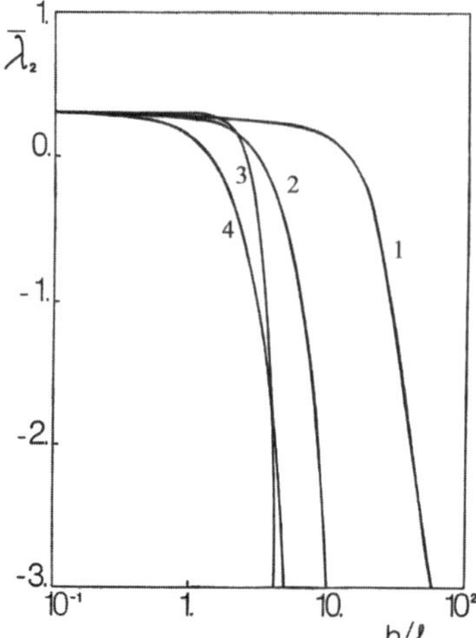

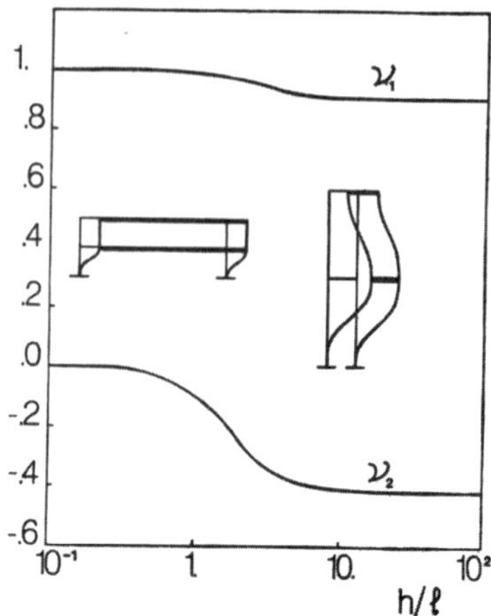

Figure 24. $\bar{\lambda}_2$ vs h/l relationship for bifurcated paths

Figure 25. Buckling mode amplitude vs h/l corresponding to curve 1 in fig 24

as illustrated in the same figure. It can immediately be shown that the slope λ_1 of the postbuckling equilibrium path is identically zero for both buckling modes (6.4) and (6.6). On the other hand, it is found that the curvature λ_2 depends on the ratio h/l. Table (6.2) furnishes a set of results which shows that the postbuckling equilibrium path is stable or unstable according to whether h/l is smaller or larger than 3.076, where h/l stands for either of the ratios h_1/l or h_2/l.

b)Multiple buckling mode

If $EI_1/h_1^2 = EI_2/h_2^2$, the two buckling loads (6.5) and (6.7) coincide and the associated buckling modes (6.4) and (6.6) occur simultaneously. A mode interaction analysis is therefore required in order to investigate the postbuckling behaviour of the frame. By applying the general theory presented in sects. 4.2, 4.3 and 4.4 its is found that there are four bifurcated paths. The results regarding $\bar{\lambda}_2 = \frac{1}{2}\frac{\lambda_2}{\lambda_c}$ which correspond to each of the bifurcated path have been plotted as a function of the aspect ratio h/l (Figure 24). Numbers 1 to 4 labelling the four curves correspond to each of the bifurcated paths. Curve 2 recovers in particular the postbuckling behaviour of a single buckling mode. The diagram shows that along the bifurcated path corresponding to curve 4 the structure is imperfection sensitive for values of h/l which are smaller then those corresponding to the single buckling mode and which fall within a range which is common among actual frames.

Figures 25, 26 and 27 show a plot of ν_1 and ν_2 against h/l. The two sketches appearing in each figure represent the buckled shape of the frame for the limit values of h/l reported on the abscissa.

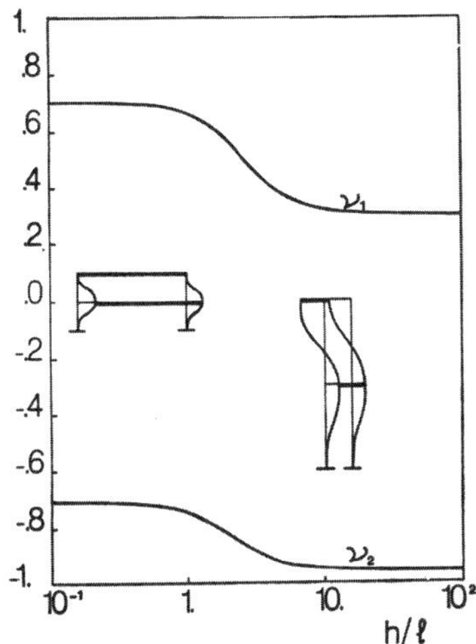

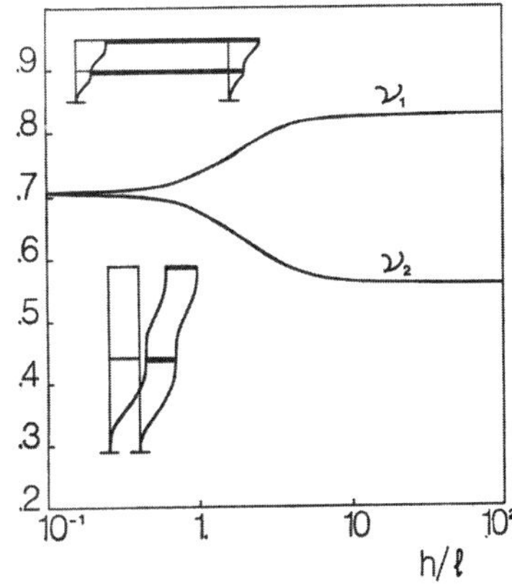

Figure 27. Buckling mode amplitude vs h/l corresponding to curve 4 in fig 24

Figure 26. Buckling mode amplitude vs h/l corresponding to curve 3 in fig 24

7 Influence of Interactive Overall Buckling on Postbuckling in Thin-Walled Compression Members

7.1 Introduction

It was already pointed out in the introduction of the previous chapter that nonlinear interaction between linearly independent (nearly) simultaneous buckling modes is largely responsible for the imperfection sensitivity exhibited by certain types of structures. The interested reader may refer to references quoted in sect. 6.1, just to mention some of the many previous contributions.

In this chapter we shall deal with the interaction between overall Euler and flexural-torsional buckling in thin-wall open cross section compression members. It will turn out that interaction is responsible for unstable postbuckling behaviour (and hence for imperfections sensitivity) even if the postbuckling behaviour of the single buckling modes, Eulerian and flexural-torsional, is stable.

Once again, the present results caution against a "naive" approach to the optimum design, consisting in the attempt to equalize the critical stresses of the individual buckling modes.

In the two next chapters, interaction between overall and local buckling in thin-walled open cross-section compression members will be investigated.

7.2 Displacement Field and Potential Energy Functional

We consider a beam with a thin-walled open cross-section (Figure 28) and assume that the axes x, y, z represent a principal reference frame. According to Vlasov theory

(Vlasov, Jerusalem 1961), the displacement of a generic point of the cross-section is given by

$$
\begin{aligned}
u(x, y, z) &= u(z) - \varphi(z)(y - y_0) \\
v(x, y, z) &= v(z) + \varphi(z)(x - x_0) \\
w(x, y, z) &= -x u'(z) - y v'(z) + \varphi'(z)\psi(x, y)
\end{aligned}
\tag{7.1}
$$

where $u(z)$, $v(z)$ denote the displacements of the shear centre C in the x and y direction, respectively, $\psi\,(x,\,y)$ is the warping function, $\varphi(z)$ the cross-section rotation around C, x_0 and y_0 the coordinates of the shear centre and a prime indicates differentiation with respect to z.

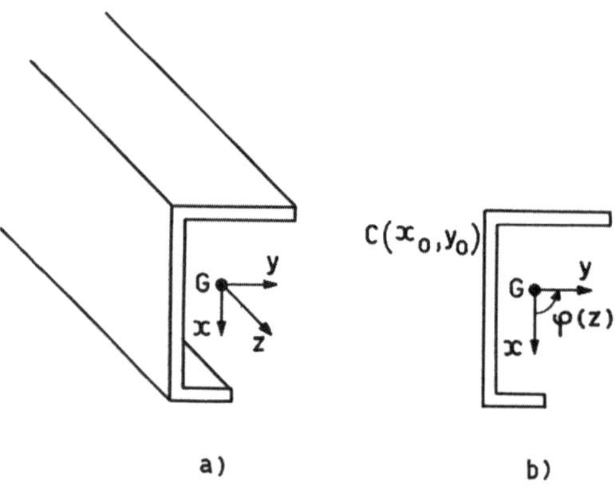

Figure 28. Thin-walled beam

By using the Green tensor for strain measure $\varepsilon_{ij} = 1/2\,(u_{i,j} + u_{j,i} + u_{h,i} u_{h,j})$ and assuming that a uniform axial load N is acting on the beam we write now the total potential energy up to third order terms. On account of eqs. (7.1) one gets

$$
\begin{aligned}
\Pi\,[\mathbf{U}, \lambda] = \frac{1}{2} EAL \int_0^1 \{\, & \theta_y U''^2 + \theta_x V''^2 + \theta_2 \Phi''^2 + \theta_1 \Phi'^2 \\
& - \lambda \theta_x \pi^2\,[\,U'^2 + V'^2 + \theta_c \Phi'^2 + 2\Phi'(c_y U' - c_x V')\,] \\
& - \Phi'^2(\theta_{cx} U'' + \theta_{cy} V'' - \theta_{c\psi}\Phi'') + 2\Phi(\theta_{bx} U' V'' - \theta_{by} U'' V')\,\} \, d\zeta
\end{aligned}
\tag{7.2}
$$

where use has been made of the dimensionless quantities

$$\begin{aligned}
\zeta &= z/L & U(\zeta) &= u(\zeta L)/L \\
V(\zeta) &= v(\zeta L)/L & \Phi(\zeta) &= (b/L)\varphi(\zeta L) \\
c_x &= x_0/b & c_y &= y_0/b \\
\theta_x &= I_x/AL^2 & \theta_y &= I_y/AL^2 \\
\theta_{bx} &= I_x/ALb & \theta_{by} &= I_y/ALb \\
\theta_1 &= C_1/EAb^2 & \theta_2 &= C_2/EAL^2b^2 \\
\theta_{cx} &= I_{cx}/ALb^2 & \theta_{cy} &= I_{cy}/ALb^2 \\
\theta_c &= I_c/Ab^2 & \theta_{c\psi} &= I_{c\psi}/ALb^3 \\
\lambda &= N/EA\theta_x\pi^2
\end{aligned} \tag{7.3}$$

which are defined in terms of

$b = $ a characteristic dimension of the cross-section

$L = $ length of the column

$A = $ cross-section area of the column

$I_y = $ moment of inertia of the cross section with respect to y axis

$I_x = $ moment of inertia of the cross section with respect to x axis

$I_c = $ polar moment of inertia with respect to the shear centre

$\Delta x = x - x_0$

$\Delta y = y - y_0$

$$I_{cx} = \int_A (\Delta x^2 + \Delta y^2) x \, dx dy$$

$$I_{cy} = \int_A (\Delta x^2 + \Delta y^2) y \, dx dy$$

$$I_{c\psi} = \int_A (\Delta x^2 + \Delta y^2) \psi(x, y) \, dx dy$$

$$I_\psi = \int_A \psi^2(x, y) \, dx dy = \text{warping constant}$$

$C_2 = EI_\psi = $ warping rigidity

$$C_1 = G \int_A [(\partial \psi/\partial x - \Delta y)^2 + (\partial \psi/\partial y + \Delta x)^2] \, dx dy = \text{torsional rigidity}$$

In eq. (7.2) the total potential energy has been denoted with Π instead of Φ as in previous chapters in order to avoid confusion with the cross-section rotation.

It is easy to verify that if the beam is simply supported at the edges, the buckling mode in vector form is

$$\bar{\mathbf{U}}\left(\zeta\right) = \left\{ \begin{array}{c} \bar{U}\, sin\, \pi\zeta \\ \bar{V}\, sin\, \pi\zeta \\ \bar{\bar{\Phi}}\, sin\, \pi\zeta \end{array} \right\} \tag{7.4}$$

all geometrical and mechanical boundary conditions being satisfied. By replacing eq. (7.4) into (7.2), the following expression of the energy in terms of the amplitude of the buckling modes is obtained

$$\Pi\left[\bar{\mathbf{U}},\lambda\right] = \frac{1}{2}EAL\theta_x\pi^4 \left\{ \; \frac{1}{2}\left[\left(\frac{\theta_y}{\theta_x} - \lambda\right)\bar{U}^2 + \left(1-\lambda\right)\bar{V}^2 \right.\right.$$
$$+ \left(\frac{\theta_2}{\theta_x} + \frac{\theta_1}{\pi^2\theta_x} - \lambda\theta_c \right)\bar{\bar{\Phi}}^2 - 2\lambda\left(c_y\bar{U} - c_x\bar{V} \right)\bar{\bar{\Phi}} \; \Big] \tag{7.5}$$
$$+ \frac{2}{3\pi}\Big[\; \Big(\frac{\theta_{cx}}{\theta_x}\bar{U} + \frac{\theta_{cy}}{\theta_x}\bar{V} \Big)\bar{\bar{\Phi}}^2 - \frac{\theta_{c\psi}}{\theta_x}\bar{\bar{\Phi}}^3 + 2\Big(\frac{\theta_{by}}{\theta_x} - \frac{\theta_{bx}}{\theta_x} \Big)\bar{U}\bar{V}\bar{\bar{\Phi}} \; \Big] \Big\}$$

When initial imperfections are present in a sinusoidal shape

$$\mathbf{U}^*\left(\zeta\right) = \left\{ \begin{array}{c} \alpha\, sin\, \pi\zeta \\ \beta\, sin\, \pi\zeta \\ \gamma\, sin\, \pi\zeta \end{array} \right\} \tag{7.6}$$

an additional term has to be added to the energy (7.5), namely

$$\Psi\left[\bar{\mathbf{U}},\mathbf{U}^*\right] = -\frac{1}{2}EAL\theta_x\pi^4\Big[\; \alpha\left(\bar{U} + c_y\bar{\bar{\Phi}}\right) + \beta\left(\bar{V} - c_x\bar{\bar{\Phi}}\right)$$
$$\gamma\left(c_y\bar{U} + c_x\bar{V} + \theta_c\bar{\bar{\Phi}}\right) \; \Big] \tag{7.7}$$

where α, β, γ are amplitude imperfections.

7.3   Buckling and Postbuckling Analysis

a) Cross-sections with two axes of symmetry

We may refer in this section to I or cruciform cross-sections which are very often encountered in practice. Since the shear centre coincides with the centre of gravity of the cross-section we have

$$c_x = 0, \qquad c_y = 0, \qquad \theta_{cx} = 0, \qquad \theta_{cy} = 0, \qquad \theta_{c\psi} = 0 \tag{7.8}$$

In this case there are three independent buckling modes: two purely flexural in the x and y direction and one purely torsional. The corresponding buckling loads coincide if

$$1 = \frac{\theta_y}{\theta_x} = \frac{\theta_2}{\theta_c\theta_x} + \frac{\theta_1}{\pi^2\theta_c\theta_x} \tag{7.9}$$

When conditions (7.8) and (7.9) are satisfied, all third order terms of the potential energy (7.5) vanish and from the general theory (4.15) we have $\lambda_1 = 0$ since all $A_{ijk}=0$.

We conclude therefore that the postbuckling behaviour is always symmetric both for simultaneous and for individual buckling modes. Fourth order terms are necessary to decide whether the postbuckling is stable or unstable.

b) Cross-sections with one axis of symmetry

Let us consider a cross-section symmetric with respect to y axis. Due to symmetry we have

$$c_x = 0, \qquad \theta_{cx} = 0, \qquad \theta_{c\psi} = 0 \tag{7.10}$$

In this case there are two independent buckling modes: one purely flexural in the y direction and one flexural-torsional. If

$$\left(\frac{\theta_y}{\theta_x} - 1 \right) \left(\frac{\theta_2}{\theta_x} + \frac{\theta_1}{\pi^2 \theta_x} - \theta_c \right) - c_y^2 = 0 \tag{7.11}$$

the corresponding buckling loads coincide.

When conditions (7.10) and (7.11) are satisfied it is easily seen that all third order terms of the energy vanish except A_{122} which is given by

$$A_{122} = \frac{1}{3\pi} EAL\theta_x \pi^4 \left[\frac{\theta_{cy}}{\theta_x} \bar{V}_1 \bar{\Phi}_2^2 + 2 \left(\frac{\theta_{by}}{\theta_x} - \frac{\theta_{bx}}{\theta_x} \right) \bar{V}_1 \bar{U}_2 \bar{\Phi}_2 \right] \tag{7.12}$$

which implies from eq. (4.15) $\lambda_1 \neq 0$. In eq. (7.12) index 1 denotes the first (Eulerian) buckling mode and index 2 the second (flexural-torsional) buckling mode. The postbuckling behaviour is asymmetric and therefore the column is sensitive to initial imperfections.

Many different types of initial imperfections may be considered according to the choice of the values of the parameters α, β, γ in eq. (7.6); for example one might consider initial imperfections in the shape of the first buckling mode, of the second buckling mode, or of a combination of them.

Figures 29 and 30 show a plot of the collapse load vs initial imperfections amplitude for an angle cross-section with equal sides and a T cross-section. Dimensions of the beams are shown in the lower left corner of the figures. Curves appearing in the diagrams refer to different choices of the initial imperfections shape. It is apparent that the worse situation occurs along the curve $\bar{\alpha} = \bar{\beta} = \bar{\gamma}$ where initial imperfections of the order 3×10^{-3} are responsible for an erosion of the buckling load around 40% for the angle section and around 25% for the T section.

c) Cross-sections with no axis of symmetry

This problem differs from the two previous ones in that linearly independent buckling modes do not occur simultaneously and therefore no interaction is possible. There is only one flexural-torsional buckling mode with flexural components in the x and y directions by correspondence with which third order terms of the energy expansion 7.5 do not vanish. Consequently, the postbuckling behaviour of these types of beams is asymmetric

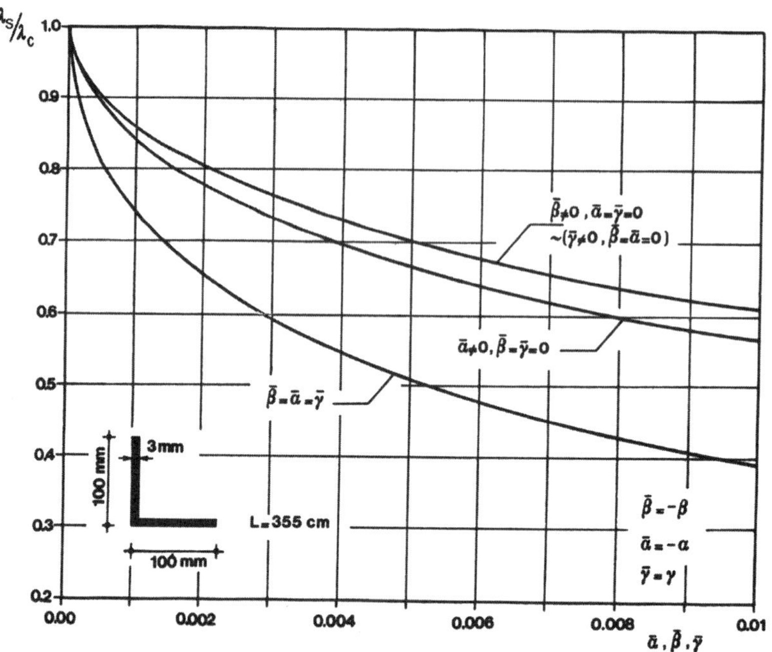

Figure 29. Sensitivity of a symmetric angle cross-section to initial geometric axis imperfection

and sensitive to initial imperfections. Figure 31 shows the dependence of the collapse load on the imperfections amplitudes α, β, γ for an angle cross-section with unequal sides when different choices of the initial imperfections are selected. As in the previous cases, the worst situation occurs along the curve $\bar{\alpha} = \bar{\beta} = \bar{\gamma}$. However, sensitivity to initial imperfection is in this case less severe since for values of the imperfections of the order 3×10^{-3} the erosion of the buckling load is around 16%.

8 Influence of Simple Interaction Between Local and Overall Buckling on Postbuckling of Thin-Walled Compression Members

8.1 Introduction

The extensive use of cold formed members in aeronautical, mechanical, naval constructions and civil engineering buildings has conferred primary importance to buckling and postbuckling analysis of these structural elements. Two basic types of buckling are possible: the local buckling assumes that the line junctions between intersecting plates remain straight and the cross-sectional shape of the member undergoes distortion; overall buckling is characterized by no distortion of the cross-section in its plane. Consequently local buckling analysis is based on shell model as a plate assemblage and overall buckling

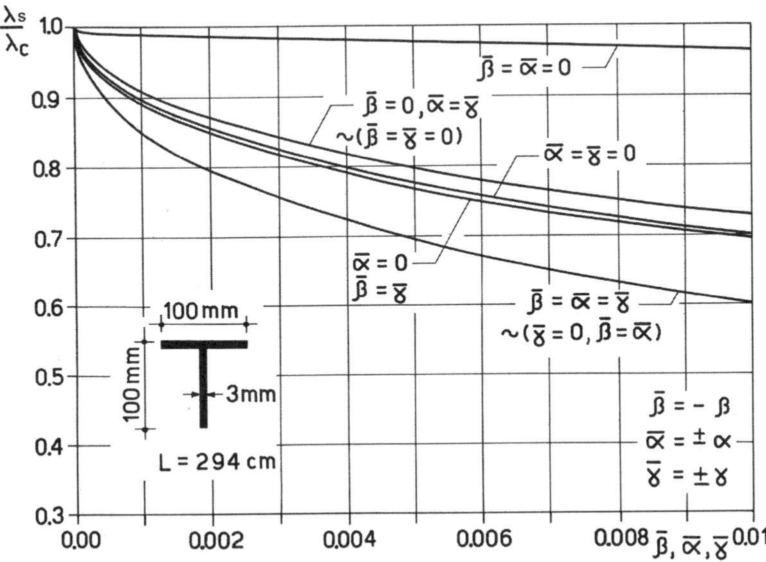

Figure 30. Sensitivity of a T cross-section to initial geometric axis imperfection

analysis on beam model (Vlasov). If two (or more) buckling modes occur simultaneously, postbuckling behaviour of the structure can be unstable even if postbuckling corresponding to each of the single buckling modes is stable. In case of interaction at the higher order terms of two buckling modes we speak of *simple interaction*; if the interactive buckling modes are more than two, we speak of *multiple interaction*. Simple and multiple interactions in thin-walled members usually regard the interaction between one overall buckling mode of the type of Euler, flexural-torsional buckling (one half-wave mode in general) and one or many local buckling modes, with many half-waves in the axial direction. Multiple interaction between several local buckling modes may also occur in shell structures. In all these cases the structure exhibits sensitivity to initial imperfections.

The interaction behaviour in thin-walled structural elements under compression has received a great deal of attention in the past years by many authors. Apart from few pioneering works which make use of analytical algorithms (Graves-Smith, 1967; Grimaldi and Pignataro, 1979; Meyer and Van der Neut, 1970; Van der Neut, Springer Verlag, Berlin 1969) or a theoretical-experimental work (Maquoi and Massonnet, 1976), more recent works have privileged numerical methods. Hundreds of papers have appeared which utilize the finite element method and we mention only very few of them (Casciaro, Garcea et al., 1998; Di Lanzo and Garcea, 1996; Garcea, 2001). However, the most competitive one for prismatic structures subjected to end loads seems to be the finite strip method. Plank and Wittrik (Plank and Wittrick, 1974) and later Graves-Smith and Sridharan (Graves-Smith and Sridharan, 1978) have determined the critical load of flat-walled structures as assemblage of plates by assuming that the fundamental path

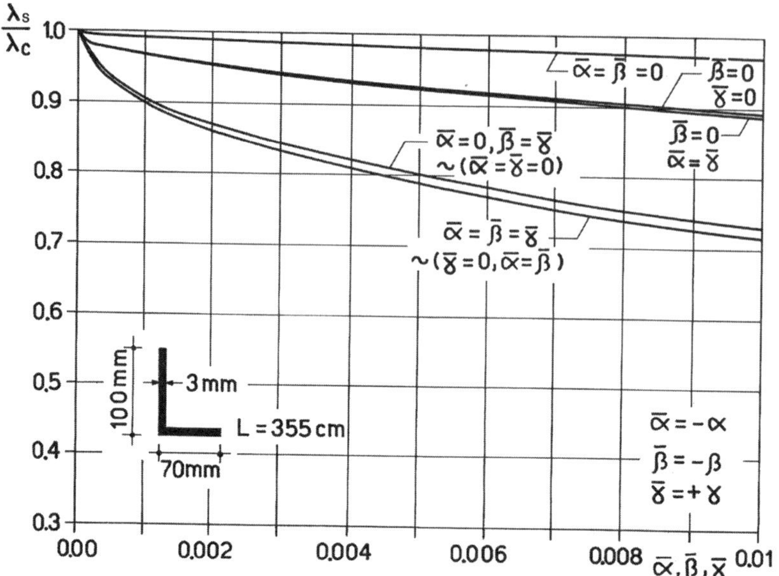

Figure 31. Sensitivity of an asymmetric angle cross-section to initial geometric axis imperfection

is linear and nonlinear, respectively. Hancock (Hancock, 1978) has analysed the local, distorsional and flexural-torsional buckling of I sections; Yoshida and Maegawa (Yoshida and Maegawa, 1978) have examined local and overall buckling of uniformly compressed I sections by taking into account the effect of residual stresses and using an elastic-plastic constitutive law. Recently Plank (Plank, 1983) has used the finite strip method to study the buckling of anisotropic plates.

A different approach has been used by Benito and Sridharan (Benito and Sridharan, 1983; Sridharan and Benito, 1984), Sridharan and Ali (Sridharan and Ali, 1985) who have employed the finite strip method within the frame of Koiter general theory of stability. They have analysed simultaneous and nearly simultaneous buckling modes for a number of sections. Along the same lines of approach are to be considered the papers by Pignataro et al. (Pignataro and Luongo, 1985,1987; Pignatro, Luongo et al.,1985), Luongo and Pignataro (Luongo and Pignataro, 1984, 1986) who have investigated interaction effect of simultaneous buckling modes on postbuckling of uniformly compressed simply supported channels.

8.2 Survey of Finite Strip Method: Plate Model

To utilize the finite strip method, a suitable plate model must be introduced first. By referring to a plate element of length l, width b and thickness t, uniformly compressed in the x (longitudinal) direction, we have for the total potential energy

$$\phi_e = \frac{Et}{2(1-\nu^2)} \int_0^b \int_0^l [\varepsilon_x^2 + \varepsilon_y^2 + 2\nu\varepsilon_x\varepsilon_y + \frac{1}{2}(1-\nu)\gamma_{xy}^2]\, dxdy$$

$$+ \frac{Et^3}{24(1-\nu^2)} \int_0^b \int_0^l [\chi_x^2 + \chi_y^2 + 2\nu\chi_x\chi_y + \frac{1}{2}(1-\nu)\chi_{xy}^2]\, dxdy \qquad (8.1)$$

$$- \lambda \int_0^b N_x[u(0,y) - u(l,y)]\, dy$$

where E is the Young's modulus and ν the Poisson's ratio. The expressions for in-plane strains and curvatures are

$$\varepsilon_x = u_{,x} + \frac{1}{2}(v_{,x}^2 + w_{,x}^2) \qquad \chi_x = w_{,xx}$$

$$\varepsilon_y = v_{,y} + \frac{1}{2}(u_{,y}^2 + w_{,y}^2) \qquad \chi_y = w_{,yy} \qquad (8.2)$$

$$\gamma_{xy} = u_{,y} + v_{,x} + w_{,x}w_{,y} \qquad \chi_{xy} = 2w_{,xy}$$

where an index preceeded by a comma denotes differentiation with respect to the corresponding variable.

For the additional energy term due to initial imperfections we take

$$\psi_e = -\frac{Et}{(1-\nu^2)} \int_0^b \int_0^l [u_{,x}\bar{u}_{,x} + \nu(u_{,x}\bar{v}_{,y} + \bar{u}_{,x}v_{,y}) + v_{,y}\bar{v}_{,y} + \frac{1-\nu}{2}(u_{,y} + v_{,x})(\bar{u}_{,y} + \bar{v}_{,x})]\, dxdy$$

$$+ \frac{Et^3}{12(1-\nu^2)} \int_0^b \int_0^l [w_{,xx}\bar{w}_{,xx} + \nu(w_{,xx}\bar{w}_{,yy} + \bar{w}_{,xx}w_{,yy}) + w_{,yy}\bar{w}_{,yy}$$

$$+ 2(1-\nu)w_{,xy}\bar{w}_{,xy}]\, dxdy$$

$$\qquad (8.3)$$

with $\bar{u}, \bar{v}, \bar{w}$ initial imperfection components.

8.3 Discrete Model

Let us consider a prismatic shell made of M plates (finite strips) rigidly connected along N lines (nodal lines) and with continuous supports at the ends. Longitudinal edges can be either free or supported. By introducing an orthogonal cartesian reference frame X, Y, Z with the X axis parallel to the nodal lines, we may describe the position of the generic n-th nodal line by assigning the displacement components $U_n(x)$, $V_n(x)$, $W_n(x)$ in the X, Y, Z direction, respectively, and the rotation $\theta_n(x)$ (Figure 32).

In order to obtain a discrete solution to our analysis it is necessary to express these functions as a series expansions suitable to satisfy the end conditions (free longitudinal displacements)

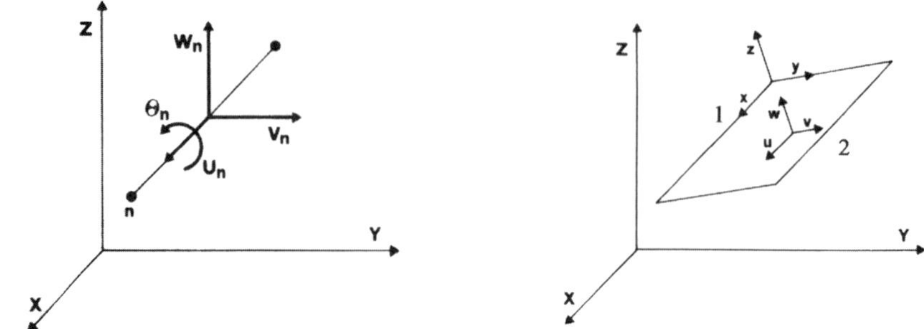

Figure 32. Nodal line and global displacement parameters

Figure 33. Finite strip and local reference frame

$$U_n(x) = \sum_{k=1}^{r} U_{nk} \cos\left(k\frac{\pi x}{l}\right) \qquad V_n(x) = \sum_{k=1}^{r} V_{nk} \sin\left(k\frac{\pi x}{l}\right)$$

$$W_n(x) = \sum_{k=1}^{r} W_{nk} \sin\left(k\frac{\pi x}{l}\right) \qquad \theta_n(x) = \sum_{k=1}^{r} \theta_{nk} \sin\left(k\frac{\pi x}{l}\right) \tag{8.4}$$

Let us now consider the generic finite strip and let us express the local displacement components as a function of the local nodal parameters through the relationships (Figure 33)

$$u^e(x,y) = \sum_{k=1}^{r} [f_1(y)u_{1k} + f_2(y)u_{2k}]\cos\left(k\frac{\pi x}{l}\right)$$

$$v^e(x,y) = \sum_{k=1}^{r} [f_1(y)v_{1k} + f_2(y)v_{2k}]\sin\left(k\frac{\pi x}{l}\right) \tag{8.5}$$

$$w^e(x,y) = \sum_{k=1}^{r} [f_3(y)w_{1k} + f_4(y)\theta_{1k} + f_5(y)w_{2k} + f_6(y)\theta_{2k}]\sin\left(k\frac{\pi x}{l}\right)$$

In eqs. (8.5) the $f_i(y)$'s ($i = 1,..., 6$) are linear or cubic polinomials and the coefficients $u_{1k}, v_{1k}, \ldots, \theta_{2k}$ represent the displacement amplitudes along the sides 1 and 2 of the strip associated with the k-th harmonic. The strip displacement parameters $\{u_k\} = \{u_{ik}, v_{ik}, w_{ik}, \theta_{ik}\}^T$ ($i = 1, 2$) are correlated with those of the adjacent nodal lines $\{U_k\} = \{U_{nk}, V_{nk}, W_{nk}, \theta_{nk}\}^T$ through the rotation of the reference system

$$\{u_k\} = [R]\{U_k\} \tag{8.6}$$

where $[R]$ is the rotation matrix. This permits to ensure compatibility between the plates edges and the nodal lines. By introducing (8.5) into (8.1) and (8.3) through the kinematic relationships (8.2) and summing the contributions from all strips, the total potential energy of the structure is obtained

$$\Phi = \sum_e [\phi_e(u^e, v^e, w^e) + \psi_e(u^e, v^e, w^e)] = \Phi(U_{nk}, V_{nk}, W_{nk}, \theta_{nk}) \tag{8.7}$$

where use has been made of eq. (8.6).

8.4 Buckling of Uniformly Compressed Stiffened and Unstiffened Channels Simply Supported at the ends

The finite strip method has been utilized to investigate the critical behaviour of channels under uniform compression simply supported at the ends. The following types of members have been analysed : a) unstiffened channels; b) stiffened web channels; c) channels with stiffened flanges; d) channels with stiffened web and flanges. Besides the range of geometric parameters which allow for interaction between two (or more) buckling modes has been investigated.

The four types of channels can exhibit an overall buckling mode (Eulerian or flexural-torsional) or a local buckling mode according to the geometry. For unstiffened and stiffened channels of constant thickness the loss of shape of the cross-section has little influence on the evaluation of the overall critical stress (Luongo and Pignataro, 1984,1986; Pignataro, Luongo et al., 1985). Therefore in these cases the analysis can be carried out with sufficient accuracy by applying the Vlasov beam theory. The curves obtained by plotting the critical stress against the length for fixed values of the other geometric parameters are qualitatively the same for the four types of channels.

To analyse the local buckling behaviour, we shall assume that the channel is sufficiently short in order to rule out overall buckling. By posing (Figure 34)

$$\alpha = \frac{l}{h}, \qquad \beta = \frac{b}{h}, \qquad \gamma = \frac{t}{h}, \qquad \delta = \frac{d}{h} \tag{8.8}$$

one gets

$$\sigma_c = k \frac{\pi^2}{12} \frac{E}{1 - \nu^2} \gamma^2 \tag{8.9}$$

where, in general, $k = k(\beta, \gamma, \delta)$. For unstiffened channels it is simply $k_0 = k_0(\beta)$.

Figure 35 shows a plot of k_0 against β (Pignataro, Luongo et al., 1985). It is observed that for $\beta \to 0$, the value $k_0 = 4$ corresponding to a plate simply supported along the longitudinal edges (with α an integer number) is obtained. By increasing β up to 0.2, approximately, it is found that k_0 increases and then decreases by recovering the value $k_0 = 4$ by correspondence with $\beta = \bar{\beta}$=0.3678. This is due to the fact that for small values of β flanges have a stabilizing effect and for $\beta > 0.2$ their stiffening effect diminishes. For $\beta > \bar{\beta}$ the destabilizing effect of flanges prevails and the curve drops rapidly down.

For channel b) a number of results have been already obtained in Luongo and Pignataro (Luongo and Pignataro, 1984). Here we present perhaps the most interesting ones (Figure 36) which shows the dependence of k/k_0 on δ for $\gamma = 0.01$. The three curves correspond to $\beta = 0$ (simply supported plate with a longitudinal stiffener at midspan), β

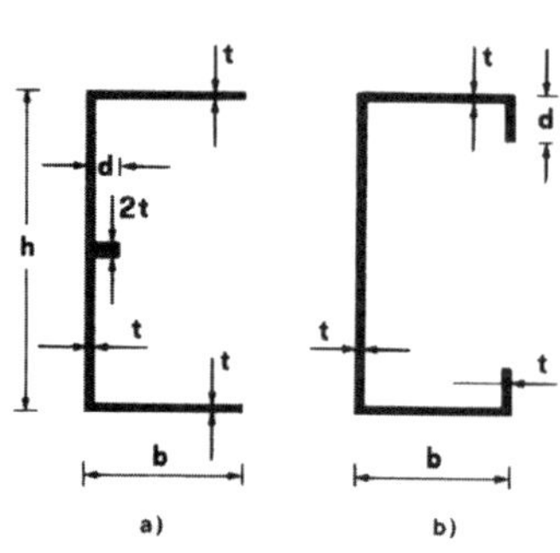

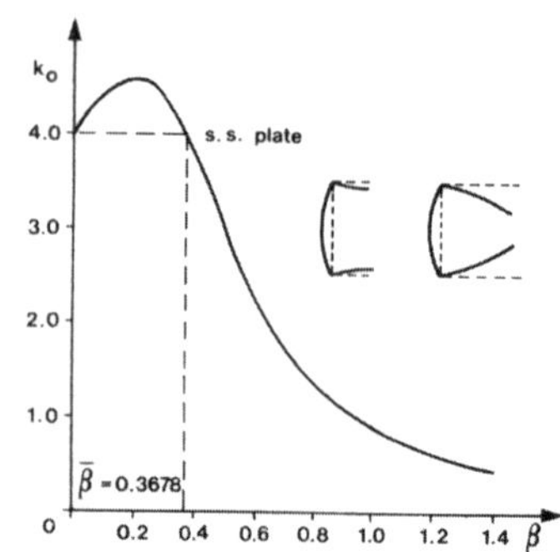

Figure 34. Channel a) with stiffened web, b) with stiffened flanges

Figure 35. Critical load of unstiffened channel against aspect ratio

$= 0.3$, $\beta = 0.6$. For $\beta = 0$ and $\beta = 0.3$, k/k_0 increases with δ up to a certain value $\delta_0 = \delta_0\,(\beta)$ to which correspond two local symmetric buckling modes, a short wave and a long wave one. For $\delta > \delta_0$, k/k_0 is practically constant and keeps the value corresponding to $\delta \to \infty$.

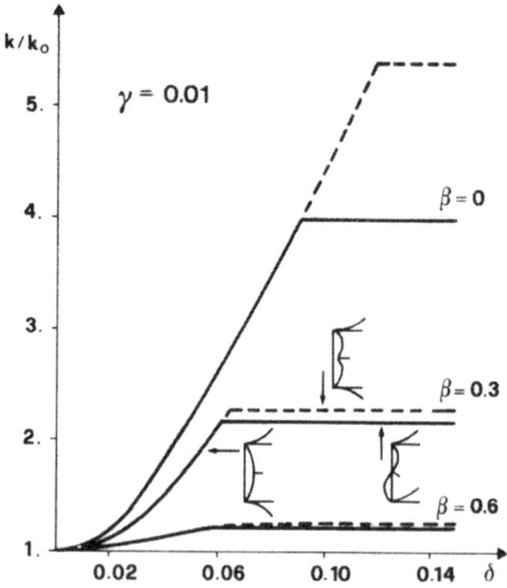

Figure 36. Web stiffened channel: stiffening effect of stringer for various aspect ratios

The intersection of the rising path and of the horizontal dashed line characterizes the geometric parameters responsible for the interaction of symmetric buckling modes. Cross-section distortions are sketched in (Figure 36). For $\beta = 0.6$ the channel has a qualitatively different behaviour since the corresponding curve passes smoothly from a

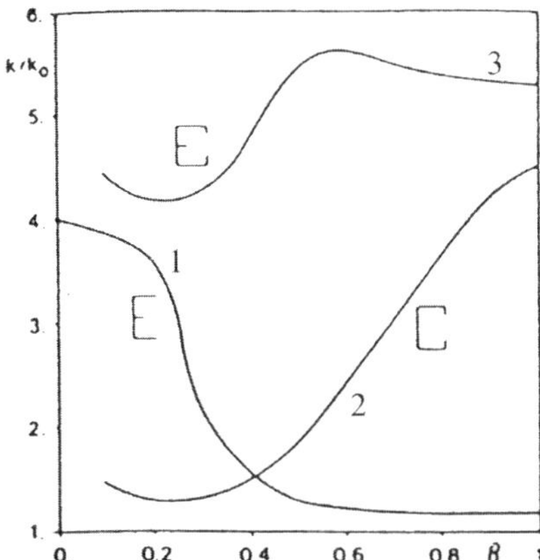

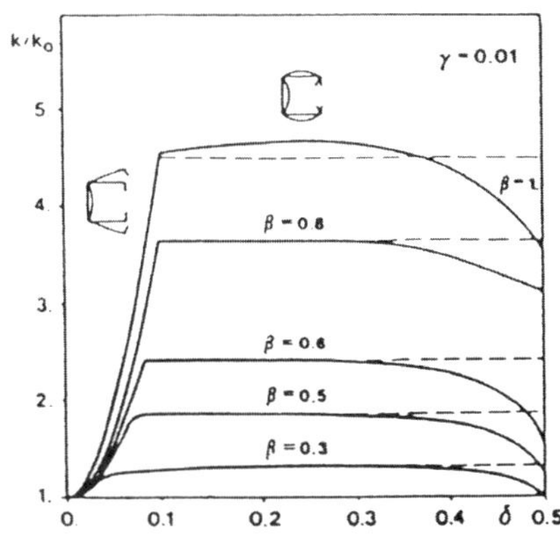

Figure 37. Limiting stiffening effects of stringers

Figure 38. Lipped channels: stiffening effect of stringers for various aspect ratios

primary ascending path to its asymptotic value. Therefore there are no points on this curve characterizing interaction between local symmetric buckling modes.

In (Figure 36) the dependence of k/k_0 on δ relative to the antisymmetric buckling mode has been also represented. Under the assumption that the stiffener has no torsional rigidity, the diagram is represented by the heavy horizontal lines. These lines lie always below the dashed horizontal ones previously described and therefore they are the only ones to deserve practical interest. By denoting with $\delta^* = \delta^*(\beta)$ the value of δ corresponding to the intersection of these lines with the rising paths we have, for each β, the following results: for $\delta < \delta^*$ a symmetric buckling mode occurs, for $\delta > \delta^*$ the antysimmetric mode is prevailing. In particular for $\delta = \delta^*$ the two buckling modes occur simultaneously. Correspondingly $k = k(\beta, \gamma, \delta)$ for $\delta < \delta^*$ and $k = k(\beta, \gamma)$ for $\delta > \delta^*$.

A global view of the maximum. stiffening effect of the stiffener is furnished by curve 1 in Figure 37 (Luongo and Pignataro, 1986) where k/k_0 against β has been plotted, k giving the value of σ_{cr} corresponding to the antisymmetric mode. One may observe that for small values of β and an appropriate choice of the stiffener, a critical stress which is about four times the one corresponding to the unstiffened channel can be attained. For increasing β the stringer stiffening effect diminishes and k/k_0 approaches an asymptotic value slightly above one for $\beta > 0.6$. Therefore, for these values of β there is little advantage in employing stiffened rather than unstiffened channels.

Let us now examine channels c), namely lipped and hat channels. Short members will again be considered in order to rule out overall buckling. Results obtained are shown in Figure 38 where k/k_0 againts δ for several values of β and fixed γ has been plotted (Luongo and Pignataro, 1986). All curves are characterized by an ascending path and a second path which is practically constant for a wide range of values of δ. The first path corresponds to a symmetric local buckling mode where the tips of the flanges displace, the second path corresponds to a buckling mode where the four corners remain fixed. The intersection points of the two paths characterize the geometric parameters responsible for the interaction of the two buckling modes.

In Figure 38 horizontal dashed straight lines can also be observed corresponding to the critical stress of unstiffened channels with hinges at the tip of the flanges. These straight lines approximate with sufficient accuracy the second path of the curves within the range of values of δ where it remains practically constant. From a mechanical point of view this can be explained by considering that for not too large values of δ, the effect of the hinges is analogous to that of the stiffeners since their torsional rigidity is negligeable. For large values of δ ($\delta > 0.3$) the stiffeners effect is destabilizing and therefore the second path of the curves drops below the horizontal lines. Along both paths of the curves the dependence of k on γ is analogous to that discussed for channels b). Note that diagram in Figure 38 is valid both for lipped and hat channels since the strain energy associated with the corresponding buckling modes is the same.

Values of k/k_0 vs β corresponding to the intersection of the ascending and of the horizontal path in Figure 38 are represented by curve 2 in Figure 37. By comparison with curve 1, one can draw the conclusion that for β less than 0.45, approximately, channels with stiffened webs are of more practical advantage, whereas for β larger than this value lipped and hat channels have to be preferred. For β around 0.45 the increment of the critical stress for channel b) or c) with respect to unstiffened channels is modest and therefore the most favourable solution in this situation is attained by employing lipped or hat channels with stiffened webs (curve 3 in fig. 37).

Before ending this section we wish to discuss in some detail the conditions which allow for the occurrence of overall and local buckling in unstiffened channels since relations in closed form are avalaible in this case. By equating two by two the Euler and flexural-torsional critical stress furnished by Vlasov theory and the local critical stress furnished by eq. (8.9) the simple geometric relations characterizing simultaneous buckling modes are obtained

$$\alpha\gamma = \phi_i(\beta) \qquad (i = 1,\ 2,\ 3) \tag{8.10}$$

The functions $\phi_i(\beta)$ have been obtained in Pignataro, Luongo et al. (Pignataro, Luongo et al., 1985) and are plotted in Figure 39. For each value of β we may select infinite pairs of values of α and γ which satisfy eq. (8.10) such that two critical stresses coincide, their actual value depending on α and γ.

From an engineering point of view it is interesting to determine in the space of the geometrical parameters the domains inside which single buckling modes occur and the boundaries of two adjacent domains along which two critical stresses coincide. After some simple manipulations it can be easily shown that the inequalities

$$\sigma_{FT} \gtreqless \sigma_E \qquad\qquad \sigma_L \gtreqless \sigma_E \qquad\qquad \sigma_L \gtreqless \sigma_{FT} \tag{8.11}$$

are satisfied if the geometrical parameters verify the relations

$$\alpha\gamma \gtreqless \phi_1(\beta) \qquad\qquad \alpha\gamma \gtreqless \phi_2(\beta) \qquad\qquad \alpha\gamma \gtreqless \phi_3(\beta) \tag{8.12}$$

In eqs. (8.11) σ_L, σ_E, σ_{FT} denote the local, the Euler and the flexural-torsional critical stress, respectively

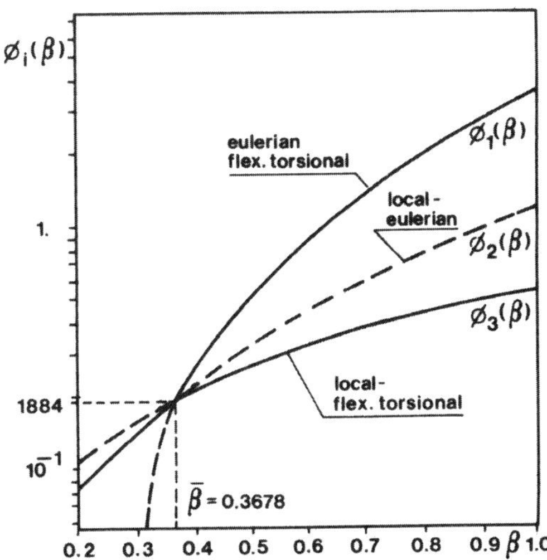

Figure 39. Functions φ_i in eq. (8.10) vs β

Inequalities (8.12) can be usefully employed to verify whether or not the critical stress corresponding to two interacting modes is the smallest possible in the column. If we consider, for instance, the interaction Eulerian/flexural-torsional, the corresponding curve $\phi_1(\beta)$ is meaningful only if $\sigma_{FT} = \sigma_E < \sigma_L$, i.e. by using eqs. (8.11) and (8.12), for $\phi_1(\beta > \phi_2(\beta)$ which is verified for $\beta > \bar{\beta} = 0.3678$. For $\beta < \bar{\beta}$ the situation is reversed and the local critical stress prevails, the Eulerian/flexural-torsional buckling interaction occurring at a higher value of the loading. In a completely similar manner it can be shown that $\sigma_E = \sigma_L < \sigma_{FT}$ if $\phi_2(\beta) < \phi_3(\beta)$ which takes place for $\beta < \bar{\beta}$ and besides $\sigma_{FT} = \sigma_L < \sigma_E$ if $\phi_3(\beta) < \phi_2(\beta)$ which occurs for $\beta > \bar{\beta}$.

In conclusion, for $\beta > \bar{\beta}$ two different types of interaction can manifest, namely the Eulerian/flexural-torsional or the local/flexural-torsional; for $\beta < \bar{\beta}$ only the local/Eulerian interaction is possible. For $\beta = \bar{\beta}$ the three buckling modes occur simultaneously. Note that the curve $\phi_1(\beta)$ is not defined for values of $\beta < \bar{\beta}^* = 0.303$ and therefore the Eulerian/flexural-torsional interaction can never exist below this value of β. In Figure 39, the meaningful branches of the curves have been represented with heavy lines; dashed branches correspond to cases of no practical interest since there is always a critical stress lower than that corresponding to the associated simultaneous buckling modes singled out by that curve.

8.5   Interaction Postbuckling Analysis

Interaction postbuckling analysis is performed on the basis of the general theory presented in sections 4,5 by restricting the investigation to the cases of two interacting buckling modes only. Before presenting numerical results we find however worthy to start with a qualitative discussion entirely based on energy considerations. Remembering that only third order terms of the total potential energy are considered in our analysis, we fix our attention on what seems to be the dominant term for the interaction effect, namely

$$\Phi_3 = \frac{Et}{2(1-\nu^2)} \int_0^l \int_C u_{,x}(v_{,x}^2 + w_{,x}^2)\, dx\, ds \tag{8.13}$$

where C is the cross-section center line.

On the basis of the general theory of stability, the postbuckling equilibrium path is symmetric if for any combination of the simultaneous buckling modes is $\Phi_3 = 0$, antisymmetric if there exists at least one combination of the simultaneous buckling modes such that $\Phi_3 \neq 0$. By remembering that the only non vanishing displacement component for the local buckling mode is w and that overall buckling mode has all displacement components different from zero, we can draw the following conclusions on the basis of geometric considerations only (Luongo and Pignataro, 1986).

In Table 2 Φ_3^{ijk} represents the integral (8.13) evaluated by correspondence with the buckling modes $\mathbf{v}_{1i}$, $\mathbf{v}_{1j}$, $\mathbf{v}_{1k}$. Indices 1 and 2 denote the overall and the local buckling mode, respectively, and n is the number of longitudinal half-waves. One can see therefore that in the interaction Eulerian/local buckling mode, postbuckling behaviour is always antisymmetric. In the interaction flexural-torsional/local, the postbuckling equilibrium path is symmetric if the local buckling mode is antisymmetric; it is antisymmetric if the local buckling mode is symmetric and n is odd. The last type of interaction has a weak effect on the postbuckling behaviour of the channel since it approaches zero as $n \to \infty$. In this case results obtained in Benito and Sridharan (Benito and Sridharan, 1983) are recovered. The presence of an adequate stiffener on the web is however desirable since it renders postbuckling behaviour symmetric.

	Eulerian/local		flexural-torsional/local	
	loc. symm.	loc. ant.	loc. symm.	loc. ant.
Φ_3^{112}	$\begin{cases} = 0 & n\ even \\ \neq 0 & n\ odd \end{cases}$	$=0$	$\begin{cases} = 0 & n\ even \\ \neq 0 & n\ odd \end{cases}$	$=0$
Φ_3^{122}	$\neq 0$	$\neq 0$	$=0$	$=0$

Table 2. Third order terms of total potential energy

For a numerical investigation of the interaction postbuckling problem, we analyse only the interaction between Euler and local buckling for perfect and imperfect members. Cross-section distortions considered are shown in the following figure where P is a particular point.

In Figures 41,42,43, the postbuckling equilibrium path relative to the interaction between Eulerian and local buckling modes has been represented for channel with reinforced flanges ($\delta = 0.08$), channel with reinforced web ($\delta = 0.07$) and unstiffened channel. In all cases $\beta = 0.3$ and $\gamma = 0.01$. Note that $\tilde{v}^P = v^P/h$ and $\tilde{\bar{u}}^P = \bar{u}^P/h$ denote the nondimensional displacement of P and the nondimensional imperfection measured at P (Figure 40).

Finally in Figure 44 the snapping load λ_s against the initial imperfections $\tilde{\bar{u}}^P$ has been plotted. It is clear that the most severe situation is determined by the simultane-

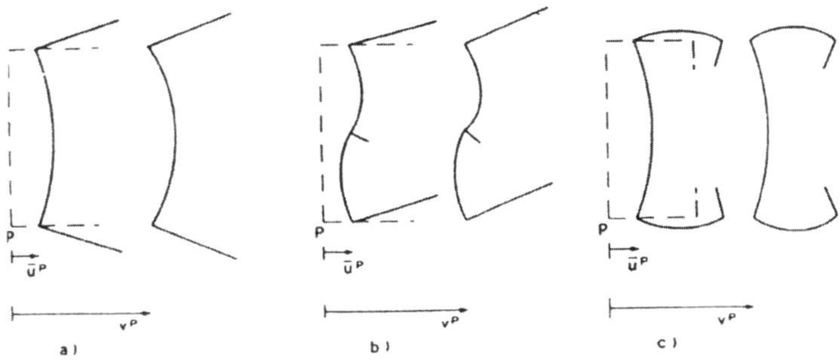

Figure 40. Cross section distorsions of imperfect channels

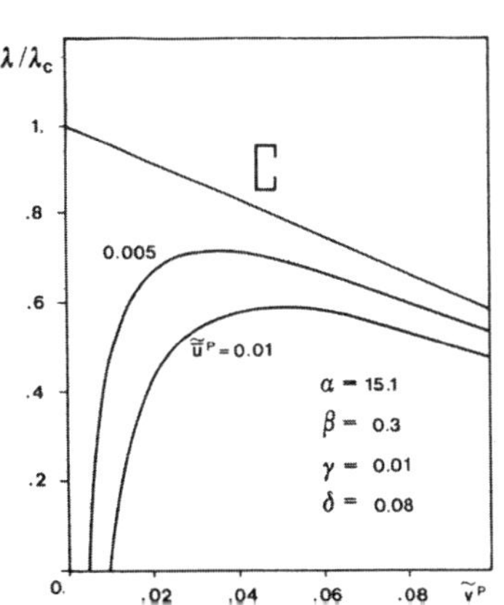

Figure 41. Postbuckling equilibrium paths for perfect and imperfect lipped channel

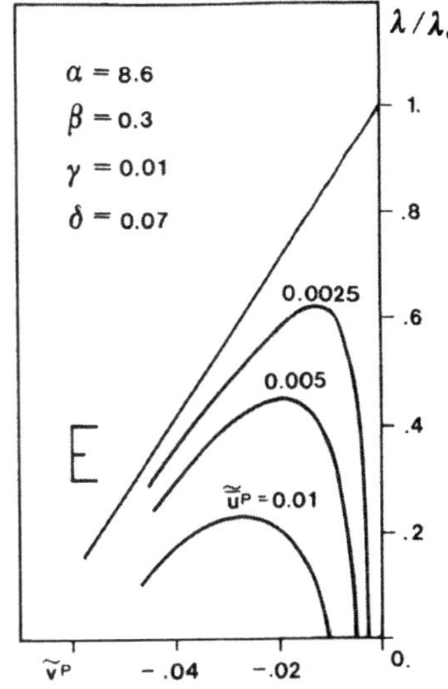

Figure 42. Postbuckling equilibrium paths for perfect and imperfect web-stiffened channel

ous occurrence of Eulerian and local (antisymmetric) buckling mode in web reinforced channels since sensitivity to initial imperfections is more pronounced.

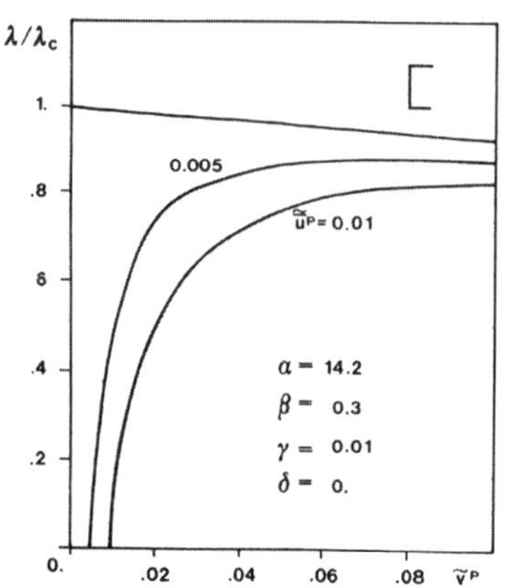

Figure 43. Postbuckling equilibrium paths for perfect and imperfect unstiffened channel

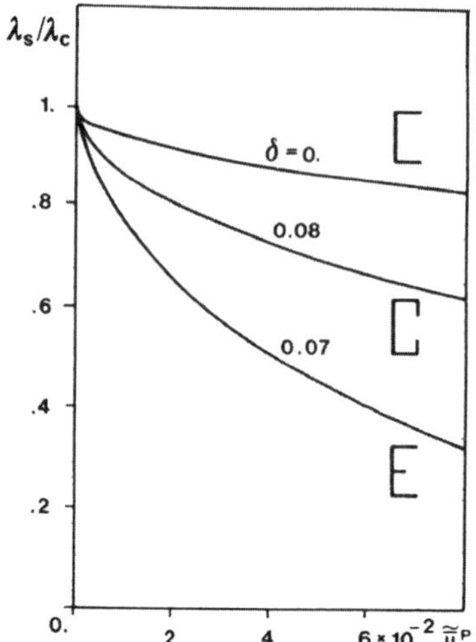

Figure 44. Snapping load against initial imperfection amplitude

9 Influence of Multiple Interactive Buckling on Postbuckling of Thin-Walled Members in Compression

9.1   Introduction

As already mentioned in previous chapters the postcritical behaviour of thin-walled-members (TWM) has been widely investigated in the last decades. In particular, the attention has been focused more recently on the effects of the interaction of simultaneous buckling modes on the postbuckling behaviour of the structure (Benito, 1983; Benito and Sridharan, 1983, 1984-1985; Byskov, 1986; Graves-Smith, 1967; Graves-Smith and Sridharan, 1978; Grimaldi and Pignataro, 1979; Hancock, 1978; Pignataro, Luongo et al., 1985; Plank 1983; Plank and Writtick, 1974; Sridharan and Ali, 1985; Sridharan and Benito, 1984; Van der Neut, 1969). Most of the results so far presented, however, regard the interaction between one overall and one local buckling mode (*single interaction*) and are therefore valid for moderately long members. For long members an investigation of the interaction must necessarily take into account the presence of more local modes which occur under (nearly) the same critical stress (*multiple interaction*).

9.2   Multiple Interaction

Let us consider the case of multiple interaction between an Eulerian buckling mode and m local modes in a simply supported beam. System (4.15) consists in this case of $m + 2$ nonlinear equations in $m+2$ unknowns ν_1 , ... , ν_{m+1}, λ_1 and its solution requires in general a great computational effort. The analysis can be however simplified by introducing the following assumptions:

i) In the local buckling modes joints remain fixed. Consequently, for the generic plate element the j-th mode is represented by the displacements

$$u_j(x, y) = 0$$
$$v_j(x, y) = 0$$
$$w_j(x, y) = a_j W_j(y) \sin n_j \pi x / l \qquad (j = 2, 3, \ldots, m + 1) \tag{9.1}$$

where n_j is a sufficiently large number.

ii) In the overall mode the TWM buckles as a shear undeformable beam with free transverse expansion and the displacement field is described by

$$u_1(x, y) = a_1 U_1(y) \cos \pi x / l$$
$$v_1(x, y) = a_1 V_1(y) \sin \pi x / l \tag{9.2}$$
$$w_1(x, y) = a_1 W_1(y) \sin \pi x / l$$

which corresponds to simply supported edges with free warping.

In eqs. (9.1), (9.2), local reference axes x and y are as shown in Figure 33. Then $A_{ijk} = 0$ for $i, j, k > 1$ and

$$A_{1jk} = \int_S Et(u_{1,x} w_{j,x} w_{k,x}) dS \qquad (j, k = 1, 2, \ldots, m + 1) \tag{9.3}$$

where S is the middle surface of the TWM and t the plate thickness.

The dominant coefficients obtained by eq. (9.3) are those for which the number of longitudinal halfwaves n_j is of the order of n_k. For $|n_j - n_k| >> 0$ the corresponding coefficients are much smaller and in particular they can be neglected if they are of the type A_{11k}, for $n_k >> 1$. Equations (4.15) then simplify into

$$A_{1jk} v_j v_k - 2 \frac{\lambda_1}{\lambda_c} v_1 = 0 \tag{9.4}$$

$$(A_{1jk} - \delta_{jk}\eta) v_k = 0 \qquad (j, k = 2, 3, \ldots, m + 1) \tag{9.5}$$

where

$$\eta = \frac{\lambda_1}{v_1 \lambda_c} \tag{9.6}$$

Note that $A_{111} = 0$ because of hypothesis ii). Results (9.5) can be extended to beams with different boundary conditions if eqs. (9.2) are replaced with appropriate equations.

Equations (9.5) are a linear eigenvalue problem whose solution furnishes m real eigenvalues. Denoting by η_h, $v_k^{(h)}$ the h-th solution, the corresponding values of $v_1^{(h)}$ and $\lambda_1^{(h)}$ can be obtained by solving eqs. (9.4) and (9.6) giving

$$v_1^{(h)} = \frac{1}{\sqrt{3}}, \qquad \frac{\lambda_1^{(h)}}{\lambda_c} = \frac{\eta_h}{\sqrt{3}} \tag{9.7}$$

As a first illustration of the theory let us consider the case of simple interaction ($m = 1$). Equation (9.5) furnishes $\eta = A_{122}$ and therefore from eqs. (9.7) and (9.3) the simple formula

$$\frac{\lambda_1}{\lambda_c} = \frac{Et}{\sqrt{3}} \int_S u_{1,x} w_{2,x}^2 \, dS \tag{9.8}$$

follows with $\nu_2 = \pm\sqrt{2/3}$.

If the cross-section has one axis of symmetry, $u_{1,x}$ is a symmetric or antisymmetric function according to whether the overall buckling is Eulerian or flexural-torsional, respectively. Consequently $\lambda_1/\lambda_c \neq 0$ in the E-local interaction and $\lambda_1/\lambda_c = 0$ in the FT-local interaction both for symmetric and antisymmetric local modes.

As a second illustration of the theory let us now examine the case of the overall-local-local buckling interaction. The eigenvalue problem (9.5) is

$$\begin{bmatrix} A_{122} - \eta & A_{123} \\ A_{123} & A_{133} - \eta \end{bmatrix} \begin{Bmatrix} \nu_2 \\ \nu_3 \end{Bmatrix} = \begin{Bmatrix} 0 \\ 0 \end{Bmatrix} \tag{9.9}$$

which leads to the following cases:

(i) $A_{123} \neq 0$, the three buckling modes are coupled and there are two postbuckling equilibrium paths;

(ii) $A_{123} = 0$, $A_{122} \neq A_{133} \neq 0$, the overall mode couples with the local mode 2 or 3 and two postbuckling paths occur with $(\lambda_1/\lambda_c)_{max} = \max(A_{122}, A_{133})/\sqrt{3}$;

(iii) $A_{123} = 0$, $A_{122} = A_{133} \neq 0$, the three buckling modes are coupled and there is an infinite number of postbuckling paths since in our approximate analysis the ratio ν_2/ν_3 is indeterminate;

(iv) $A_{123} = A_{122} = A_{133} = 0$, third-order analysis furnishes $(\lambda_1/\lambda_c) = 0$ and postbuckling paths can be determined only by fourth-order analysis.

Overall modes	*Local modes*			
	any	*symm.-symm.*	*symm.-ant.*	*ant.-ant.*
Eulerian	$A_{122} \neq 0$ $A_{133} \neq 0$	$A_{123} \neq 0$	$A_{123} = 0$	$A_{123} \neq 0$
Flex.-tors.	$A_{122} = 0$ $A_{133} = 0$	$A_{123} = 0$	$A_{123} \neq 0$	$A_{123} = 0$

Table 3. Coefficients A_{ijk} for overall-local interaction

If the cross-section has one axis of symmetry, one can easily show, by using eq. (9.9) and invoking geometric considerations, that the coefficients A_{ijk} take the values displayed in Table 3. It is apparent that Euler buckling can interact with two local modes both symmetric or antisymmetric. On the other hand the flexural-torsional mode interacts only with a symmetric and an antisymmetric local mode. In this case (λ_1/λ_c)

Mathematical Modelling of Instability Phenomena 67

$= \pm A_{123}/\sqrt{3}$ and $\nu_3 = \pm \nu_2 = \pm 1/\sqrt{3}$. It should be borne in mind, however, that A_{123} is very small if $|n_2 - n_3| >> 0$ and consequently the coupling is weak.

Table in Figure 45 furnishes the coefficients A_{ijk} relative to the interaction between the Euler and two local symmetric buckling modes of a lipped channel sketched in the picture. It is seen that the two off-diagonal terms are very small in comparison with the main diagonal ones, due to the fact that the wavelengths of the two local modes are very different. This implies that the three modes interaction is weak in comparison with the two modes interaction. Note that parameter $\alpha, \beta, \gamma, \delta$ are defined by (8.8).

Table in Figure 46 collects the A_{ijk}'s relative to the interaction between the Euler and three local modes sketched in the picture of a web-reinforced channel.

The term A_{123} is much smaller than the main diagonal ones for the reasons illustrated above. Terms A_{124} and A_{134} vanish since interaction between a symmetric and an antisymmetric mode is involved. As a result there exists a single interaction between the Euler and the antisymmetric mode and a weak multiple interaction between the Euler and the two symmetric local modes. Both interactions are characterised by an approximately equal slope of the bifurcated paths.

j	A_{1jk}		Local mode	c/h
2	0·052	0·19 10^{-4}		3·41
3	0·19 10^{-4}	−0·038	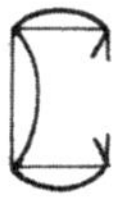	0·88

Figure 45. Lipped channel: E-local symm.-local symm. interaction ($\alpha = 30.7$, $\beta = 0.6$, $\gamma = 0.01$, $\delta = 0.065$)

Results relative to FT-local symmetric-local antisymmetric interaction for a web-reinforced channel are given in the Table in Figure 47. In contrast with the two previous results the dominant terms are the off-diagonal ones so that only multiple interaction occurs. This is due to the fact that the two local modes have the same wavelength. The slope of the bifurcated path is of the order of the largest ones obtained in the E-local interaction and therefore represents the same level of risk for the TWM.

The preceding analysis can be further simplified if the transverse shape of the local critical mode is independent of the number of longitudinal halfwaves. This is true to a good approximation whenever single plates can be considered simply supported along longitudinal edges. In this case, by normalizing the local critical modes in terms of the second order elastic energy term, it is found that the corresponding amplitudes a_i are proportional to l/n_i. In view of this results, eqs. (9.5) become

$$(c_{1jk} - \delta_{jk}\eta^*)\nu_k = 0 \tag{9.10}$$

j	A_{1jk}		Local mode	c/h
2	0·018 0	−0·000 9	0	 2·18
3		0·092 3	0	 0·73
4	Symm.		0·092 1	 0·73

Figure 46. Web-stiffened channel: E-local symm.-local antysymm. interaction ($\alpha = 28.6$, $\beta = 0.6$, $\gamma = 0.01$, $\delta = 0.08$)

where

$$c_{1jk} = \frac{1}{1 - (n_j - n_k)^2} \qquad (j, k = 2, 3, \ldots, m + 1) \tag{9.11}$$

and $\eta^* = \eta/\eta_0$ is the ratio between the eigenvalues of multiple and single interaction.

Eqs. (9.10) are independent of geometrical and mechanical properties of the structure and their solution depends on the number m of the local buckling modes only. They can be extended to beams with different boundary conditions if eqs. (9.2) are adequately modified.

Figure 48 shows that for increasing m, η^* approaches rapidly the asymptotic value 1.56 for simply supported beams and 3.60 for fixed-supported beams. The analysis of the problem at hand can therefore be carried out by determining η_0 from the problem of single interaction and multiplying the result by the *amplifying factor* η^*.

j	A_{1jk}		Local mode	c/h
2	0	0·094		1·24
3	0·094	0		1·24

Figure 47. Web-stiffened channel: FT-local symm.-local antisymm. interaction ($\alpha = 28.6$, $\beta = 0.6$, $\gamma = 0.01$, $\delta = 0.08$)

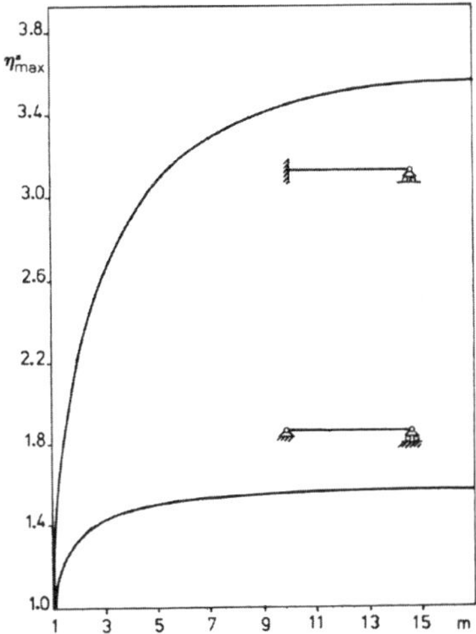

Figure 48. Amplifying factor vs number of local buckling modes

9.3   Numerical Results

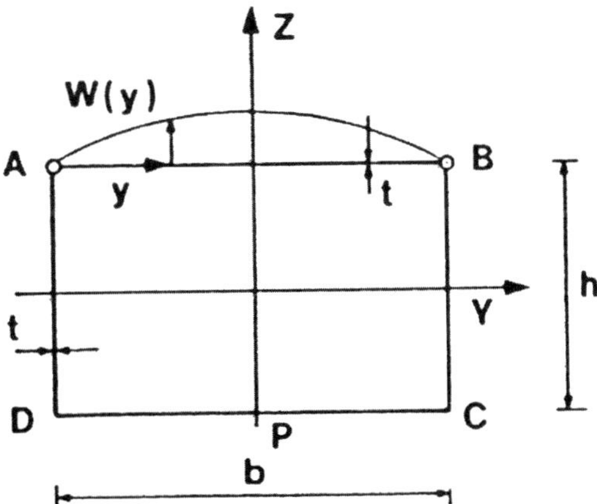

Figure 49. Cross-section of a box girder

As a further application of the theory presented (eqs. (9.5) and eqs. (9.10)) we analyse the postbuckling behaviour of the box girder simply supported at the ends whose cross-section is shown in Figure 49. In the local buckling only the plate AB simply supported along the longitudinal edges displaces according to a sinusoidal law. Therefore the transverse shape $W(y)$ is independent of the number of longitudinal halfwaves and eq. (9.10) can be applied.

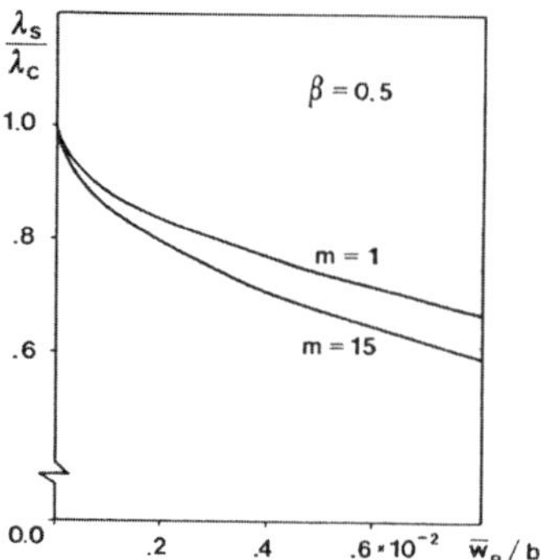

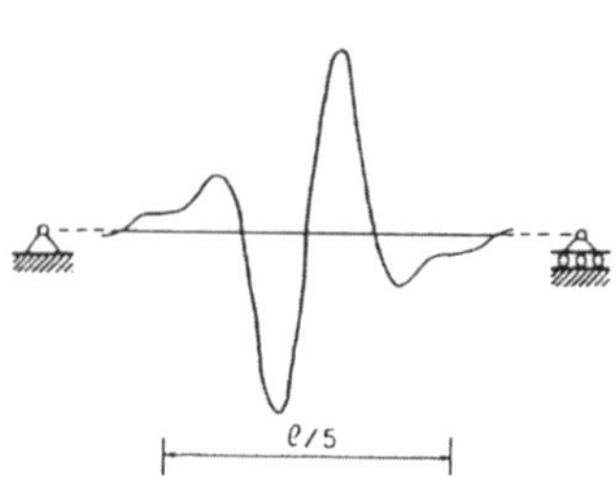

Figure 50. Limit load as a function of the imperfections amplitude for simple and multiple interaction

Figure 51. Localization phenomenon

Figure 50 shows the dependence of the snapping load on the initial imperfections for a cross-section aspect ratio $\beta = h/b = 0.5$ obtained by applying eq. (5.74) which in this case reads

$$\frac{\lambda_s}{\lambda_c} = 1 - 2\sqrt{\mu_0 \frac{\bar{W}^P}{b}} \tag{9.12}$$

being $\rho = 1$, $(\lambda_1/\lambda_c) = \mu_0 = \eta_0 b/a_1 = (12/\pi)(1+\beta)/\beta(3+\beta)$, and $\bar{\xi} = \bar{W}^P/b$, with $\bar{W}^P$ the initial displacement of a generic point P of the cross-section Figure (49). The solution to multiple interactive problems is obtained by replacing in (9.12) μ_0 with

$$\mu = \eta^* \mu_0 \tag{9.13}$$

where η^* is the amplifying factor defined in (9.10). In Figure 50 the sensitivity to initial imperfections both for single ($m = 1$) and multiple interaction ($m = 15$) is displayed. In the latter case the sensitivity increases although not drammatically. Figure 51 shows the axial dependence of the resulting lateral displacement due to all local buckling modes and evidences that the cross-section distorsion is concentrated in a narrow band of length $l/5$, approximately, around midspan. This phenomenon is known in the literature as *localization phenomenon* of lateral displacements. Further numerical results of this phenomenon are shown in Figures 52,53, where lateral displacements have been denoted with $\phi(x)$.

For different boundary conditions, by adequately modifying eqs. (9.2), the associated local modal shapes for a fixed-supported beam are obtained and illustrated in Figures 54,55.

Finally, for fixed-sliding beam we have the results depicted in Figure 56.

As a second example we consider a lipped channel shown in the Figure 57, simply supported at the ends.

Due to the difficulties in determining in closed form the local critical stress, the numerical approach of finite strip method has been employed. The coefficients A_{ijk} have

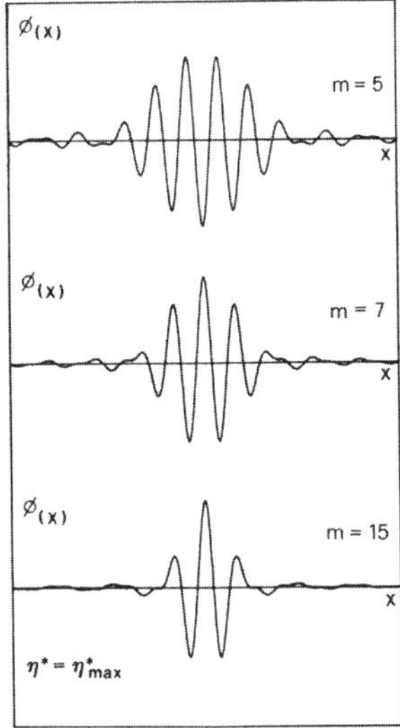

Figure 52. Localization along path of steepest descent (simply supported beam)

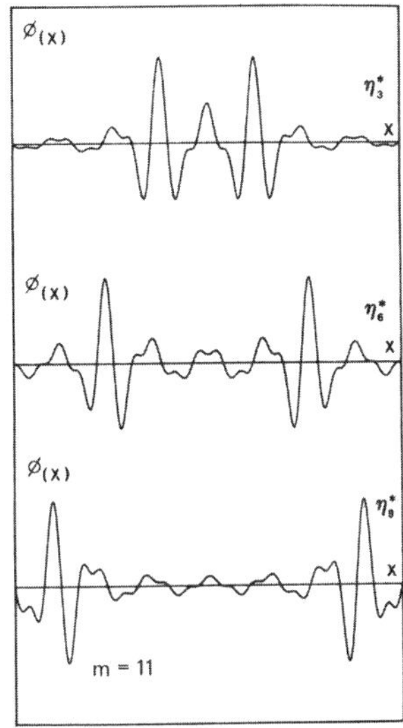

Figure 53. Localization along different bifurcated paths (simply supported beam)

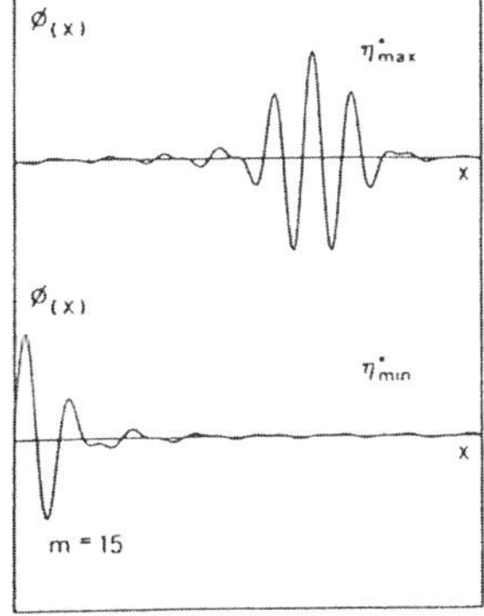

Figure 54. Localization along paths of maximum positive and negative slope (fixed-supported beam)

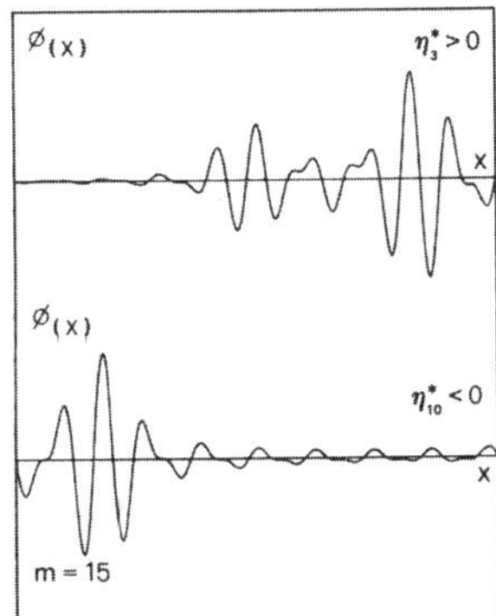

Figure 55. Localization along different bifurcated paths (fixed-supported beam)

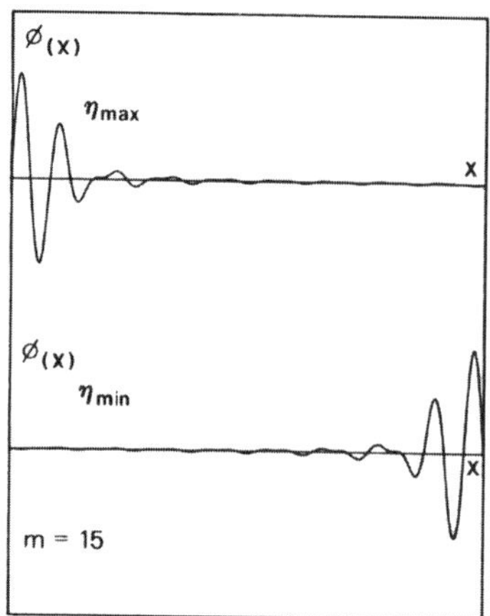

Figure 56. Localization corresponding to the extreme values of η (fixed-sliding beam)

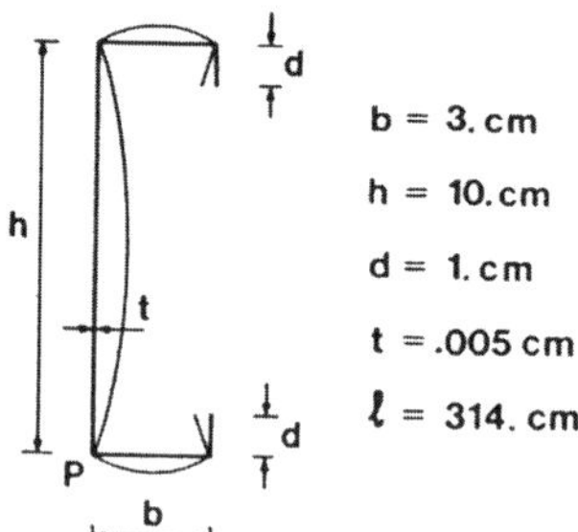

Figure 57. Lipped channel and local buckling mode

then been evaluated by removing hypotheses (9.1) and (9.2). Among these coefficients, the A_{1jk} $(j, k = 2, 3, \ldots, m+1)$ are at least one hundred time larger than the others, so that the reduction of nonlinear problem (4.15) to the linear eigenvalue problem (9.5) appears justified.

The eigenvalue problem has been solved for single and multiple interaction including different numbers of local modes. The Table 4 summarizes the results for η. For η^*, "exact" values through finite strip method and eigenvalue problem (9.2.10)(9.10) have been obtained. It should be noticed on one hand the rapid convergence of η, and on the other hand the excellent agreement between the "exact" η^*'s and the approximate ones.

m	n_i	η	$\eta^*=\eta/\eta_0$	
			f.strip	*approx*
1	1,42	-0.207	1.	1.
3	1,40,42,44	-0.297	1.436	1.439
5	1,38,40,42,44,46	-0.312	1.507	1.514
7	1,36,38,40,42,44,46,48	-0.316	1.529	1.539
9	1,34,36,38,40,42,44,46,48,50	-0.318	1.537	1.551

Table 4. Lipped channel: E-Local symm. interaction

10 A Qualitative Approach to Coupled Instabilities in Thin-Walled Beams

10.1 Introduction

Thin-walled beams are the basic ingredient of the realization of metal constructions in civil, naval and aeronautical engineering. A lot of papers have been dedicated in the last decades to the analysis of postbuckling behaviour when two or more buckling modes manifest themselves by correspondence with the same critical stress. Large account of the results achieved has been given in chapts. 7, 8 and 9, where analytical and numerical methods have been employed. In some cases the Vlasov beam model has been adopted (overall buckling), in some other cases the shell model has been utilized and the beam has been treated as a plate assemblage (local buckling).

It seems desirable to perform a qualitative analysis capable of furnishing information on postbuckling behaviour of beams and framed structures. In contrast with analyses of previous chapters, a direct one-dimensional beam model will be utilized in this chapter. A model of this type was first introduced in Epstein (Epstein, 1979), while in Reissner (Reissner, 1983) this model was derived from a three-dimensional one. More recently, in Simo and Vu-Quoc (Simo and Vu-Quoc, 1991) it has been suggested a direct one-dimensional beam model with a sophisticated kinematical description accounting for warping, but simple linear elastic constitutive relations have been assumed.

A direct one-dimensional beam model which accounts for nonlinear hyperelastic constitutive relations has been introduced and utilized in Pignataro and Ruta (Pignataro and Ruta, 2002); in Rizzi and Tatone (Rizzi and Tatone, 1996) and in Tatone and Rizzi (Tatone and Rizzi, 1991). Qualitative results obtained in these papers are in agreement with those stemming from different beam model analyses.

10.2 One-Dimensional Beam Model

The beam configurations are described by the position of the axis $\mathbf{p}(\rho;\tau)$, the rotation of the sections $\mathbf{R}(\rho;\tau)$ and a scalar parameter $\alpha(\rho;\tau)$ roughly accounting for warping, provided ρ is the axial abscissa and τ an evolution parameter. The deformation is defined as difference between a generic motion and a rigid one

$$\mathbf{e} := \mathbf{R}^{\mathbf{T}}\mathbf{p}' - \mathbf{q}', \qquad \mathbf{E} := \mathbf{R}^{\mathbf{T}}\mathbf{R}', \qquad \alpha, \qquad \eta := \alpha' \qquad (10.1)$$

where $\mathbf{q}(\rho;\tau)$ denotes the position of the axis in the reference configuration, $\mathbf{e}$ is the strain, $\mathbf{E}$ is the curvature and a prime indicates differentation with respect to ρ.

By writing the power of the external and internal dynamical actions and requesting the balance of power (Di Carlo, 1996; Germain, 1973a,b), balance equations are derived by standard theorems

$$\mathbf{t'} + \mathbf{b} = \mathbf{0}, \qquad \mathbf{T'} + \mathbf{p'} \wedge \mathbf{t} + \mathbf{B} = \mathbf{0}, \qquad \gamma_0 = \beta + \gamma_1' \tag{10.2}$$

where $\mathbf{t}$, $\mathbf{T}$ represent contact forces and couples, $\mathbf{b}$ and $\mathbf{B}$ body forces and couples, respectively, γ_0, γ_1 are interpreted as bi-shear and bi-moment, respectively, and β is the body action spending power on the warping; besides $\wedge$ denotes the external product.

Some inner constraints are supposed to hold: the warping is assumed to linearly depend on the twist and the beam is assumed shear-undeformable. Due to these constraints, suitable nonlinear hyperelastic constitutive relations are

$$
\begin{aligned}
Q_1 &= a\varepsilon_1 + \frac{1}{2}d\kappa_1^2, \\[4pt]
M_1 &= (c + d\varepsilon_1 + f_2\kappa_2 + f_3\kappa_3)\kappa_1 - h\xi^2\kappa_1'', \\[4pt]
M_2 &= b_2\kappa_2 + \frac{1}{2}f_2\kappa_1^2, \\[4pt]
M_3 &= b_3\kappa_3 + \frac{1}{2}f_3\kappa_1^2, \\[4pt]
\gamma_1 &= h\xi\kappa_1' + \frac{1}{2}g\kappa_1^2,
\end{aligned}
\tag{10.3}
$$

where: ε_1, κ_1, κ_2, κ_3 are the axial strain, the twist and the two bending curvatures, respectively, provided the beam axis is initially parallel to the first element of an orthonormal basis $(\mathbf{i}_1, \mathbf{i}_2, \mathbf{i}_3)$; Q_1, M_1, M_2, M_3 are the axial force, the twisting and the bending couples, respectively; ξ is the linear coefficient relating warping and twist (Reissner, 1983; Rizzi and Tatone, 1996; Simo and Vu-Quoc, 1991; Tatone and Rizzi 1991; Vlasov, 1961) the coefficients a, b_j, c, h stand for the axial, bending, torsion and warping stiffnesses, respectively; d, f_j, g take into account the coupling between torsion and extension (Poynting's effect, see Truesdell and Noll, 1965), torsion and bending, torsion and warping (Mollmann, 1986), respectively.

By comparing eq. $(V.1.10)_3$ in Vlasov (Vlasov, 1961) with the corresponding one which can be obtained from this analysis, under a suitable choice of the co-ordinate frame one has $a = EA$, $f_j = EI_{f_j}$, $b_j = EI_j$, $d = EI_d$, $c = GI_c$, $h\xi^2 = EI_\omega$, where E, G are the Young and shear moduli, A is the area of the cross-section, I_j are the principal moments of inertia, I_c, I_d the torsion moment of inertia and the polar moment of inertia with respect to the shear centre, respectively, I_ω is the warping rigidity, and $I_{f_j} = \int_R x_j r^2 dA$, with x_j the co-ordinates of a point of the cross-section with respect to the centroid and r its distance from the shear centre. It is remarkable how, with this identification, the following results for the critical loads are qualitatively the same as those in Timoshenko and Gere (Timoshenko and Gere, 1961).

By introducing the constitutive relations into balance and compatibility equations, the problems for evaluating the critical load(s) and the associated buckling mode(s) are stated. By means of adequate choices of the constitutive constants, i.e., of the stiffnesses of the beam, single or coupled buckling modes can be searched, making use of the static perturbation analysis developed in Budiansky (Budiansky, 1974).

10.3 Buckling and Postbuckling Analysis

Let us consider a beam axially compressed by a dead force of magnitude λ. The fundamental solution is given by the axial strain only $-\lambda/a$ and the axial force $-\lambda$, all the other relevant quantities vanishing. Compatibility, balance and constitutive equations can be written in terms of differences between the bifurcated and the fundamental path.

If we consider a beam of length l clamped at $x_1 = 0$ and with a roller at $x_1 = l$, the corresponding eigensolutions are

$$
\begin{aligned}
\bar{u}_1 &= 0, \\
\bar{u}_2 &= U_2[\cos(\frac{\pi\,x_1}{2l}) - 1], \\
\bar{u}_3 &= U_3 \left[\sin\left(\frac{A_0}{l}x_1\right) - A_0 \cos\left(\frac{A_0}{l}x_1\right) + \frac{A_0}{l}(l - x_1) \right], \\
\bar{\varphi}_1 &= \Phi_1 \left[\sin\left(\frac{A_0}{l}x_1\right) - A_0 \cos\left(\frac{A_0}{l}x_1\right) + \frac{A_0}{l}(l - x_1) \right], \\
\lambda_{c1} &= \frac{ac}{d}\left(1 + \frac{A_0^2}{l^2}\frac{h\xi^2}{c}\right), \\
\lambda_{c2} &= \frac{a}{2}\left(1 - \sqrt{1 - \frac{\pi^2 b_3}{al^2}}\right), \\
\lambda_{c3} &= \frac{a}{2}\left(1 - \sqrt{1 - \frac{4A_0^2 b_2}{al^2}}\right),
\end{aligned}
\tag{10.4}
$$

where U_2, U_3, Φ_1 are arbitrary constants and A_0 is the smallest solution of the characteristic equation $\tan(x) = x$ (see for instance (Timoshenko and Gere, 1961)). The bending buckling loads depend on both bending and axial rigidities, while the torsional buckling load depends on axial, torsion, warping and coupled torsion-axial rigidities. When $a \to \infty$, eqs. $(10.4)_{6,7}$ yield the usual buckling loads provided, for instance, in Timoshenko and Gere (Timoshenko and Gere, 1961). Remark, however, that in this case no torsional buckling occurs.

We shall consider interaction between the torsional mode $\bar{\varphi}_1$ and the bending mode $\bar{u}_2$, neglecting the other quantities. The second-order perturbation field equations $(4.8)_2$ for u_2, φ_1 read

$$
\begin{aligned}
\bar{\bar{u}}_2^{IV} + \frac{\pi^2}{4l^2}\bar{\bar{u}}_2'' &= -2\frac{a - 2\lambda_c}{ab_3}\bar{\lambda}_c\bar{\kappa}_3 - \frac{2f_3}{b_3}[(\bar{\kappa}_1')^2 + \bar{\kappa}_1\bar{\kappa}_1''], \\
\bar{\bar{\varphi}}_1^{IV} + \frac{A_0^2}{l^2}\bar{\bar{\varphi}}_1'' &= -\frac{2d}{2ah\xi^2}\lambda_c\bar{\kappa}_1' + \frac{2f_3}{h\xi^2}(\bar{\kappa}_1\bar{\kappa}_3)'.
\end{aligned}
\tag{10.5}
$$

Fredholm compatibility conditions involving the right hand sides of eqs. (10.5) and $\bar{u}_2, \bar{\varphi}_1$, in addition to the normalization condition $\Phi_1^2 + U_2^2 = 1$ yield a system of three nonlinear algebraic equations in the unknowns $\bar{\lambda}_c, \Phi_1, U_2$, whose solutions are

$$\bar{\lambda}_c = 0, \qquad U_2 = \pm 1, \qquad \Phi_1 = 0,$$
$$\bar{\lambda}_c = 0, \qquad U_2 = 0, \qquad \Phi_1 = \pm 1,$$
$$U_2 = \pm \frac{50.04\sqrt{dl}}{\sqrt{(50.04)^2 ad + 2.47(a - 2\lambda_c)b_3}},$$
$$\Phi_1 = \mp \sqrt{\frac{2.47(a - 2\lambda_c)b_3}{(50.04)^2 ad + 2.47(a - 2\lambda_c)b_3}},$$
$$\bar{\lambda}_c = \mp \frac{56.15\sqrt{b_3}f_3}{\sqrt{dl^2}\sqrt{(50.04)^2 ad + 2.47(a - 2\lambda_c)b_3}},$$

$$(10.6)$$

under no restrictions, provided $a > 2\lambda_c$ because of $(10.4)_6$.

According to Bezout's theorem (Van der Waerden, 1950), in addition to the two buckling modes occurring separately, there are two actual solutions, depending on the sign combinations of U_2 and Φ_1. Remark that, due to the normalization condition, eq. $(10.6)_3$ can be equivalently written as a linear function in terms of U_2.

The slope of the bifurcated coupled paths is not zero and depends on many mechanical parameters, among which d and f_3 play a crucial role. The postbuckling equilibrium path is thus non-symmetric and the beam is imperfection sensitive.

When the cross-section exhibits symmetry with respect to x_2 axis, $f_3 = 0$ and no coupling occurs between bending and torsion; hence $\bar{\lambda}_c = 0$ and the bifurcated path becomes symmetric. In particular, this happens when the section has two axes of symmetry, thus confirming the results previously achieved in Grimaldi and Pignataro (Grimaldi and Pignataro, 1979). On the other hand, when for instance x_3 is an axis of symmetry for the section, $f_3 \neq 0$ and the post-buckling behaviour is asymmetric.

11 Multiple Interactive Modes in Postbuckling of Corrugated Panels

11.1 Introduction

Quite recently, trapezoidal profiles have been largely investigated because of their increasing importance in the industry. Profiled panels are widely used for girders, roofing, decking, wall cladding and off-shore structures. Due to the remarkable stiffening provided by corrugations, thin metal sheets are used in the manufacturing process of the panels. This implies that fabrication costs for elements with corrugated panels are generally lower than those with stiffened plates. Because of their high slenderness ratio, buckling is of primary importance in the design of elements with trapezoidally corrugated profiles; some design codes have been published for the design of profiled sheetings (AISI, 1980; Eurocode 8, 1992).

Relevant experimental investigations have been performed in the past (Bernard, Bridge et al., 1993,1995) and a design method has been proposed (Bernard,Bridge et al., 1996).

Numerical analyses of corrugated panels have been carried out under different hedge loading conditions (Leiva-Aravena, 1987; Luo and Edlund, 1994; Papangelis and Hancock, 1995). They make use in general of finite strip method and are limited to linear analysis. Non linear interaction of buckling modes is therefore not considered. Experimental and numerical results are often compared in determining the critical stress and the agreement is found to be satisfactory. More recently, postbuckling behaviour of a simply supported trapezoidal profiles under compression in the presence of interaction between one overall and several local buckling modes has been investigated by means of finite strip method (Pignataro, Pasca et al. 2000). Results confirm that interaction is responsible for unstable postbuckling behaviour and for sensitivity to initial imperfections.

11.2 Postbuckling Analysis and Localization Phenomena

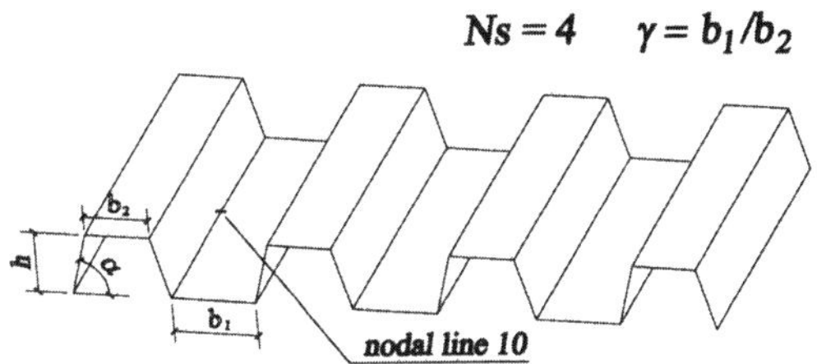

Figure 58. Geometric description of a panel element

The corrugated panels analyzed are of the type shown in Figure 58). The panel is assumed to have length L and to be simply supported at the transversal edges where a uniform axial load is applied. Longitudinal edges are supposed to be free. In the following, we shall refer (improperly) to *stiffener* as to that portion of the profile of length L and trapezoidal cross-section described by two inclined plates of width $h/\sin\alpha$ and one horizontal plate of width b_2; $\gamma = b_1/b_2$ is defined as the inverse of the corrugation density. Three classes of profiles have been considered: *i*) profiles with four stiffeners (number of stiffeners $N_s= 4$), *ii*) eight stiffeners, *iii*) twelve stiffeners.

From examination of the catalogues of industrial profile producers it can be seen that the first class of profiles ($N_s= 4$) well describes the basic module of corrugated sheets available on the market. The other two classes have been selected in order to investigate behaviour changes when more modules are assembled together as often requested in technical realizations.

Postbuckling behaviour of corrugated panels has been analysed for different values of α, γ and of the sheet thickness t taken from producer catalogues; in particular $\alpha = 60^\circ$, 75°, 90°, $\gamma = 1, 2, 3$ and $t = 0.5, 1.0$ mm have been chosen. By a suitable choice of the panel length, overall buckling and several local buckling modes manifest themselves simultaneously.

The next two figures show the postbuckling deformation of four panels all having thickness $t= 1$ mm. It is apparent from Figure 59(a) that when $\gamma =1, \alpha = 90^\circ, N_s = 4$, the panel local deflections start at the free longitudinal edges which are weaker while the

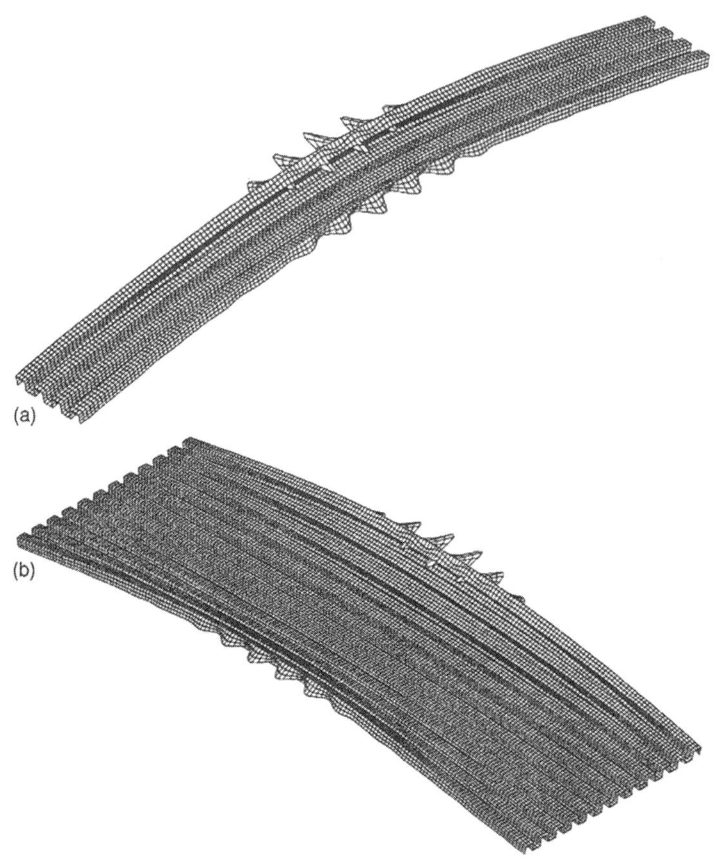

Figure 59. Postbuckling deformations: (a) $\gamma = 1$, $\alpha = 90$, $N_s = 4$; (b) $\gamma = 1$, $\alpha = 90$, $N_s = 12$

central part of the profile, having a higher stiffness, practically undergoes only overall deflection.

The changes of deformations from $N_s = 4$ to $N_s = 12$ are hardly noticeable, the deformation pattern being qualitatively the same with local buckling confined at the free longitudinal edges. However, the changes induced by different values of α are more evident as obtained from numerical results. A decrease of the angle α reflects on an increase of the free longitudinal edges distortions.

Figure 60 shows the postbuckling deformations of two panels having $\gamma = 3$ and the same values for α and N_s as in the previous case. One can see that localization appears here clearly concentrated at midspan as it was found in chapter 9 for simply supported thin-walled compression members.

The sensitivity of corrugated panels to initial imperfections is illustrated in figure 61

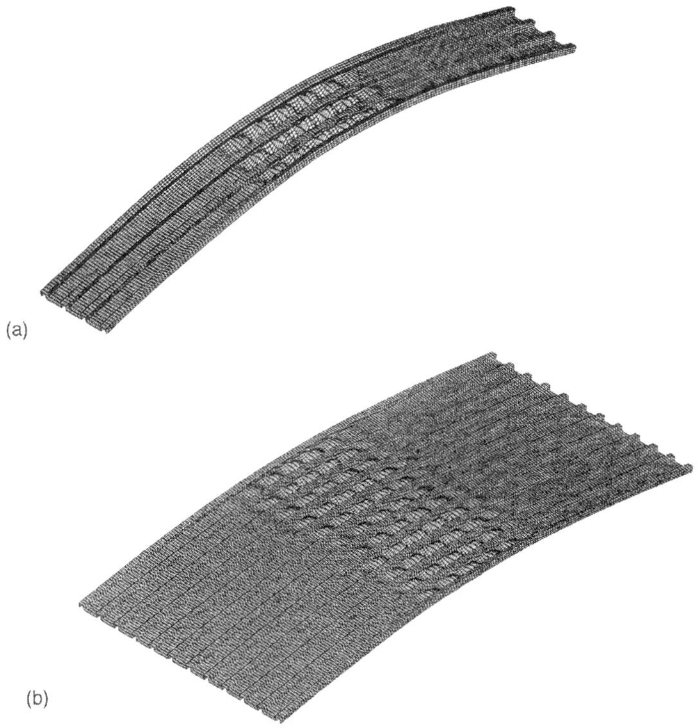

Figure 60. Postbuckling deformations: (a) $\gamma = 3$, $\alpha = 90$, $N_s = 4$; (b) $\gamma = 3$, $\alpha = 90$, $N_s = 12$

where the collapse load λ_s/λ_c vs the overall imperfection amplitude $\bar{u}/t$ has been plotted. For industrial panels where initial imperfections are of the order of magnitude of the plate thickness the collapse load ranges from 50% to 65% of the critical stress approximately, according to the value of γ, both for $N_s = 4$ and $N_s = 12$. The most detrimental effect is observed by correspondence with $\gamma = 1$ and decreases for increasing values of γ. This is easily explainable since for large values of γ, postbuckling behaviour of corrugated sheets tends towards that of flat plates which are insensitive to initial imperfections. The erosion of the critical stress in the present analysis due to initial imperfections is such that we may speak of strong interaction according to Gioncu's terminology (Gioncu, 1994).

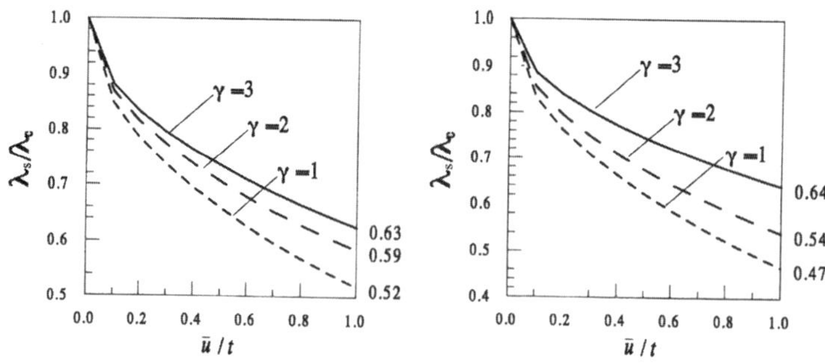

Figure 61. Collapse load for different imperfection values: $N_s = 4$; $N_s = 12$

Bibliography

S.S. Antman. Bifurcation problems for nonlinear elastic structures. In P.H. Rabinowitz, editor, *Applications of Bifurcation Theory*. Academic Press, New York, 1977.

R. Benito. *Static and Dynamic Buckling of Plate Assemblies*. PhD thesis, Washington University in St Louis, 1983.

R. Benito and S. Sridharan. Mode interaction in thin-walled structural members. *J. Struct. Mech.*, 12:517–542, 1984-1985.

R. Benito and S. Sridharan. Interactive buckling: a novel approach and some new results. In *Third Intern. Colloq. on Stability of Metal Structures*, pages 91–111, Toronto 1983.

E. S. Bernard, R. Q. Bridge, and G.J. Hancock. Design method for profiled steel decks with intermediate stiffeners. *J. Constr. Steel Res.*, 38:61–68, 1996.

E. S. Bernard, R.Q. Bridge, and G. J. Hancock. Test on profiled steel decks with v-stiffeners. *J. Struct. Engng, ASCE*, 119:2277–2293, 1993.

E.S. Bernard, R.Q. Bridge, and G.J. Hancock. Test on profiled steel decks with flat hat stiffeners. *J.Struct. Engng, ASCE*, 121:1175–1182, 1995.

M.A. Bradford and G.J. Hancock. Elastic interaction of local and lateral buckling in beams. *Thin-Walled Structures*, 2:1–25, 1984.

B. Budiansky. Theory of buckling and post-buckling behaviour of elastic structures. In *Advances in Applied MechanicsAdvances in Applied Mechanics*. Chia-Shun Yih editor, Vol. 14 1-65 Academic Press, New York 1974.

E. Byskov. Elastic buckling problem with infinitely many local modes. Technical Report 327, Technical University of Denmark, Lyngby, 1986.

E. Byskov and J.W. Hutchinson. Mode interaction in axially stiffened cylindrical shells. *AIAA J.*, 15:941–948, 1977.

R. Casciaro, A. Di Carlo, and Pignataro M. A finite element technique for bifurcation analysis. Technical Report II-192, Istituto di Scienza della Costruzioni, University of Rome, 1976.

R. Casciaro, G. Garcea, G. Attanasio, and F. Giordano. Perturbation approach to elastic post-buckling analysis. *Comput. & Struct.*, 66/5:585–595, 1998.

A. Di Carlo. A non-standard format for continuum mechanics. In Batra R. C. and Beatty M. F., editors, *Contemporary Research in the Mechanics and Mathematics of Materials*. CIMNE,Barcelona, 1996.

A. Di Carlo, M. Pignataro, and N. Rizzi. On the proper treatment of axial and shear undeformability constraints in post-buckling analysis of beams. *Int. J. Non-Linear Mechanics*, 16 (2):221–229, 1981.

A. Di Lanzo and G. Garcea. Koiter's analysis of thin-walled structures by a finite element approch. *Int. J. Numer. Methods Engng.*, 39:3007–3031, 1996.

M. Epstein. Thin-walled beams as directed curves. *Acta Mechanica*, 33:229–242, 1979.

F. Gantmacher. *Lectures in Analytical Mechanics*. MIR, Moscow, 1970.

G. Garcea. Mixed formulation in koiter analysis of thin-walled beams. *Comput. Methods Appl. Mech. Engng.*, 190:3369–3399, 2001.

P. Germain. La méthode des puissances virtuelles en mécanique des milieux continus, 1ère partie: La théorie du second gradient. *Journal de Mécanique*, 12:235–274, 1973a.

P. Germain. The method of virtual power in continuum mechanics, part II, application to continuous media with microstructure. *SIAM Journal of Applied Mathematics*, 23: 556–575, 1973b.

V. Gioncu. General theory of elastic stability. *Thin-Walled Structures*, 19:81–127, 1994.

T.R. Graves-Smith. The ultimate strength of locally buckled columns of arbitrary length. In K.C. Rochey and H.V. Hill, editors, *Thin-Walled Steel Constructions*. Crosy Lockwood, London, 1967.

T.R. Graves-Smith and S. Sridharan. A finite strip method for the buckling of plate structures under arbitrary loading. *Int. J. Mech. Sci.*, 20:685–693, 1978.

A. Grimaldi and M. Pignataro. Postbuckling behaviour of thin-walled open cross-section compression members. *J. Struct. Mech.*, 7:143–159, 1979.

G.J. Hancock. Local, distorsional and lateral buckling of i-beams. In *Proc. ASCE, 104,ST11*, pages 1787–1798, 1978.

G.J. Hancock. Nonlinear analysis of thin sections in compression. In *Proc. ASCE, 107 ST3*, pages 455–471, 1981.

American Iron and Steel Institute. *Specifications for the Design of Cold-Formed Steel Structural Members*, Washington: AISI, 1980.

W. T. Koiter. Post-buckling analysis of a simple two-bar frame. In *The Folke Odkvist volume*. Almqvist and Wiksell, Stockholm, John Wiley and Sons, 1967.

W.T. Koiter. *On the Stability of Elastic Equilibrium (in Dutch)*. PhD thesis, H.J. Paris, Amsterdam, 1945, English translation as NASA TT F-10,833,1967 and AFFDL report TR 70-25, 1970.

W.T. Koiter and G. D. C. Kuiken. The interaction between local buckling and overall buckling on the behaviour of built-up columns. WTHD Report 23, Delft, 1971.

W.T. Koiter and M. Pignataro. A general theory of the interaction between local and overall buckling of stiffened panels. WTHD report 83, Delft, 1976. Technical report.

W.T. Koiter and M. Pignataro. An alternative approach to the interaction between local and overall buckling in stiffened panels. In B. Budiansky, editor, *Buckling of structures*. Springer-Verlag, New York, 133-148, 1976.

W.T. Koiter and M. Skaloud. Interventión sur le comportement postcritique des plaques utilisées en constructions metalliques. *Mem. Soc. R. Sci. Liège*, [5] 8(5):64–68,103,104, 1963.

M. Kotelko and Z. Kolakowski. Coupled instabilty related collapse behaviour of channel-section beam-column. In *Proc. CIMS 2000*, Lisbon, (2000), 205-212.

M. Królak, Z. Kolakowski, and M. Kotelko. Influence of load-nonuniformity and eccentricity on the stability and load carrying of orthotropic tubular columns of regular hexagonal cross-sections. *Thin-Walled Structures*, 39:483–498, 2001.

J. La Salle and S. Lefschetz. *Stability by Liapunov's Direct Method with Applications.* Academic Press, New York, 1961.

L. Leiva-Aravena. Trapezoidally corrugated panels. buckling behaviour under axial compression and shear. *Chalmers Univ. of Technology at Gothenburg*, Sweden, Publ. 87,1, 1987.

A.M. Liapunov. *Problème genéral de la stabilité du mouvement (in Russian).* ,Karkov, 1892; French traslation in Ann. Fac. Sci. Univ. Tolouse,9,1907; English traslation : Stability of motion, Academic Press, New York, 1966.

R. Luo and B. Edlund. Buckling analysis of trapazeoidally corrugated panels using spline finite strip method. *Thin-Walled Structures*, 18:209–224, 1994.

A. Luongo and M. Pignataro. On the use of finite strip method in the analysis of nontraditional problems of thin-walled members. In *Proc. VII Italian Corgress of Theoretical and Applied Mechanics AIMETA (in italian)*, pages 193–204, 1984.

A. Luongo and M. Pignataro. Buckling and post-buckling analysis of stiffened channels under uniform compression. *Costruzioni Metalliche*, 4:242–249, 1986.

A. Luongo and M. Pignataro. Multiple interaction and localization phenomenon in post-buckling of compressed thin-walled members. *AIAA J.*, 26:1395–1400, 1988.

R. Maquoi and C. Massonnet. Interaction between local plate buckling and overall buckling in thin-walled compression members. theories and experiments. In B. Budiansky, editor, *Buckling of structures*, pages 350–382. Springer-Verlag New York, 1976.

C.M. Menken, W.J. Groot, and Stallenberg G.A.J. Interactive buckling of beams in bending. *Thin-Walled Structures*, 12:415–434, 1991.

J.J. Meyer and A. Van der Neut. The interaction of local buckling and column failure of imperfect thin-walled compression members. Technical Report 160, WTHD, report 160,Delft, 1970.

H. Mollmann. Theory of thin-walled beams with finite displacements. In W. Pietraszkiewicz, editor, *EUROMECH Colloquium 197.* Springer-Verlag, New York, 195-209, 1986.

K.E. Moxham. Buckling tests on individual welded steel plates in compressions. Technical Report CUED/C-Structures/TR3, Cambridge University Engineering Department, 1971.

Commission of the European Community. *Eurocode 8, Design of Steel Structures Part 1.3 Brussels*, 1992.

J.P. Papangelis and G. J. Hancock. Computer analysis of thin-walled structural members. *Computer Struct.*, 56:157–176, 1995.

M. Pignataro, A. Di Carlo, and R. Casciaro. On the non-linear beam models from the point of view of computational post-buckling analysis. *Int. J. Solids Struct.*, 18 (4): 327–347, 1982.

M. Pignataro and A. Luongo. Asymmetric interactive buckling of thin-walled columns with initial imperfections. *Thin-Walled Structures*, 5:365–386, 1987.

M. Pignataro and A. Luongo. Simultaneous buckling modes and imperfection sensitivity of channels in compression. In J. Szabó, editor, *EUROMECH Colloquium 200*, 233-252 1985.

M. Pignataro, A. Luongo, and N. Rizzi. On the effect of the local-overall interaction on the postbuckling of uniformly compressed channels. *Thin-Walled Structures*, 3: 293–321, 1985.

M. Pignataro, M. Pasca, and P. Franchin. Post-buckling analysis of corrugated panels in the presence of multiple interacting modes. *Thin-Walled Structures*, 36:47–66, 2000.

M. Pignataro and N. Rizzi. On the interaction between local and overall buckling of an asymmetric portal frame. *Meccanica*, 18:92–96, 1983.

M. Pignataro and G. C. Ruta. Coupled instabilities in thin-walled beams: a qualitative approach. *European Journal of Mechanics A/Solids*, 22:139–149, 2002.

R.J. Plank. Developments in the finite strip method for the buckling analysis of compression members. In *Trosième Colloque International "Stabilité des Structures Metalliques"*, Paris. 1983.

R.J. Plank and W.H. Wittrick. Buckling under combined loading of thin-flat-walled structures by a complex finite strip method. *Int. J. Num. Meth. Engng.*, 8:323–339, 1974.

H. Poincaré. Sur l'equilibre d'une masse fluide animée d'un mouvement de rotation. *Acta Math.*, 7:259, 1885.

L. Pontriaguine. *Equations Differentielles Ordinaires*. MIR, Moscow, 1969.

M. Potier-Ferry. Wavelength selection and pattern localization in buckling problems. In J.E. Wesfreid and S. Zalesky, editors, *Cellular Structures in Instability Problems*. Lecture Notes in Phisics, Springer-Verlag Berlin, 1984.

E. Reissner. On a simple variational analysis of small finite deformations of prismatical beams. *ZAMP*, 34:642–648, 1983.

N. Rizzi and M. Pignataro. The effect of multiple buckling modes on the postbuckling behaviour of plane elastic frames. part I: Symmetric frames, part II: Asymmetric frames. *J. Struct. Mech.*, 10:437–458 and 459–474, 1982.

N. Rizzi and A. Tatone. Nonstandard models for thin-walled beams with a view to applications. *J. Appl. Mech.*, 63:399–403, 1996.

J. Roorda. Stability of structures with small imperfections. *J. Eng. Mech. Div. ASCE*, 91, No. EM1,:Proc. paper 4230,87, 1965.

J. C. Simo and L. Vu-Quoc. A geometrically exact rod model incorporating shear and torsion-warping deformation. *Int. J. Solids Struct.*, 27:371–393, 1991.

S. Sridharan. Doubly symmetric interactive buckling of plate structures. *Int. J. Solids Struct.*, 19:625–641, 1983.

S. Sridharan and A. Ali. Interactive buckling in thin-walled beam-columns. *ASCE*, 111, EM12:1470–1486, 1985.

S. Sridharan and R. Benito. Columns: static and dynamic interactive buckling. *ASCE*, 110 EM1:49–65, 1984.

A. Tatone and N. Rizzi. A one-dimensional model for thin-walled beams. In H. Troger W. Schneider and F. Ziegler, editors, *Trends in Applications of Mathematics to Mechanics*. Longman, Avon: 312-320, 1991.

J. M. T. Thompson, J. D. Tulk, and A. C. Walker. An experimental study of imperfection-sensitivity in the interactive buckling of stiffened plates. In B. Budiansky, editor, *Buckling of structures*. Springer-Verlag, New York, 149-159, 1976.

S. P. Timoshenko and J. M. Gere. *Theory of Elastic Stability*. McGraw-Hill New York, 1961.

C. Truesdell and W. Noll. The non-linear field theories of mechanics. In *Handbuch der Physik III/3*. Springer-Verlag, New York, 1965.

V. Tvergaard. Imperfection sensitivity of a wide integrally stiffened panel under compression. *Int. J. Solids Struct.*, 9:177–192, 1973.

V. Tvergaard and A. Needleman. On the localization of buckling patterns. *J. Appl. Mech.*, 47:613–619, 1980.

A. Van Der Neut. Mode interaction with stiffened panels. In B.Budiansky, editor, *Buckling of Structures*. Springer-Verlag, New York, 117-132, 1976.

A. Van der Neut. The interaction of local buckling and column failure of thin-walled compression members. In *Proc. 12th International Congress of Applied Mechanics*, pages 389–399, Springer Verlag, Berlin 1969.

B. L. Van der Waerden. *Modern Algebra, Vol. II*. Ungar, 1950.

V.I. Vlasov. Thin-walled elastic beams, 2nd ed. the Israel Program for Scientific Translations, Jerusalem 1961.

H. Yoshida and K. Maegawa. Local and member buckling of h-columns. *J. Struct. Mech.*, 6:1–27, 1978.

Phenomenological Modelling of Instability

Victor Gioncu

Department of Architecture "Politehnica" University of Timişoara, Romania

Abstract. The lecture presents the background of structure instability: main research directions, phenomena, instability types, classifications, etc. The author gives the main aspects of instability in the light of theories developed for evolving systems: Synergetics, Dissipative Systems, and Catastrophe Theory. A phenomenological methodology for instability design, based on stable and unstable components of critical load, is also presented.

1 Introduction

Improving the efficiency of constructions by reduction its weight and material consumption is a major trends in structural field. This aim can be achieved by using new materials and new structure types, or by optimization the existing solutions. The permanent trend to optimize the solutions leads to intensive using of the light weight and thin-walled structures. The using of new structure types is also closed linked to the rapid development of digital computers which provides in the hands of the progressive designer the possibility of analysis of complex structures and the correct assessment of their behaviour. At the same time, the designers are facing new structural problems created by reducing the amount of material required for the construction. As the structures are becoming lighter, i.e. as their cross-section dimensions are decreasing, the structures are influenced by the factors which had not been significant previously:

a) Due to the reduction of weight the structure slenderness increases and the importance of strength checking decreases, being replaced by the problems of instability checking.

b) Another consequence is the increasing of importance of dynamic actives in comparison with dead loads, as live movable, time variable, or suddenly applied loads.

Therefore, for the contemporary structures the nonlinear analysis and checks for static and dynamic stability is a very important problem, playing the role of the prima-dona among of structural analysis. Despite a remarkable progress made in the last years, many problems are still unsolved, theoretical researches and design procedures have to be developed or improved. The increasing number of publications and conferences strongly indicates the interest of scientists and engineers in this subject. A careful examination of these publications shows that they tend to fall into theoretical studies, aiming at mathematical modeling of phenomena, and practical studies to provide solutions in order to prepare specifications or codes.

The purpose of the present lectures is to transfer the very complex theoretical achievements to practice in order to fill the gap existing between the accumulated theoretical knowledge and practical applications.

2 Structural Stability Today

2.1 Survey of Present Problems

Although the stability of bars was first studied over 250 years ago (Euler's paper was published in 1744), adequate solutions are still not available for many problems in structural stability. So much has been and being studied and written in the field of structural stability, that one may well wonder why, after such intellectual and financial efforts, there are no definite solutions to these problems. With the help of large computers and the FE method it is possible to calculate the stress distributions with a great accuracy for very complex structures. Why, when the structure is buckling-critical, then even today, one will encounter in many cases great difficulties in finding a reliable value for the buckling load? Because determining the load under a structure collapses due to the loss of stability is still one of the most difficult problems of structural designing. Its difficulties are essentially due to the following factors:

a) The loss of stability depends on numerous factors, some of which are very difficult to control. This is confirmed by a number of recent structure accidents (Gioncu, 2003). Faulty design and execution, overstressing or the use of inadequate materials have been shown to be mainly responsible for these accidents. It should be noted that these accidents practically cover the entire range of structures. Today, only a specialist can carry out stability checks in complete agreement with the actual behaviour of the structure.

b) Instability occurs in a region with strong geometrical and material nonlinearities. For the pre-critical range an extensive literature of effective solution exists. For the post-critical range, only after remarkable progress in the field of electronic computing equipment, non-linear analysis using FE and some special numerical techniques in the neighborhood of the limit point, has made it possible to correctly describe the behaviour of structure, shortly before failure. However such analysis requires rather complex programs and large-capacity computers, which are not available for many designers.

c) In no other field of structural mechanics is the influence of imperfections due to the execution so significant as in the field of instability. In strength analysis the stress-strain state is determined by means of an idealized scheme of the structure, neglecting the geometrical and mechanical imperfections, compared to the actual structure, differences are relatively small. In the case of instability, on the other hand, loads on the actual structure may be only 20% at worst of those of the ideal structure.

d) Checking the buckling of structure experimentally is very difficult, because it is impossible to test the actual structure just until it collapse. In strength analysis, the reduced model tests are used for checking the validity of theoretical values. In stability analysis, testing on reduced models is irrelevant in most cases, because a correct modelling of the effect of imperfections is practically impossible.

e) There is a wealth of information available in numerous papers dealing with the stability and instability of structures, but design codes and standards with modern conceptions of stability checking are scarce. In this situation structural designers may commit grave errors in the structure instability checking.

2.2   Principal Directions of Research Works

The behaviour of a structure is known if is well defined the three sequence of load-deformation curve: pre-critical, critical, and post-critical paths (Fig .1). In function of considered type of behaviour, today conceptions and trends in the structural stability shows that the investigations have tended to fall into one of the following directions:

a) *Practical design direction*, in which methods of analysis, specific for stability problems related to some structural forms, are developed. These studies are invaluable since they aim is to provide solutions to practical problems, to supply designers with data useful for design and to prepare norms, specifications or codes. In this field of structural stability work the engineers who believe especially in the physical sense of the phenomena than in the mathematical results. All the researches in this direction are based on *Euler's concept*, which considers that an analysis for stability is restricted to determining the buckling load and all the structural members have the same sensitivity to the imperfections as the standard bar hinged at both ends, and the difference between buckling and bifurcation loads is due to the influence of mechanical and geometrical imperfections.

b) *Theoretical direction*, in which the main problem is the study of phenomena with aid of theoretical methods. This direction is concerned with a more profound understanding of the phenomenon of instability, especially in the post-critical range. This approach, which has led to the *Theory of Post-critical Behaviour of the Structures* is less familiar to designers. A wide range of research has gone into these theories: mathematicians and mechanical engineers specialized in applied mathematics and mechanics, engineers who have been working in the field of space and naval constructions rather in building and industrial constructions, have all contributed to it. The start of this research direction was the Koiter's doctoral thesis (1967). Koiter showed that, beside Euler's type of bifurcation, there are other types of structural behaviour at the critical point, connected with the post-critical curve and that the collapse load of the actual structure, with its geometrical imperfections, is in direct connection with the post-critical behaviour. Hence *Koiter's concept* demands that, besides determining the critical load , the post-critical behaviour

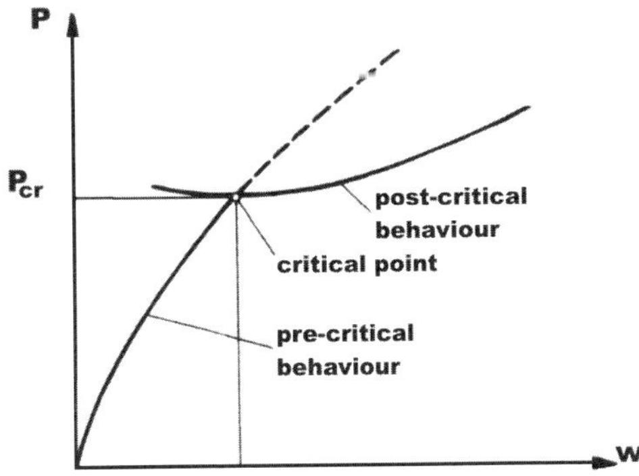

Figure 1. Stability and instability of structures.

of the structure should be studied so as to determine to what extent the structure is sensitive to imperfections. This concept, applied to struts, plates and shells, explained their fundamental different response. It seems fair to say that the Koiter's post-buckling theory represents one of most important contributions in the field of stability of elastic systems since the initial investigations of Euler.

c) *Instability of evolving systems.* For systems governed by a gradiental potential, Thom (1972) developed a very important theory, the well known *Theory of Catastrophes.* Other two theories, which deal with stability and instability of systems, were developed in the same time. The concept of *Dissipative Systems* elaborated by Prigogine (Prigogine and Stengers, 1979) covers all natural phenomena which occur far from its initial equilibrium as they can appear beyond a critical value, in the domain of non-linear processes. The second theory, developed by Haken, (1978) and known as the *Synergetics*, deals with the co-operation of the individual parts of a system, both in the equilibrium and instability ranges. These theories are interdisciplinary fields of research works, which aid the better understanding of the instability phenomena. Therefore, the instability theory interacted vigorously for the last years with these general theories.

3 New Phenomenological Models of Evolving Systems

3.1 Models for Continuous Phenomena

After the Newton's development of mechanics principles, the evolution was considered as a succession of continuous phenomena. According to mechanics laws, if the present state of a system is known, the evolution is also well defined. The world of systems is governed by universal, logical, causal and deterministic laws. The nature acts as a clock which moves complying with a mechanism. If somebody knows the evolution law, he can influence the future behaviour of the system. There are not place for spontaneous and unexpected events. Newton, Laplace and Leibnitz have developed the mathematics of continuous phenomena, based on the hypothesis that small continuous changing of variables produces continuous small effects on the system. The essence of mechanistic model of evolution is condensed in the Leibnitz's aphorism that *Natura non facit saltus.* But this image of a harmonious world evolution was changed in the last time due to knowledge accumulations about discontinuous phenomena in nature.

3.2 Models for Discontinuous Phenomena

In contrast with the mechanist models of evolution ordered by exact laws which consider continuous movements, in the actual world one can observe sudden changes of states which denies the image of an ideal world. Besides of regular movement of planets on the periodical trajectories one can see explosions or collapse in black holes of some stars. The relief of a zone remains unchanged for years, but a strong earthquake, volcano eruption or landslide can produce a sudden modification of the relief. Some species remain unmodified for thousand years and suddenly they exhibit a genetic changing or completely disappear. Even the Darwin's theory about the species evolution is now contested due to the detection of some genetic explosions. The same sudden changing can be observed in meteorology, economics, social sciences, etc. So, the famous Leibnitz's aphorism must be changed in the *Nature facit saltus*. In these conditions, the question is: *why for us the perception of evolution is more a continuous phenomenon than a discontinuous one?* The answer is very simple: our existence is very briefly in comparison with the universe life and the chance to participate at more jumps is very small. Therefore these changes are considered only some accidents of a harmonious and continuous evolution, not a rule of the actual world. The evolution of a complex system (Fig.2) is characterized by a cascade of bifurcation points. One can see that the evolution path is composed by fragments with slow growing, followed by fragments with very fast growing. After these jumps a new continuous development is recorded. For an observer situated in the point A the start of evolution is the point O', ignoring that the actual start was the origin O, and to arrive in point O' the system undergoes some jumps in development. If only the evolution from point O' is considered, the model of continuous evolution can be used. But, if the complete evolution is studied, this model is out of availability and the model of discontinuous evolution must be developed. For these approaches one must appeal to some new theories elaborated in the last few decades for evolving systems. These theories deal with the stability and instability of global systems in nature and science and the structural mechanics can takes advantage of their developments.

a) *Synergetics* developed by Haken (1978) deals with the co-operation between the individual parts of a system. During evolution some sub-systems dramatically change their behaviour, leading to new systems with different properties instead of the old ones.

b) *Dissipative Systems* elaborated by Prigogine (Prigogine and Stengers, 1979), covers all natural phenomena which occur far from initial equilibrium state in the domain of nonlinear processes. This theory introduces the notions of being near and far from

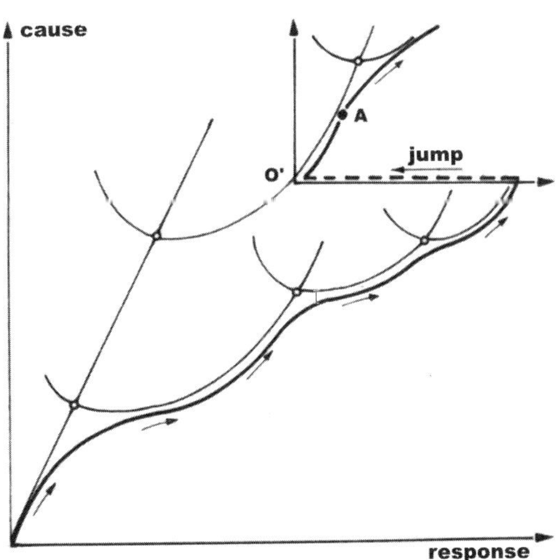

Figure 2. Behaviour of the complex systems.

primary equilibrium, the influence of fluctuations and explains the essential changes in phase transformations.

c) *Catastrophe Theory* elaborated by Thom (1972) which is a mathematical study of the jumps in evolving systems. It provides an universal mathematical method for the study of jump translations, discontinuities and sudden qualitative changes.

d) *Theory of Hidden Asymmetry* developed by Hunt (1986) which deals with loss of symmetrical behaviour of a class of systems

If one attempt to characterize these general theories, one can find that they answer to the following questions:

Why some very important changes are produced in the evolution of the systems? Synergetics explains why some variables of systems lead, during the evolution, to important changes of system behaviours.

What happens after these changes of systems? The Theories of Dissipative Systems and Hidden Asymmetry explain the essential changes in the far from equilibrium phases.

How much are these changes of systems? The Catastrophe Theory offers the mathematical support of the questions asked by the above mentioned theories.

3.3 Synergetics and the Domination Principle

The evolving systems presenting jumps come from very different disciplines: mathematics, physics, astrophysics, chemistry, biology, sociology, etc. Accordingly their topics seem to be quite different, but there are many behavioural aspects which bind together these disciplines. *Synergetics* is a general science referring to all these complex systems, formed by sub-systems, which are self-organized in order to have a good response to different external factors. The word synergetics is composed by two Greek words meaning working together. The theory focuses its attention to those situations when the systems undergo dramatic changes in the global behaviour. Particularly, the theory investigates how one or two sub-systems can produce these changes in an entirely self-organized system. In the first stage (Fig. 3) all N sub-systems co-operate to the overall behaviour of the system, practically in the same hierarchy, and this behaviour is the sum of all sub-system contributions. In the following step, during the evolution of the system, some altered sub-systems begin to have a particular behaviour. So, the influence of the sub-systems i, j came to be out of the general rule in comparison with the other ones. The non-linear behaviours of these sub-systems change the overall behaviour of system and the contribution of the unaltered sub-systems begin to be more and more reduced. In the last step, when the system behaviour is dramatically changed, only the variables corresponding to altered sub-systems govern the system behaviour, the other variables practically disappearing from behaviour of the overall system. Based of this observation on can define the *Domination Principle, DOMP*, which attests the reduced number of variable (generally one or two) which produced the system collapse.

3.4 Theory of Dissipative Systems

The fundamental characteristic of dissipative systems is that they occur far from equilibrium, as they can appear only beyond a critical phase in the field of non-linear processes.

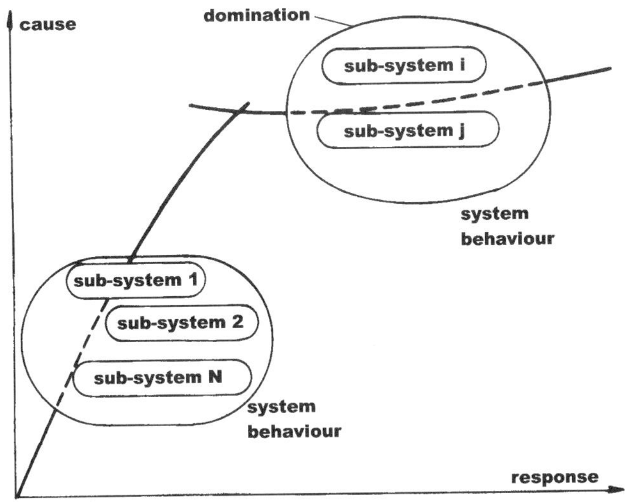

Figure 3. Co-operation of sub-systems.

They are than maintained only by external constrains and have need to be permanently supplied by external energy. For these systems an important part plays the bifurcations and fluctuations. Two states are characteristic for the dissipative systems: near to equilibrium and far from equilibrium (Fig. 4a). The change from first state to the second one is produced by instability of the system, when a very important role play the fluctuations, representing the deviations of the actual system from the ideal one. If in the phase near to equilibrium these fluctuations do not change very much the system behaviour, in the far from equilibrium state these fluctuations have a very important influence on structure behaviour. A right fluctuation produces an acceleration of the development, while a left one generates a delay of this development (Fig. 4b). Therefore, differences in fluctuations, unimportant in the equilibrium state, generate very important behavioural changes in the far from equilibrium states. Another very important characteristic of these systems is the possibility to generate equilibrium jumps (Fig. 5). If the system have a left fluctuation, for instance, due to some external factors, a jump to the right fluctuation can occur.

3.5 Catastrophe Theory and the Determination and Perturbation Principles

About the Theory of Catastrophes a special Session is present in our course, that in this lecture only some general aspects are shown. There are two different ways in which the Thom's works can be understood:

a) *Catastrophe Theorem* classifies the jumps effects of systems governed by a potential. From mathematical point of view this theorem studies how the qualitative nature of the equation solutions depends on the parameters that appear in the equations. From

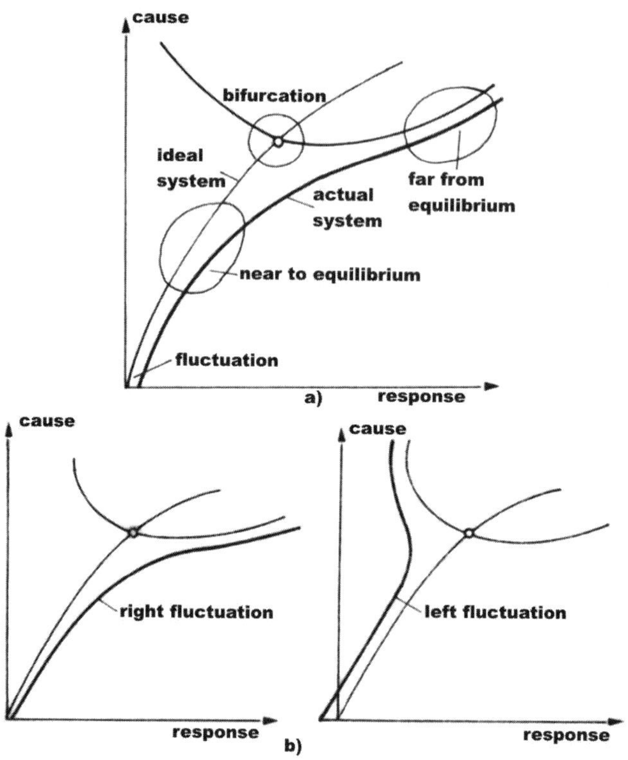

Figure 4. Bifurcations and fluctuations.

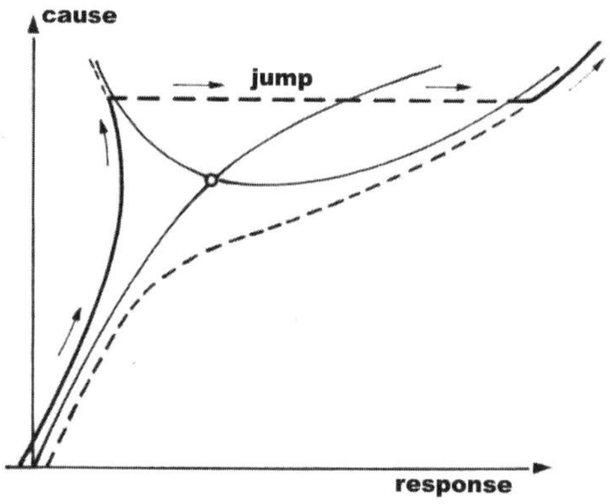

Figure 5. Equilibrium jump.

phenomenological point of view, the theorem allows to study the changes in behaviour of evolving systems. The program of Catastrophe Theorem is to construct the canonical form of the Elementary Catastrophes:

$$\text{Catastrophe} = \text{Germ} + \text{Perturbation}.$$

The *Germ* depends of *state variables* and the *Perturbation* is function of *state variables* and *control parameters*. The Catastrophe Theorem determines, for one or maximum two state variables the form of catastrophe germ and the perturbation function. So the basis theorem of Catastrophe Theory is the *Classification Theorem* which specifies the possible catastrophe types, the number of state variables and control parameters which defines the phenomenon.. This theorem is based on:

  - *Determination principle, DETP*, which determines the form of catastrophe germ;

  - *Perturbation Principle, PERT*, which determines, for each form of catastrophe germ, the canonical form of perturbation and the control parameters which govern the phenomenon.

The feature of Classification Theorem is the determination of the germs sensible to some form of perturbations. Figure 6a shows a germ without important behaviour changing if a perturbation exists. Figure 6b presents a germ for which the perturbation produces an important change in behaviour. Due to this fact, it is possible to have some discontinuous change in system behaviour if some traces are covered (Fig. 7).

b) *Catastrophe Theory* which is more a philosophy than a mathematics for evolving systems. The father of this theory, Thom, says that catastrophe theory is not a theory, but a language or a method that can be used in modeling natural phenomena in order to describe reality. He insists on the on the qualitative aspect of the catastrophe theory, imagining the situation presented in Fig. 8. As result of a very careful experimental works the curve *a* is obtained. By using a statistical methodology results the curve *b*, which presents the minimum deviation from the experimental curve. On the basis of this analysis a theory was elaborated, limited to the field of used parameters. In the

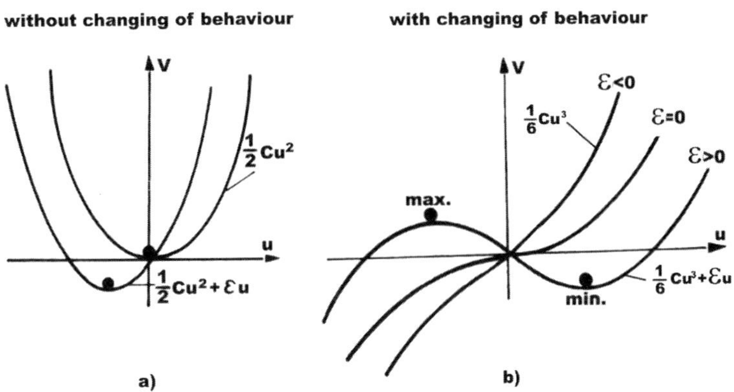

Figure 6. Influence of perturbations.

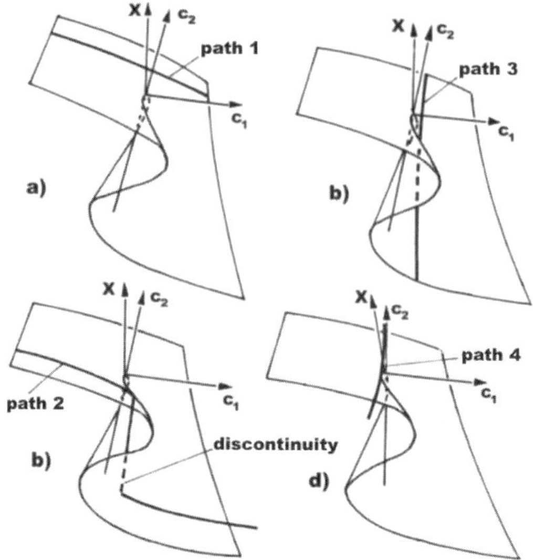

Figure 7. Different paths on the catastrophe surface.

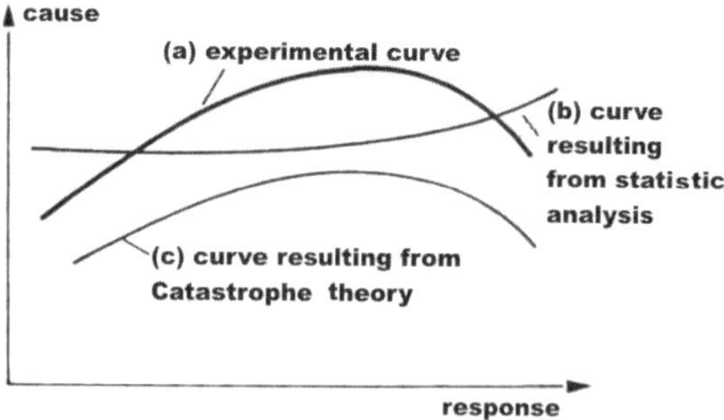

Figure 8. Theories and experience agreements.

same time one can see that the curvatures of experimental and theoretical curves are opposite. Using the Catastrophe Theory another curve c is obtained, which presents some deviations from experimental results, but the curvature and the development tendency is the same as the experimental values. An experimenter researcher decides to use curve b, while for the theoretical researcher the curve c is more adequate because describes more correct the development tendency of phenomenon. Therefore the results obtained using the Catastrophe Theory serve to widen the understanding capacity of pathological situations during the evolution of the systems, when some jumps in behaviour occur.

3.6   Hidden Asymmetry Theory

The scientists prefer to retain as much system symmetry as possible in order to simplify the calculations and the drawings. But the system symmetry is only a hypothesis which is contradicted by the reality and the near-symmetry or asymmetry is more frequently than the symmetry. One of more complicated situation results when the structure is inherent symmetric, but during the behaviour some hidden asymmetries appear. To ignore this aspect is to miss a very important feature of the systems, while to consider only the asymmetrical systems makes the theories too complicated. For these cases, Hunt (1986) developed a symmetry test in which the conditions in which the hidden asymmetry must be considered in analysis.

3.7   Application of these Theories to Structural Instability

These theories concerning the evolving systems introduced some principles which allow the elaboration of the *Coherent Theory of Structure Instability* (Gioncu, 2003), with a mathematical rigorous support based on the following principles:

a) *Domination Principle, DOMP,* which identifies and establishes the active coordinates that dominate the structure behaviour near the buckling. Among the great number of coordinates, this principle determines the one or maximum two active coordinates, the other being eliminate from the analysis. This operation represents a very important simplification of design process.

b) *Determination Principle, DETP,* which states precisely the instability type that assure a proper description of the structure critical and post-critical behaviour. This instability type depends of structure and loads configurations.

c) *Perturbation Principle, PERP,* which identifies and establishes the number of control parameters (loads and imperfections) that are essential for a complete description of structure behaviour. This principle allows the determination of the main imperfection types which must be included in the analysis.

d) *Symmetry Test, ST,* which establishes the conditions for the loss of behaviour symmetry during the structure instability.

The application refers to the roof-truss presented in Fig. 9. This truss is a system composed by some sub-systems (9 members, 4 joints, 2 supports) which co-operate to assure a good overall behaviour. In the same time the truss is an evolving system because the vertical loads increase from zero to the collapse loads. Therefore, the above mentioned theories can be used to describe the truss behaviour. In the pre-critical state,

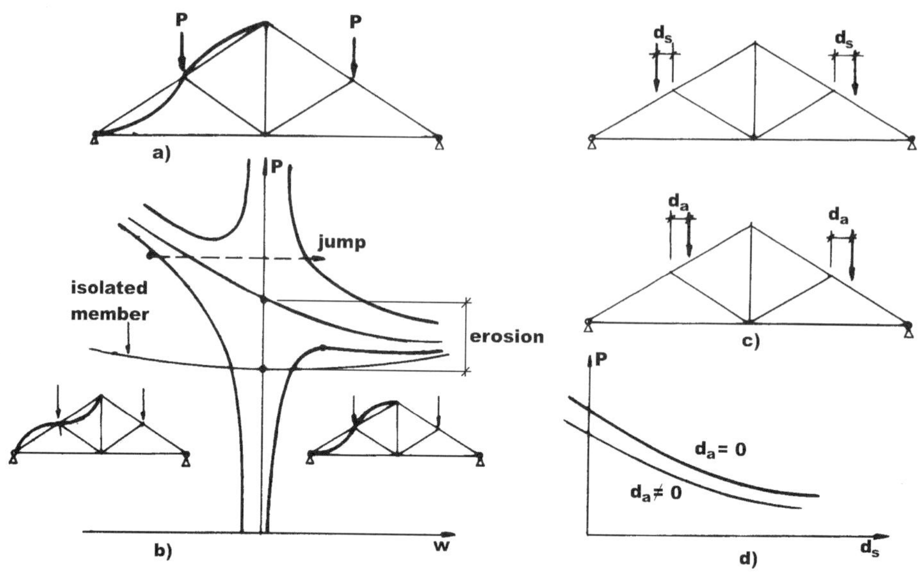

Figure 9. Roof-truss behaviour.

the truss behaviour is determined by all components, but in the critical and post-critical behaviour, only the axial shortening of one member dominates the truss behaviour. Considering the *DOMP*, the other shortenings can be neglected from the truss analysis. Only if the improve of results is intended, the effect of surrounding members can be considered. The domination principle assures that this simplified methodology is correct one, in the sense that the structure instability is produced by only one member component. The second principle, *DETP*, allows the establishing of the instability type (see the next Section). If an isolated member is considered, the behaviour of this member is a symmetric stable point of bifurcation. In structure, due to the effect of elastic supports of surrounding members, the bifurcation critical load increases, but considering the asymmetry of member behaviour in the structure, the instability type is changed. The loss of stability is produced by an asymmetric point of bifurcation (Fig. 9b). The third principle, *PERP*, considers the influence of fluctuations, which for the examined truss are the geometrical imperfections. When the imperfection favor the unstable mode of post-critical, a progressively increase of external loads leads to a peak value. But when the imperfections favor the stable mode of deformation, peak loads do not results from the progressive loading. However, in such situations, a jump to the associated unstable mode may be achieved quite easily. The behaviour of member in structure at large deformations may deviate appreciably from the initial post-critical behaviour. Due to the fact that the truss behaviour is influenced by many imperfection types, the *PERP* allows the determination the major and minor imperfections. For the studied truss, the eccentricity of applied loads can be symmetrically as well anti-symmetrical ones (Fig. 9c). Clearly the symmetrical eccentricity is the major imperfection, while the anti-symmetrical one is the minor one and can be neglected in analysis.

4   Instability Types

4.1   Ideal or Actual Structures

Actual structures are affected by imperfections which can be either of geometrical or mechanical nature (Fig. 10).

Geometrical imperfections result from the non-coincidence between the prescribed form and that constructed and can affect the whole structure or only some portions of it. Mechanical imperfections are produced by some modifications of mechanical properties (initial tensions, variation of yield and strength stresses, etc). An important concept in the theory of structures is the definition of *ideal structures* by disregarding the imperfections of the actual structures. This schematization allows the definition of the criterion for the classification of different structures according to manner in which they lose their stability.

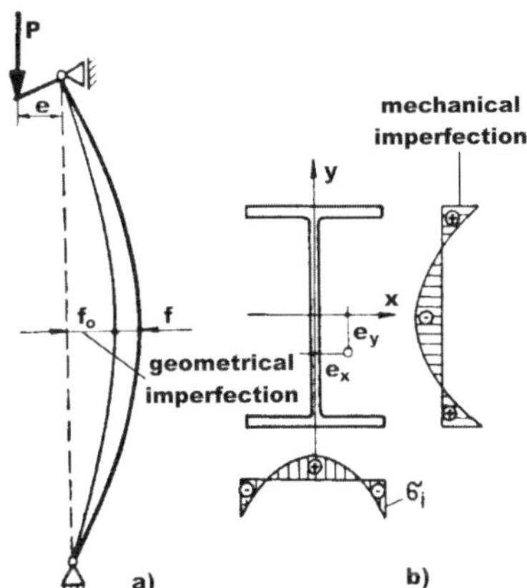

Figure 10. Geometrical and mechanical imperfections.

4.2   Instability by Bifurcation or Limitation

The complete behaviour of a structure is characterized by its pre-critical curve, critical point and post-critical curve. There are two types of critical point in which the structure loses its stability:

a) *Bifurcation point* (Fig. 11a) which results in the case that, at some load level, there are two equilibrium states corresponding to the pre-critical and post-critical paths.

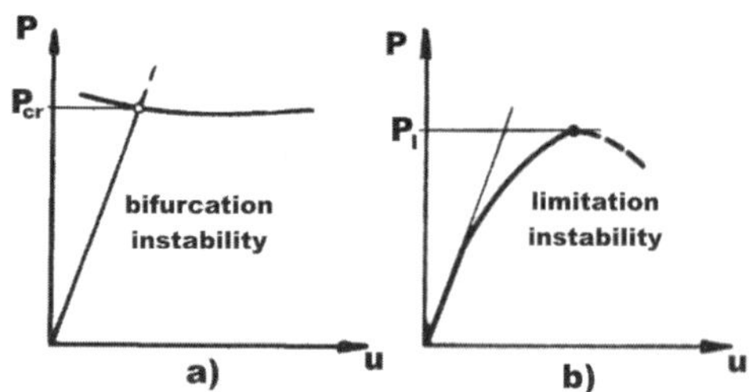

Figure 11. Bifurcation and limitation points.

b) *Limitation point* (Fig. 11b) where the post-critical path is an extension of the pre-critical one and the load-deformation curve presents a maximum value.

Corresponding to these critical points it is possible the definition of two instability types:

a) *Instability by bifurcation*, produced when:

- the pre-critical deformations do not correspond with the instability deformations. For instance, the portal frame (Fig. 12a) has a symmetrical pre-critical deformation, while the post-critical deformation is asymmetrical one:

- the structure is an ideal one, without geometrical imperfections, as in case of compression bar (Fig. 12 b), when the bar in the pre-critical state is straight, while the post-critical form is bended;

- the structure is an actual one with geometrical imperfections, but these are not similar with the post-critical deformations. As example, the arch can have a symmetrical

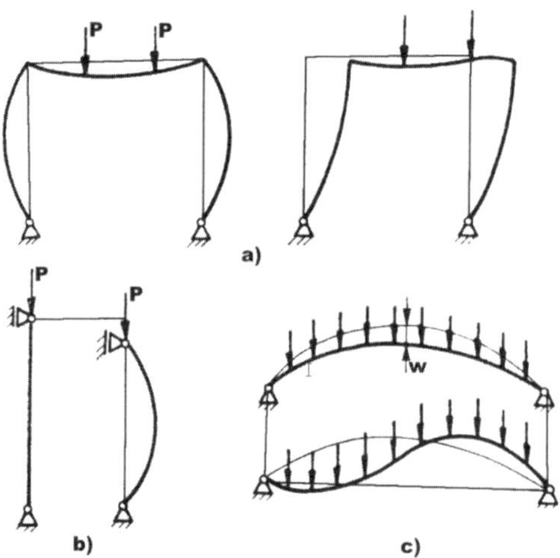

Figure 12. Instability by bifurcation.

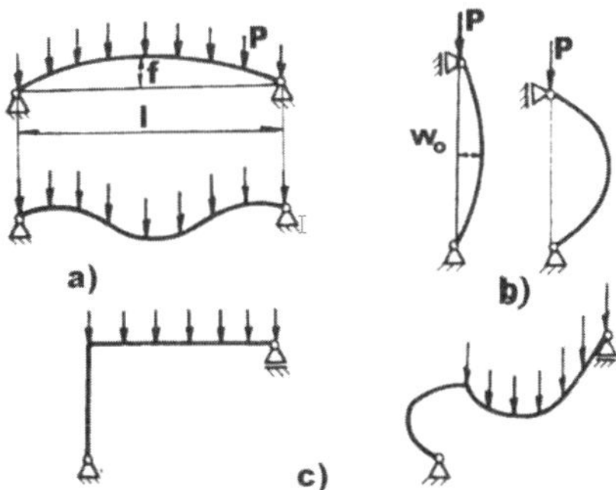

Figure 13. Instability by limitation.

deviation from the designed form, but the post-critical deformation is asymmetrical one (Fig. 12c).

a) *Instability by limitation*, produced for:

- the shallow structures with transversal loads, as the shallow arch (Fig. 13a);

- the actual structure with geometrical imperfections with the same form as the post-critical deformations. For instance, the actual elastic-plastic compressed bar (Fig.13b);

- the structure for which the pre-critical deformations contain as well the instability deformations. The two-bar frame from Fig. 13c is an example for this case.

4.3 Bifurcation Instability Types

a) *Modification of member stiffness in the post-critical range*. Due to the nonlinear deformations during the post-critical behaviour, there are two stiffness modification types (Fig. 14):

- *increasing stiffness*, when the nonlinear deformations have stabilizes effect. For instance, the beam with tension stresses presents an increasing stiffness during the post-critical deformation;

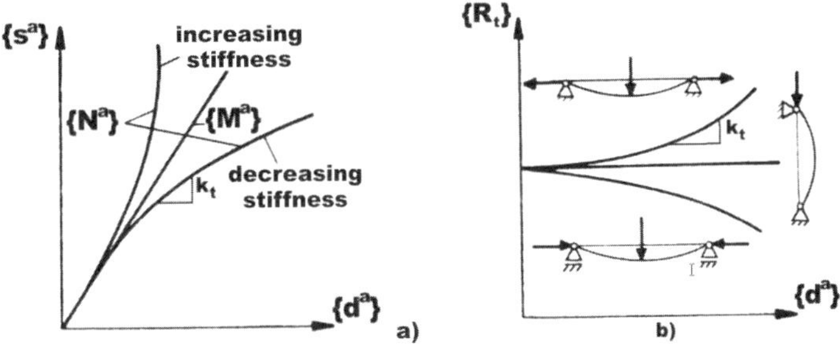

Figure 14. Stiffness modifications.

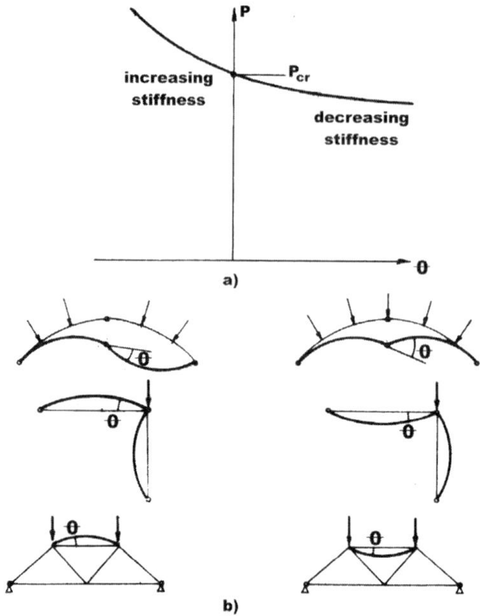

Figure 15. Asymmetric bifurcations.

- *decreasing stiffness*, when the nonlinear deformations have unstabilizes effect. For instance, the beam with compression stresses presents a decreasing stiffness during the post-critical deformation.

b) *Simple bifurcation instability types.* These instability types are characteristic for the cases when the structure presents a single mode of bifurcation. There are three distinct simple bifurcation types:

- *asymmetric bifurcation* (Fig. 15a) occurs when for positive or negative deformations the stiffness of supporting members have contrary effects, as in cases of three-hinged arch, two bars frame, or bridge truss (Fig. 15b);

- *symmetric stable bifurcation* (Fig 16a) occurs when for positive or negative deformations the stiffness of supporting members increases, as in the cases of portal frame or ring (Fig. 16b);

- *symmetric unstable bifurcation* (Fig. 17a) occurs then for positive or negative deformations the stiffness of supporting members decreases, as in the cases of two-hinged arch and compressed beam on elastic foundation (Fig. 17b).

c) *Coupled bifurcation instability types.* These instability types are characteristic for the structure for which, at the same critical load, there are possible two or more instability modes. In these cases some additional unstable paths corresponding to the coupled instability occur. There are two main coupled bifurcation types:

- *the semi-symmetric coupled bifurcation*, characteristic for the structures presenting almost one simple asymmetric bifurcation. There are three semi-symmetrical coupled

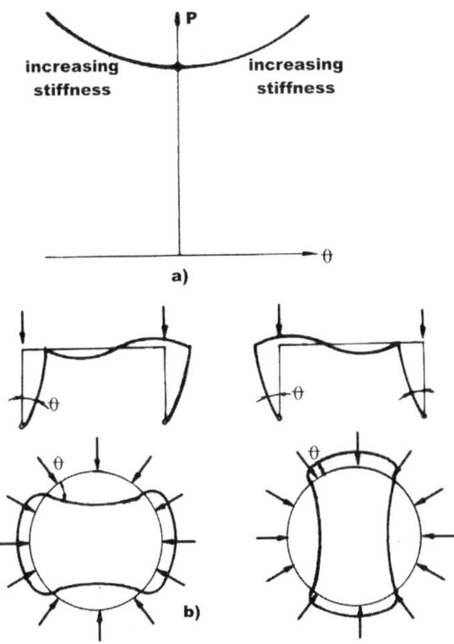

Figure 16. Symmetric stable bifurcations.

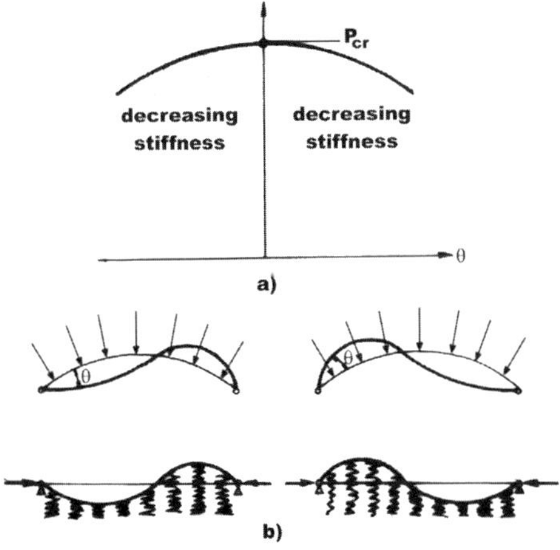

Figure 17. Symmetric unstable bifurcation.

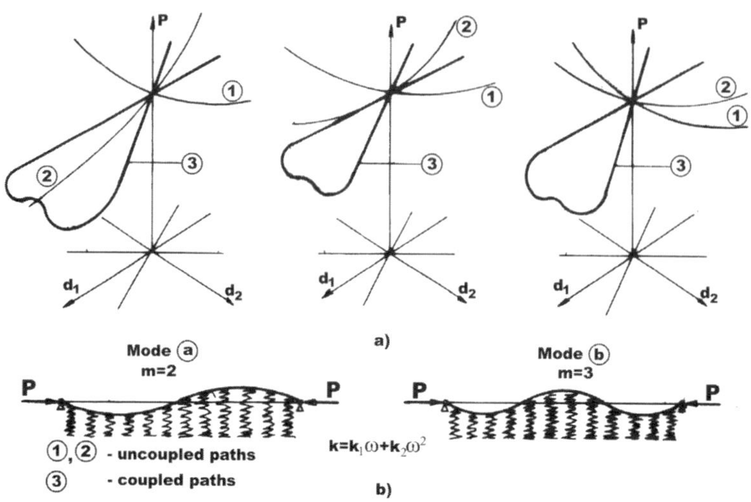

Figure 18. Semi-symmetric coupled bifurcation.

bifurcations: homeoclinal, paraclinal and anticlinal, in function of the slopes of coupled and uncoupled paths (Fig. 18a). The example refers to the compressed beam on elastic foundation (Fig. 18b);

- *the double-symmetrical coupled bifurcation*, in the case when the two uncoupled modes are symmetrical ones. The coupled paths result to be unstable, even if the uncoupled modes are stable. There are two double-symmetrical coupled bifurcations: anticlastical and sinclastical ones (Fig.19a). The most interesting example refers to the pitched-roof frame (Fig. 19b).

4.4 Limitation Instability Types

a) *Limitation due to geometrical imperfections.* Structural systems can never be built precisely as planned and inevitable contain imperfections associated to geometrical errors and material defects. These imperfections can drastically change the response of the structure. The effects of these imperfections in a simple bifurcation instability types are shown in Fig. 20. An interesting feature of these instability types is the existence of some complementary paths to the natural paths in the vicinity of the bifurcation points. For the *asymmetric bifurcation* (Fig. 20a) a negative imperfections yields a stable natural path, but an unstable complementary path. In contrast, a positive imperfection yields a natural path which loses its initial stability at limit point, at a buckling load significantly lower than of the perfect system. For the *stable symmetric bifurcation* (Fig. 20b), positive and negative imperfections have essentially similar effects, yielding continuously stable natural equilibrium path. Only the complimentary paths present limit points. For the *unstable symmetric bifurcation* (Fig. 20c) the effects of imperfections are similarly, but

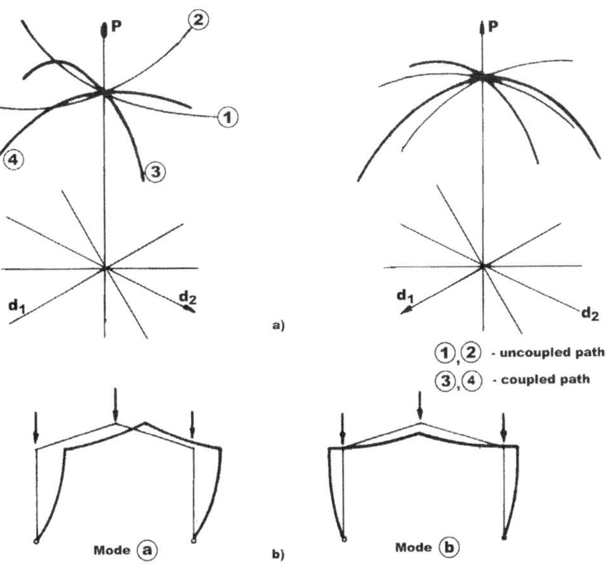

Figure 19. Double-symmetric coupled bifurcation.

in this case natural paths show that the stability losses are produced by limitation at considerably reduced values of the load. In exchange the complimentary paths have not limitation points. The imperfection sensitivity for coupled instabilities is presented in Fig. 21. The lower surfaces correspond to the natural paths, while the upper ones are produced by complimentary paths.

b) *Limitation due to nonlinear deformations*. In a number of structural forms under applied load the change of geometry becomes significant and the structure whilst remaining elastic exhibits a nonlinear load-deflection relationship. A loss of stiffness and the possibility of structure collapse by limitation results. The load-deformation curve is defined in both pre-critical and post-buckling ranges in Fig. 22a. This is the cases of shallow two bar frame and shallow arch (Fig. 22b).

4.5 Dynamic Instability

The transition of a structure from a stable state into an unstable one is a dynamic process and should be analyzed accordingly. There are, however, a number of cases for which the investigation of stability problems is possible using the static theory only. As such cases are numerous in practice, the development of a static theory of stability was elaborated and the dynamic characteristics of the equilibrium stability were forgotten. But there are some cases when this methodology is not adequate:

- the loads are dynamic characteristics, as step, impulsive or periodic loads (Fig. 23);

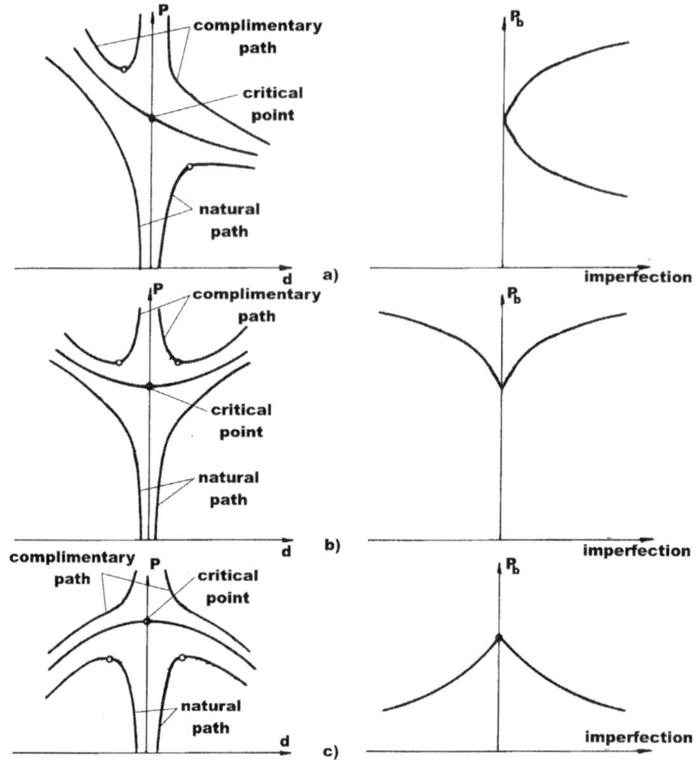

Figure 20. Limitation instability due to imperfections.

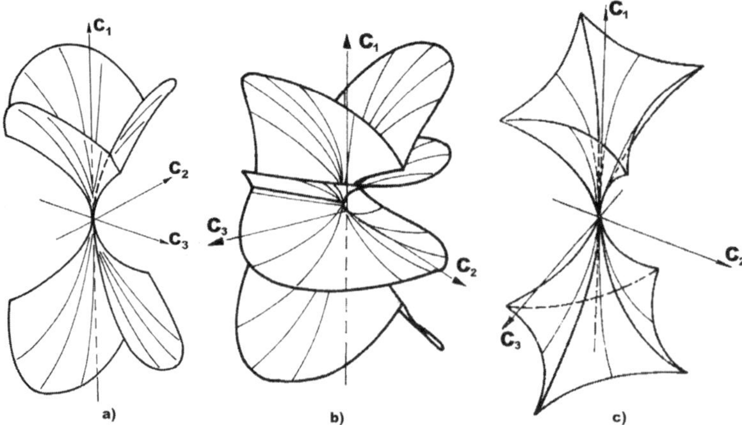

Figure 21. Influence of imperfections for coupled instabilities.

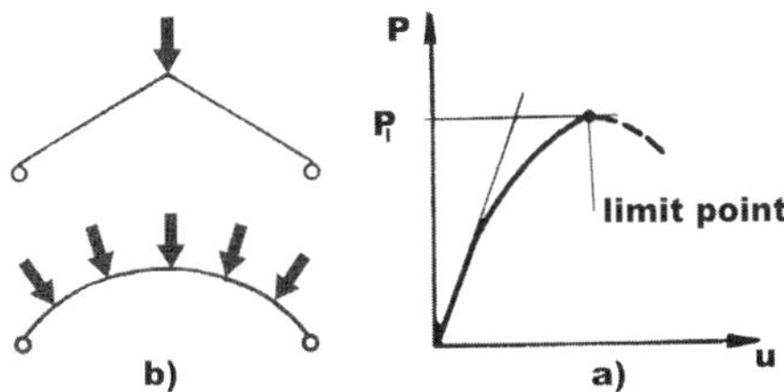

Figure 22. Limitation instability due to nonlinear deformations.

- the lose of stability has a very violent characteristic, so it transformation into a static phenomenon is a mistake. This is the case of snap-through instability and progressive buckling.

a) *Dynamic instability under different load types.* There are some different dynamic loads for which the critical loads are different in comparison with the static critical load.

- *Step loads.* The problem is studied by Hutchinson and Budiansky (1966). The buckling under suddenly applied loads (Fig. 24a) that are maintained at the constant magnitude for an imperfection-sensitive structure is examined. The aim of the study is to compare the influence of imperfections under dynamic loads with the case of static one. A simple model, an archetypal one for the unstable behaviour is shown in Fig. 24b. The ratio between static and dynamic buckling load depends on the ratio between actual and ideal structure. Thus, the dynamic load is smaller than the static one, but the reduction is always at least 3/4.

- *Impulsive loads.* This is a step loading but with finite duration, which involves the vibration period of structure (Fig. 25a). If the loading time is shortly in comparison with natural vibration period, it is possible to obtain dynamic buckling loads several times larger than the static one. This increase is due to the fact that the detrimental effect of secondary order deformations has no time to act. The theoretical and experimental research works performed by Ravinger and Sokol (1994) on the stable post-critical

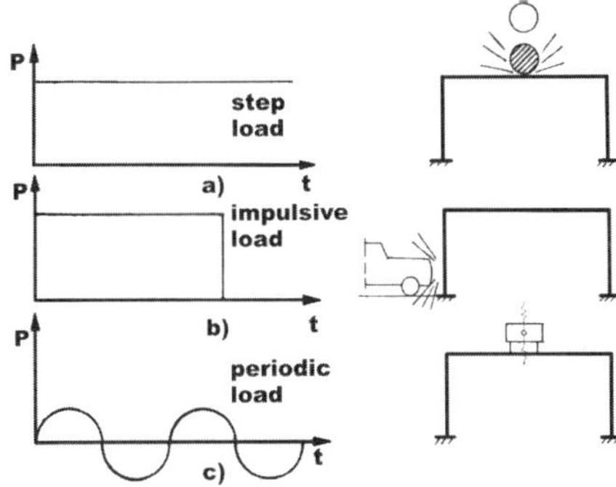

Figure 23. Dynamic loads.

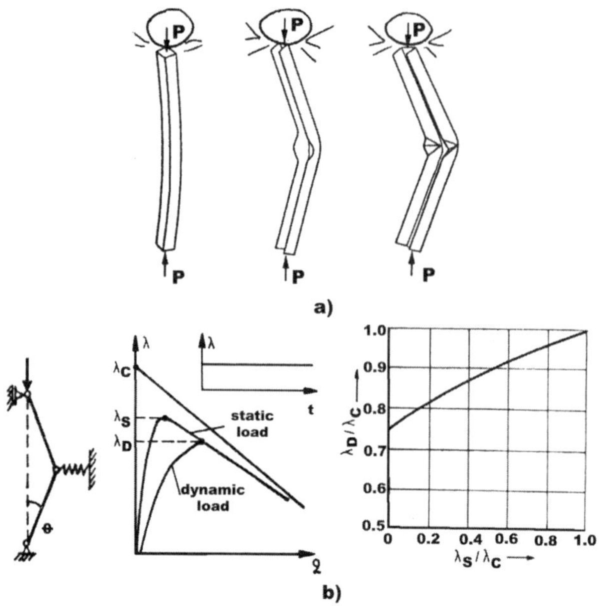

Figure 24. Influence of step loads.

behaviour structures, as actual columns and plates, have shown the effects of dynamic buckling response to impulse load are not significant. But if the structure is imperfection-sensitive, the effect is the increasing of buckling loads (Fig. 25b), especially for triangular loads (Hutchinson and Budiansky, 1966).

- *Periodic loads.* The time-dependent loads may cause of stability by flutter or divergence. A very important factor is the *natural period*, influenced by axial forces and geometrical imperfections. The loading-frequency relationship in the presence of geometrical imperfections is studied by Souza (1994) and Ravinger (1994). Fig.26 summarizes these results, both for stable and unstable post-buckling curves. For the force corresponding to critical load, the natural frequency is zero. In the case of imperfect structures, for systems exhibiting stable post-critical behaviour, the influence of initial imperfections is to raise the level of natural frequencies as compared to the ideal system. In exchange, for systems that exhibit unstable post-critical behaviour, the influence of initial imperfections consists in lowering the natural frequency.

b) *Snap-through instability.* There are structures which present more forms of stability for the same level of load, the first on the pre-critical path and the second far from this position. A model for this type loss of stability is the broken beam shown in Fig. 27a, consisting of two bars, hinged at the ends and supported at mid-span on elastic spring. Fig. 27b shows the behaviour of this model, in which the mirrored position with respect to the initial shape is the most important aspect. For the case of loss of stability by bifurcation or limitation, the curves of post-critical behaviour along which the loading must decrease in order to obtain stable equilibrium positions are shown in Fig. 27c. This

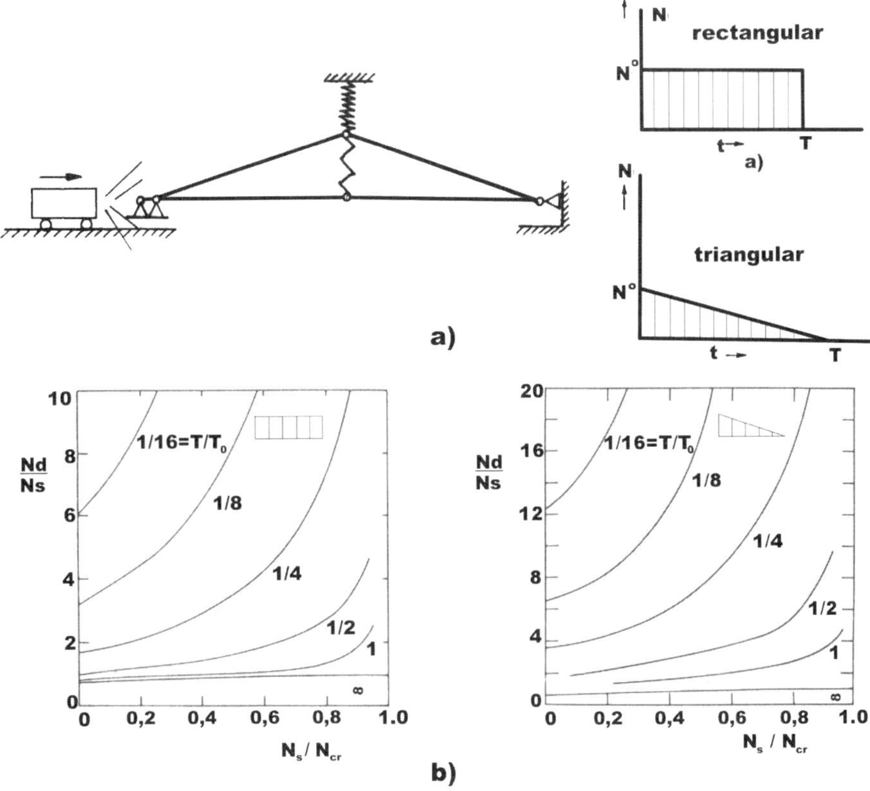

Figure 25. Influence of impulsive loads.

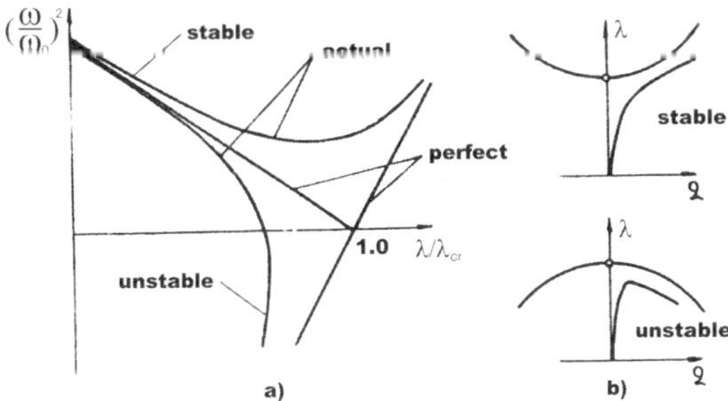

Figure 26. Influence of periodic loads on structure frequency.

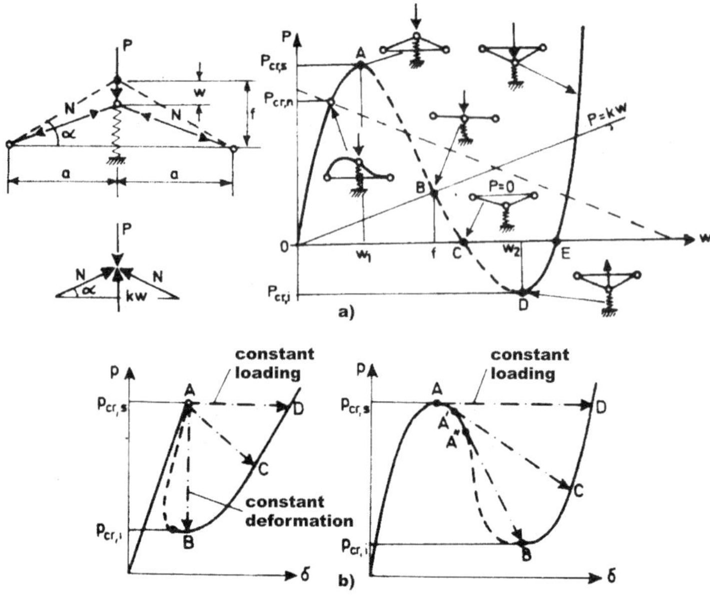

Figure 27. Snap-through instability.

decreasing must correspond to a certain load-deformation law. But, in many cases, the loads are independently of structure displacements (e.g. own weight and snow) and remain constant all along the time in which the loss of stability appears. Therefore, from the point A the structure suddenly passes to the position defined by point D, corresponding to another position of stable equilibrium loading. This passing from a stable equilibrium position to another position, which also provides stable equilibrium is produced by very large deformations and is called *equilibrium jump* or *snap-through instability*. The rate at which the equilibrium jump takes place is very high and large inertial forces appear. When the external load P reach the limit value (Fig. 28) the geometry suddenly changes with dynamic effects. The segment L represents the gap between the applied load and the possible elastic reaction during the jump. The area B represents the kinematical energy gained by the system, which is transformed in an impulsive action. Fig.29 shows how the behaviour of the truss varies when the rigidity of spring increases. Increasing this rigidity results the decreasing of the kinematical energy. The Table 1 gives the maximum acceleration values obtained as function of K. One can see a very important acceleration obtained during this equilibrium jump, which characterizes the violence of the snap-through instability.

c) *Change in buckling patterns.* Some interaction between different post-critical modes can occur, and a jump in stability from the first mode to the second one is produced. During this jump a great velocities and accelerations occur with a very important dynamic effect. For some structures this pattern changing represents only an aspect changing, but for others, this changing produced a very dangerous collapse.

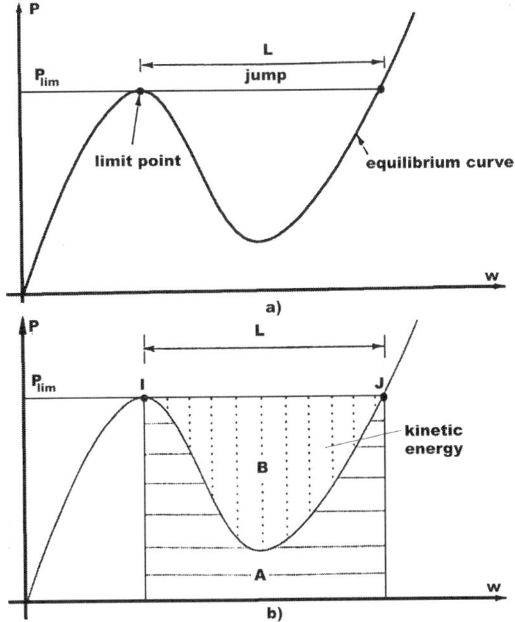

Figure 28. Energetic behaviour during the jump.

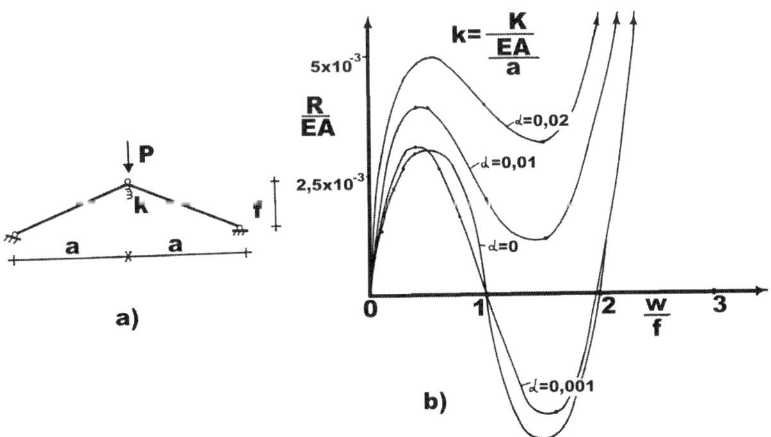

Figure 29. Influence of spring rigidity.

Table 1. Accelerations during the snap-through.

K/ EA/a	A/g
0	7.62
0. 001	7.13
0.01	3.77
0.02	1.58

- *Rectangular plate.* It is well known that an elastic plate subjected to in-plane compressive loading undergoes a primary buckling, which is a stable symmetric bifurcation (Fig.30a). In this primary post-critical range, the plate is able to carry a greater load till this primary path intersects a secondary post-critical path. Due to the fact that the secondary bifurcation is unstable, a snap-through is produced. For instance, from a shape with three half-waves, the pattern changes suddenly in five half-waves (Nakamura and Uetani, 1979). This phenomenon is very well known in the plate buckling, where the pattern changing is associated with intense noise.

- *Cylindrical shell.* The unstable post-critical paths for cylindrical shell are characterized by sequentially changing of circumference patterns (Fig. 31) (Sheinman and Simitses, 1982). The successive passing from a post-critical path to other by snap-through produces a dynamic effect. The local instability starts with 14 half-wave and ends with 8 half-wave.

d) *Progresssive instability,* which can be defined as widespread propagation of instability following damage to a large portion of structure or over the whole of it.

- *Pipeline,* the phenomenon being observed for the first time in the early seventies. Under some conditions the local geometric degradation of the pipe (local dent or buckling, Fig. 32), results a flattening of complete length of the pipe, produced with high velocity. Recently it has been observed that under external pressure a similar propagation can also occur in the case of pipe embedded in a confining medium (Fig. 33). A local dent can initiate a buckle that propagates in a characteristic U mode, collapsing the pipe inside the confinement.

- *Double-layer grids.* As this structure type consists of a very large number of members and is redundant, a conception exists that local collapse does not affect overall safety, the structure having adequate redistribution capability. But a number of failures have shown that this is not the reality. With tension members failure the forces keep constant and even benefit by hardening effects as well. In return, with compression members, after the critical load is reached, a carrying capacity sharp reduction occurs, in the end the member being able to take only 40 - 50 % of the initial load. Under these conditions a much more drastic redistribution takes place in comparison with tension member failure where only additional load has to be redistributed. If the members adjacent to the buckled member are able to take over supplementary loads, the phenomenon keeps local. Otherwise, if the carrying capacity of these adjacent members is exceeded, the buckling of members propagates and progressive instability takes place (Fig. 34). This phenomenon is known as

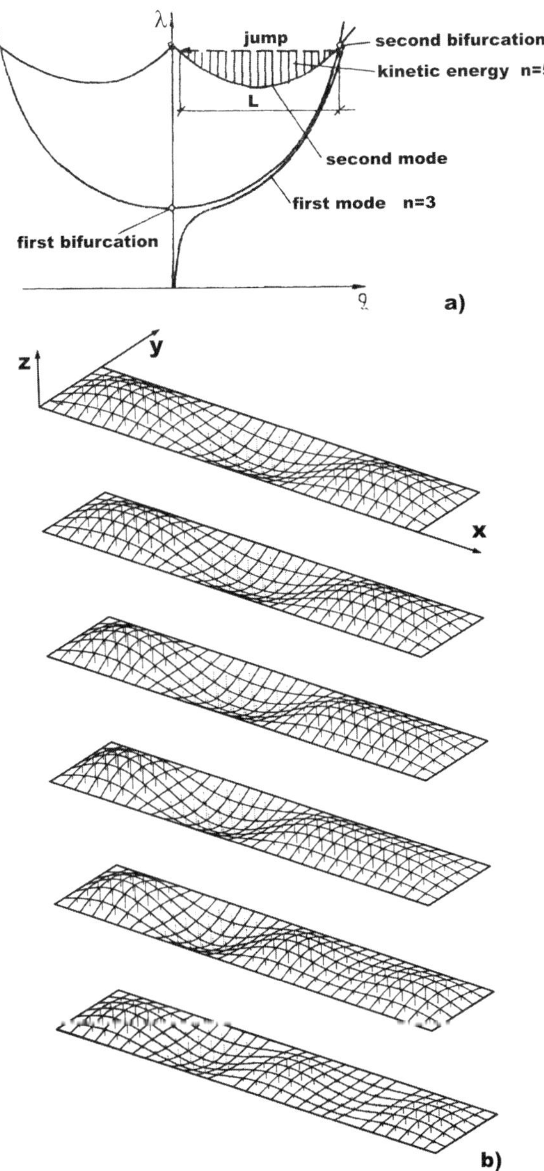

Figure 30. Pattern change for plate.

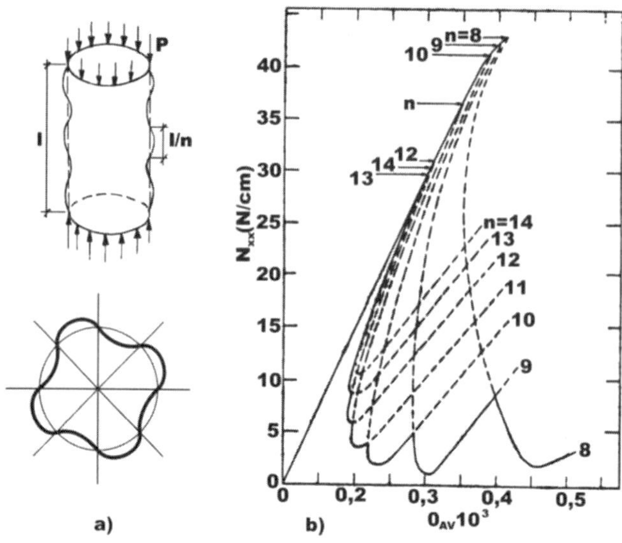

Figure 31. Pattern change for cylindrical shell.

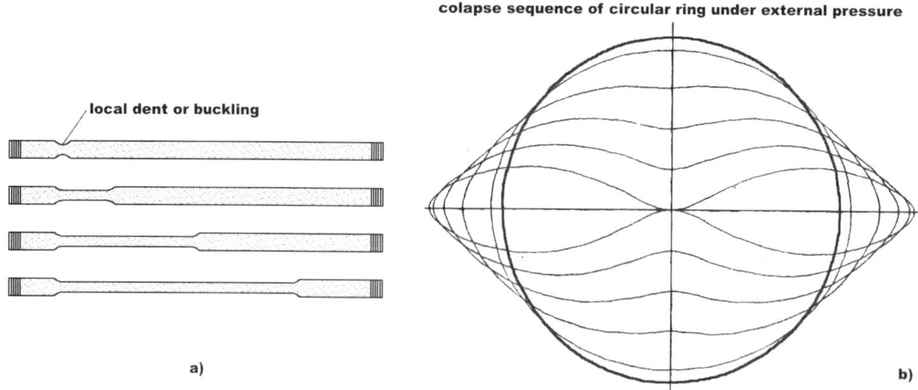

Figure 32. Progressive instability for pipeline.

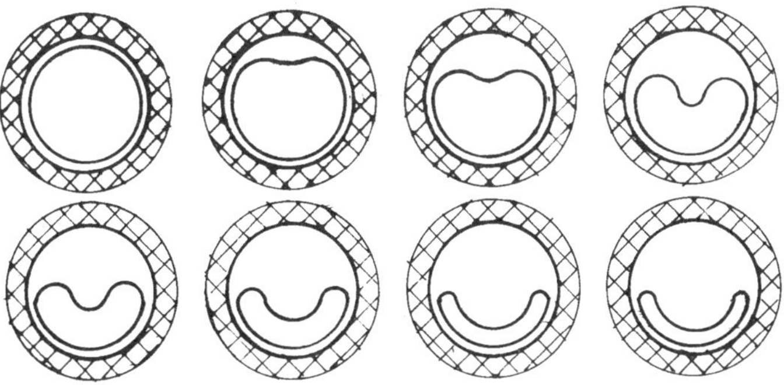

Figure 33. Progressive instability for embedded pipeline.

domino effect (Gioncu and Dinculescu, 1996). The propagation has an accented dynamic character and the space structure seldom escapes an overall collapse.

- *Single-layer reticulated shells.* The local instability of a node is one of the most dangerous forms of instability in the case of reticulated shells (Gioncu and Lenza, 1993, Abedi and Parke, 1996, Ivan and Gioncu, 2000). This phenomenon can be initiated by a variety of causes such material defects, fabrication errors, impact, deviations in the geometry of the ideal form or abnormal concentrated forces in nodes. The nodal snap-trough failure affects a small portion of the structure initially, but has potential for propagating to other parts of the structure and may cause total collapse of the structure. This mechanism of failure requires an inversion in the local curvature and so a violent dynamic effect has to be considered. The final position of the structure in the mirrored form represents a stable equilibrium state. Figure 35 presents the case of domes, where

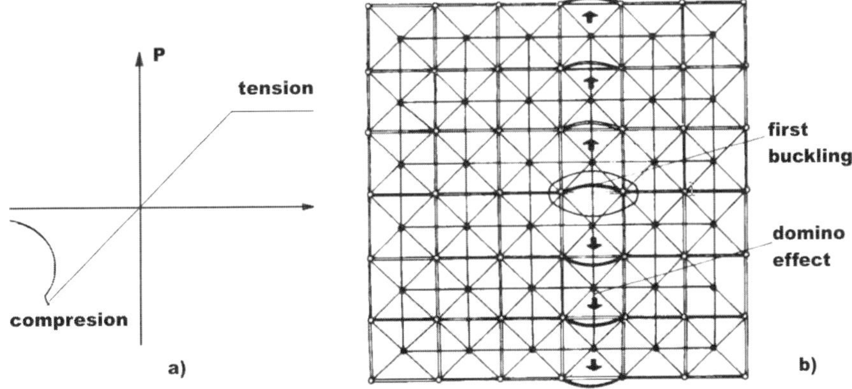

Figure 34. Domino effect for double-layer grids.

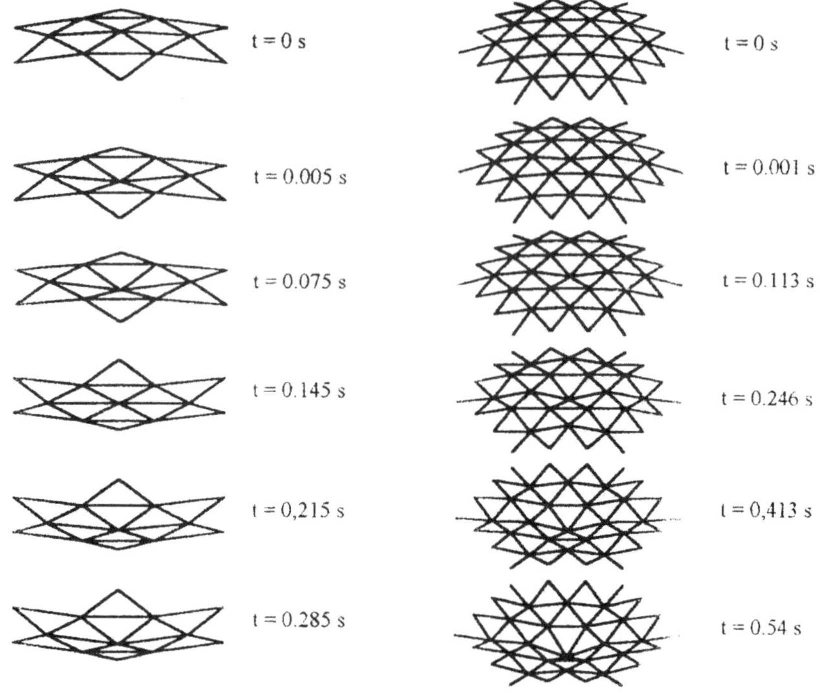

Figure 35. Instability propagation for single-layer reticulated shells.

the instability is started by the snap-through of the central node and the instability propagates over whole surface until the domes arrives in a mirrored position.

5 Phenomenological Methodology for Instability Design

Although the application of post-critical theory to various types of structures has provided remarkable results, this has not yet gained ground in design practice. There some reasons for this situation. The analysis is extremely complex and the equations of nonlinear theory are not readily used by engineers. It is very important for the designer to understand the phenomenon. He needs to understand why it is that in some these nonlinear deformations produce a stable post-critical behaviour, and in others an unstable behaviour. Also, he wants explanations that should be based not on complicated mathematical calculations, but on simple physical considerations concerning the post-critical behaviour of structures. Therefore the results of post-critical theory will find its way into design practice if simple but coherent engineering concept, going right to the very essence of the phenomenon, are developed.

5.1   Erosion of the Critical Bifurcation Load

An ideal structure loses its stability through bifurcation. The critical bifurcation load is reached at the point of intersection between the pre-critical curve and the post-critical one. An actual structure, i.e. one with imperfections, loses stability through limit point equilibrium (Fig. 36). Normally, since structures inevitably have imperfections, the limit load ought to be determined. But this task is very difficult even with very powerful computers due to the geometrical and physical nonlinearities are very pronounced when the limit load is reached. Computation of the limit load by means of complicated and time-consuming computer programs will remain the task of research works only. In practice of structure design the critical bifurcation can by determined by means of much simpler methods than the ones required for determining the limit load. Therefore, for future, the computation of bifurcation load remains the methodology used in practice. The main problem is the correction of this bifurcation load considering the influence of geometrical and mechanical imperfections.

The difference between the critical bifurcation load and the limit load is the *erosion of the critical load*. This erosion may be more or less pronounced, depending on the type of structure and the size of the imperfections. Thus if one accept the critical bifurcation load as the basic value, the main problem will be the determination of this erosion. Since this erosion is little in some structures and significant in others, the idea is now that the critical bifurcation load consists of two components, a stable and an unstable one, the latter being the only one that is eroded by imperfections (Fig. 37). The degree of the erosion depends on the imperfection size and on the ratio of the unstable to the stable component. So, the two aspects of instability, the critical bifurcation load and the post-critical behaviour, can be considered as a result of the sum of the two components (Fig. 38):

Critical bifurcation load = Stable component + Unstable component

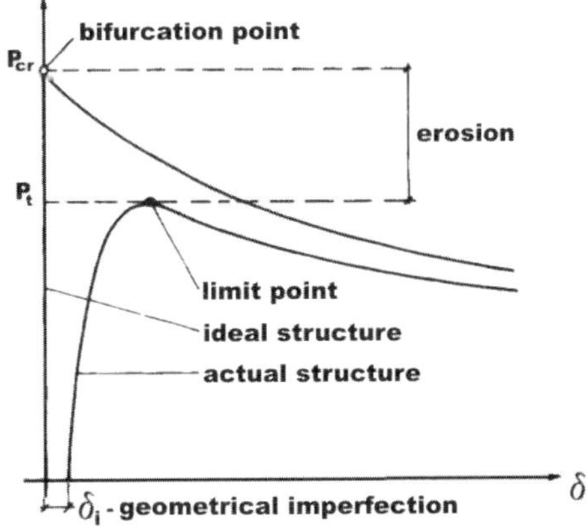

Figure 36. Erosion of critical bifurcation load.

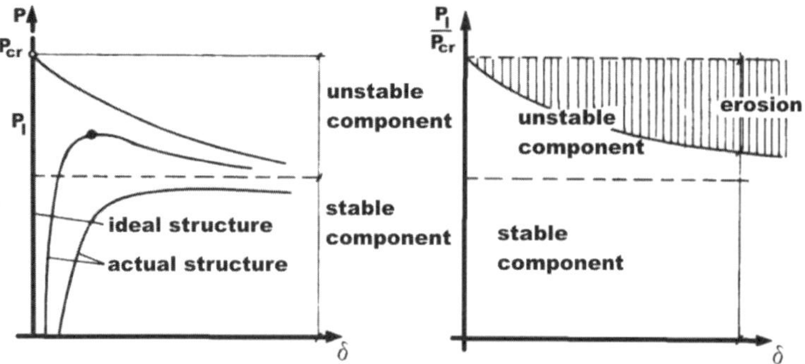

Figure 37. Stable and unstable components.

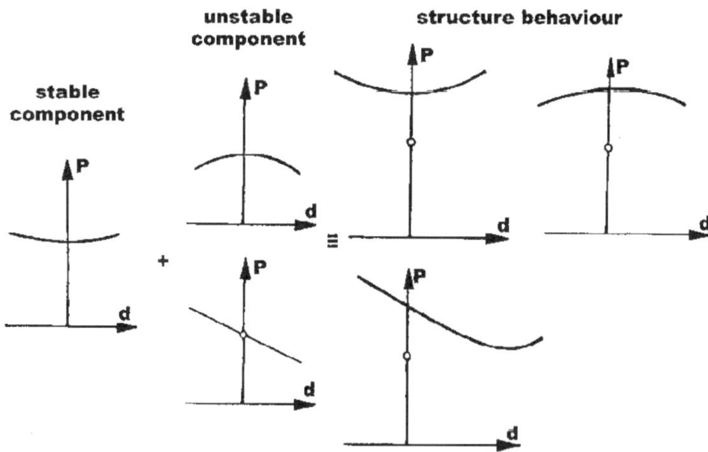

Figure 38. Post-critical behaviour as result of the sum of the two components.

The stable component presents a stable symmetric post-critical behaviour, while the unstable component can show an unstable symmetric or asymmetric post-critical behaviour. By superposing these two components, result the post-critical behaviour of structure. If the both components are symmetrical ones, the post-critical behaviour results symmetrical too, but in function of the characteristics of the two components, the behaviour can be stable or unstable. If the unstable component presents an asymmetric post-critical behaviour, the result of the superposing is always an unstable asymmetric post-critical behaviour.

This idea is not really new. The isolated bar has been analyzed on the similar principle, the European buckling curves representing, in fact, the erosion of the bifurcation load as a function of the bar slenderness, which introduces the size of unstable component due to plastic deformations, eroded by geometrical and mechanical imperfections. The concept of a component eroded by imperfections has also been used in the analysis of shells. Thus, Croll (1981) and Batista (1979) introduced the method of reduced the method of reduced axial stiffness for determining the lower limit of the buckling load. The studies are based on the idea that circumferential stresses introduce unstable components eroded by geometrical imperfections. The concept of reduced components is also used in the analysis of thin-walled members. Thus, the bearing capacity is considerably reduced by coupling the buckling of walls with the global buckling of the bar. A number of methods account for this effect, such as the well-known active width, which suggests that the stiffness of the bar be reduced by eliminating part of the bar walls as an effect of their buckling.

An analysis of these studies shows that the idea of taking into account the unstable component eroded by imperfections has introduced satisfactory results for many types of structures. Since each of these studies is rather particular in character, a more general approach to the problem is now required to provide a unitary concept for all types of structures. In the first analysis it is necessary to determine the factors that account for unstable components in the critical bifurcation load. An analysis carried out by Gioncu (1986) has established that the stable component is introduced by bending stresses, while the unstable component is produced by the following factors:

- extensional deformations;
- elastic lateral supports,
- coupled instabilities;
- plastic deformations.

Once these factors are known, a method applicable in design practice must be found to eliminate the actual difficulties of the stability analysis of structures. In the followings, a number of results selected from literature will be presented in such manner as to prove the existence of unstable components in the value of the critical bifurcation load.

5.2 Effect of Extensional Deformations

a) *Simple supported bar* and *circular ring* are presented in Figs. 39 and 40, after El Nashie (1975, 1977) and Gioncu and Ivan (1983). It can be seen that the critical load and the post critical curves, for the both elements, consist in two components, the first corresponding to the bending stresses and the second to the axial stresses. But the last

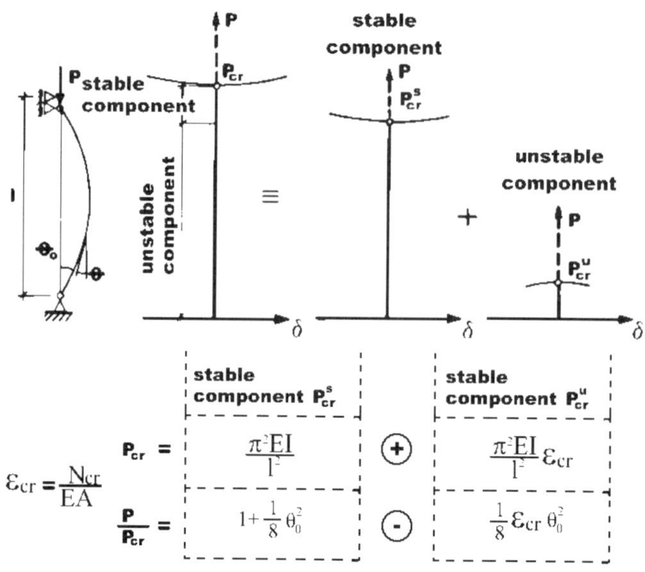

Figure 39. Simple supported bar.

ones are very small in comparison with the firsts and for this reason the extensional deformations are commonly neglected in the determination of critical bifurcation load.

b) *Double-hinged circular arch* with constant direction pressure (Fig. 41a) is analyzed by Dym (1974) and Gioncu and Ivan (1983). The main behavioural parameter is the central angle. The critical load is composed by a stable component resulted from bending stresses and an unstable component produced by axial stresses. If this angle is 90, one can obtain the critical load of ring, composed mainly by stable component. In the interval 90 and 64 the stable component dominates the arch behaviour, but for angles less than 64 (shallow arches) the situation is changed, the arch showing an unstable behaviour, due to the domination of unstable component on the post-critical behaviour (Fig. 41b).

c) *Compressed cylindrical panel* (Fig. 42 a) is studied by Koiter (1956)., in which longitudinal axial deformations were neglected, but the transversal one, due to panel curvature, was taken into account. The critical stress consists of two terms, the first corresponding to the compressed plate, while the second represents the load increment due to the curvature of panel, resulting from the effect of the extensional deformations. The load incremental is obviously due to the favourable effect of the membrane circumferential stresses which prevent the premature buckling of the panel. If, on the other hand, the post-critical behaviour is analyzed, the component due to the plate is seen to correspond to the stable post-critical behaviour, while the one corresponding from the panel curvature has an unstable post-critical behaviour. Hence, it is proved once more that unstable component is introduced by extensional deformations, but in the case of cylindrical panel, this component is very important. One can see that the favourable ef-

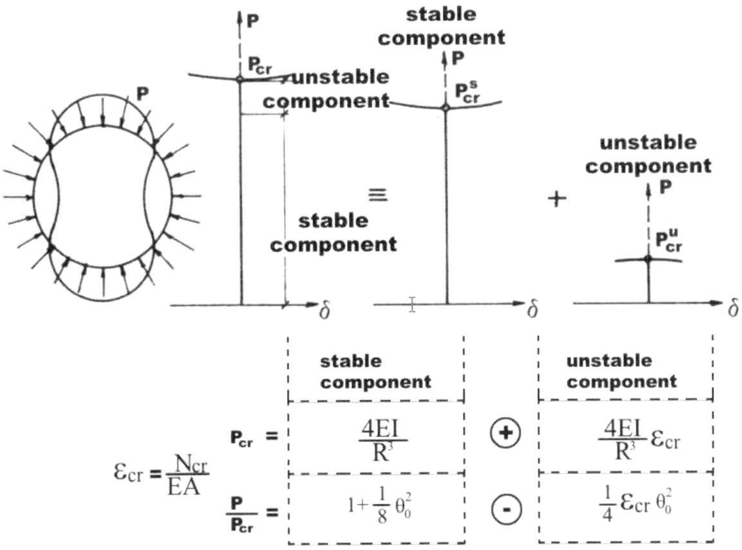

Figure 40. Circular ring.

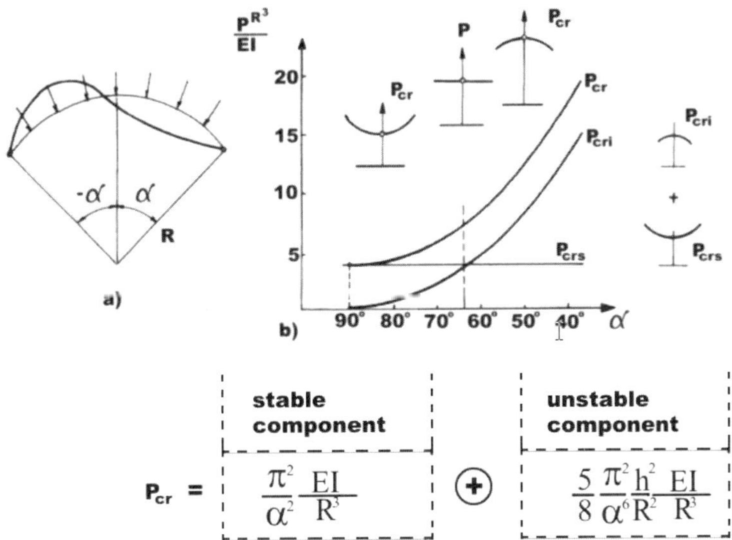

Figure 41. Double-hinged arch.

fect of the curvature which increases the critical load is partial cancelled by the unstable component that may be eroded by geometrical imperfections.

5.3 Effect of Lateral Elastic Support

The second factor responsible for the introduction of unstable components into the critical load is the elastic lateral supports. It may be of two kinds: continuous along the bar or isolated supports.

a) *Compressed beam on elastic foundation* (Fig. 43a), studied by Gioncu and Ivan (1983). Just like in the above studied cases, it can be seen that the critical load is composed of two terms, the first corresponding to the bending stresses and the second, to the supporting effect of elastic medium. An examination of the post-critical behaviour shows that the first term provides a stable component, while the second one, an unstable component that may be eroded by imperfections. The proportion of stable and unstable components for the first three mode of instability is given in Fig. 43b. One can be see that, even if the two components are equal, the unstable component shows an unstable post-critical behaviour that predominates over that of the stable component, thus producing an unstable post-critical behaviour. Therefore, the favourable effect of elastic lateral support, which leads to an increase in the critical load as compared to the isolated bar, is partially eroded , because the load increment is produced by an unstable component.

b) *Two-bars frame* (Fig. 44a) have been studied by Koiter (1967), Rizzi et all (1980) and Gioncu and Ivan (1983). The critical load of this frame increases in comparison with the simple-supported bar is due to presence of horizontal bar. This increase is due to the end moment (producing inextensional deformations) and axial force (producing extensional deformations), only the last effect introduces unstable component. Depending on the slope of force P the stress in horizontal bar may by either a tensile or a compressive stress. Tensile stresses are seen to lead to an increase of the critical load, while compressive stresses produce a drop, due to the increase or decrease of the stiffness of the horizontal bar as a result of the effect of axial force. For the vertical load, there are two possible way of losing stability, one with a clockwise rotation and the other with an anti-clockwise rotation. In the first case the shearing forces produced in buckled bar introduce tensile stresses in the horizontal bar. In the second case, compressive stresses appear in the bar. Hence, buckling with a clockwise rotation increases the stiffness of the horizontal bar, while buckling with an anti-clockwise rotation decreases it. Thus it can be noticed that elastic lateral support introduces an unstable component that may be eroded by geometrical imperfections.

c) *Guyed mast* (Fig. 45), studied by Haftka and Nachbar (1970). The critical load is composed by a stable component corresponding to the cantilever bar and an unstable one, introduced by the guy support.

5.4 Effect of Coupled Instability

The erosion of unstable components produced by extensional deformation, elastic lateral support and plastic deformations is the *primary erosion* (Fig 46a). In the case of coupled instability a *secondary erosion* occurs (Fig. 46b). One can see that the effect of geometrical

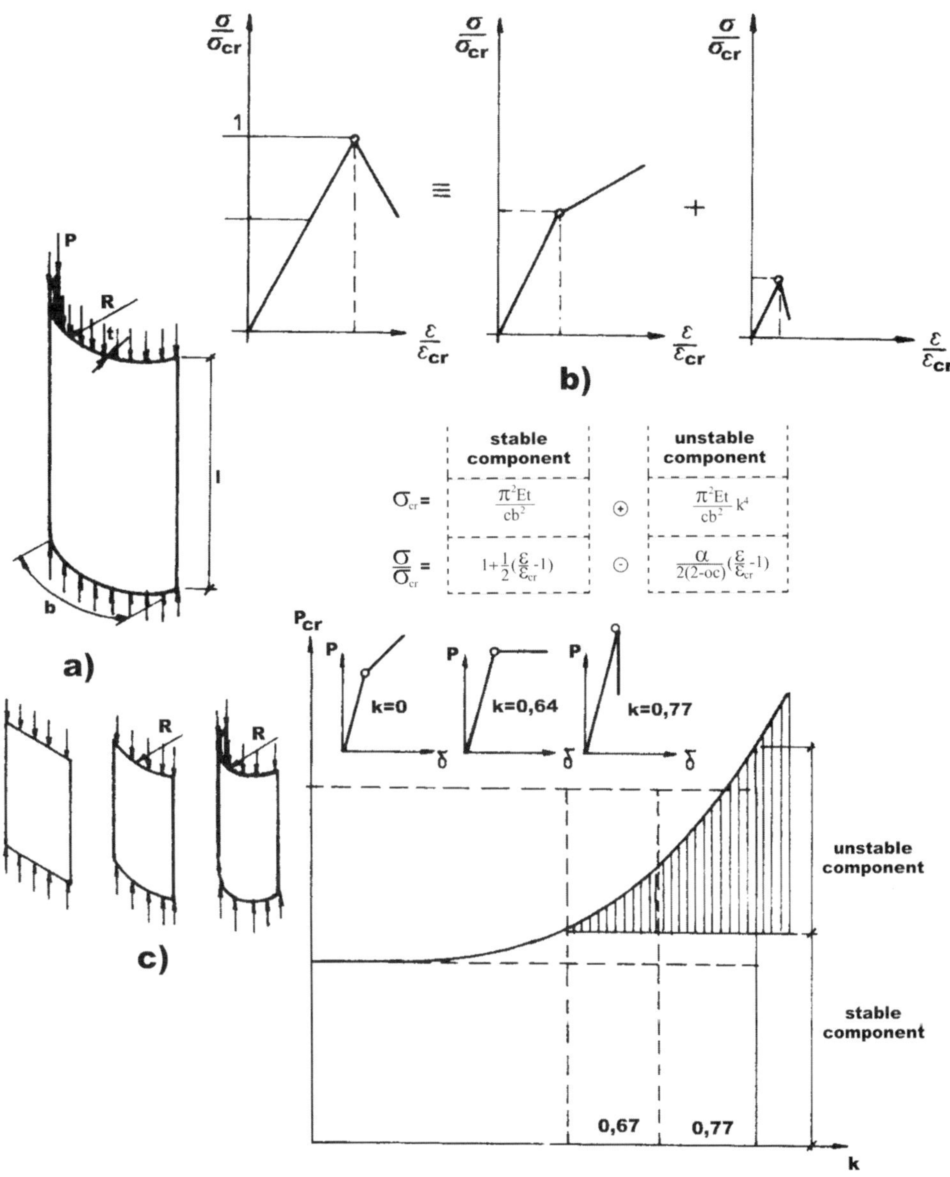

Figure 42. Cylindrical panel.

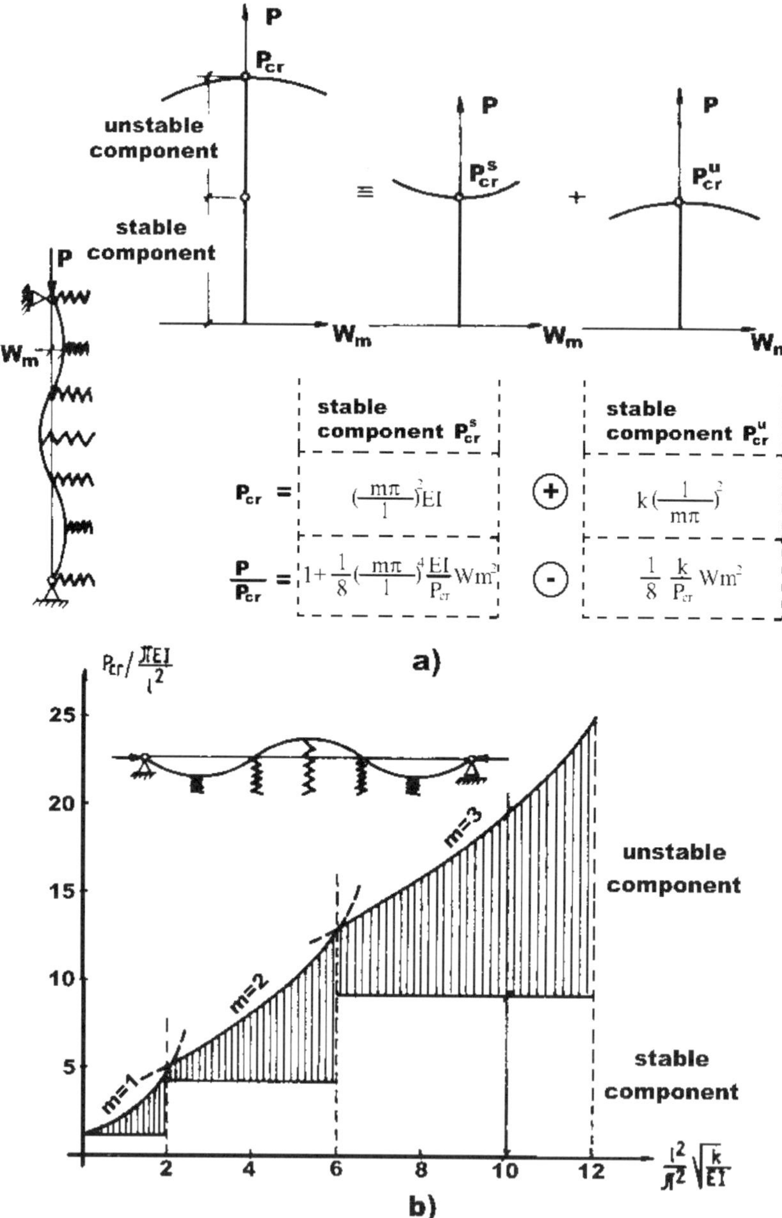

Figure 43. Compressed beam on elastic foundation.

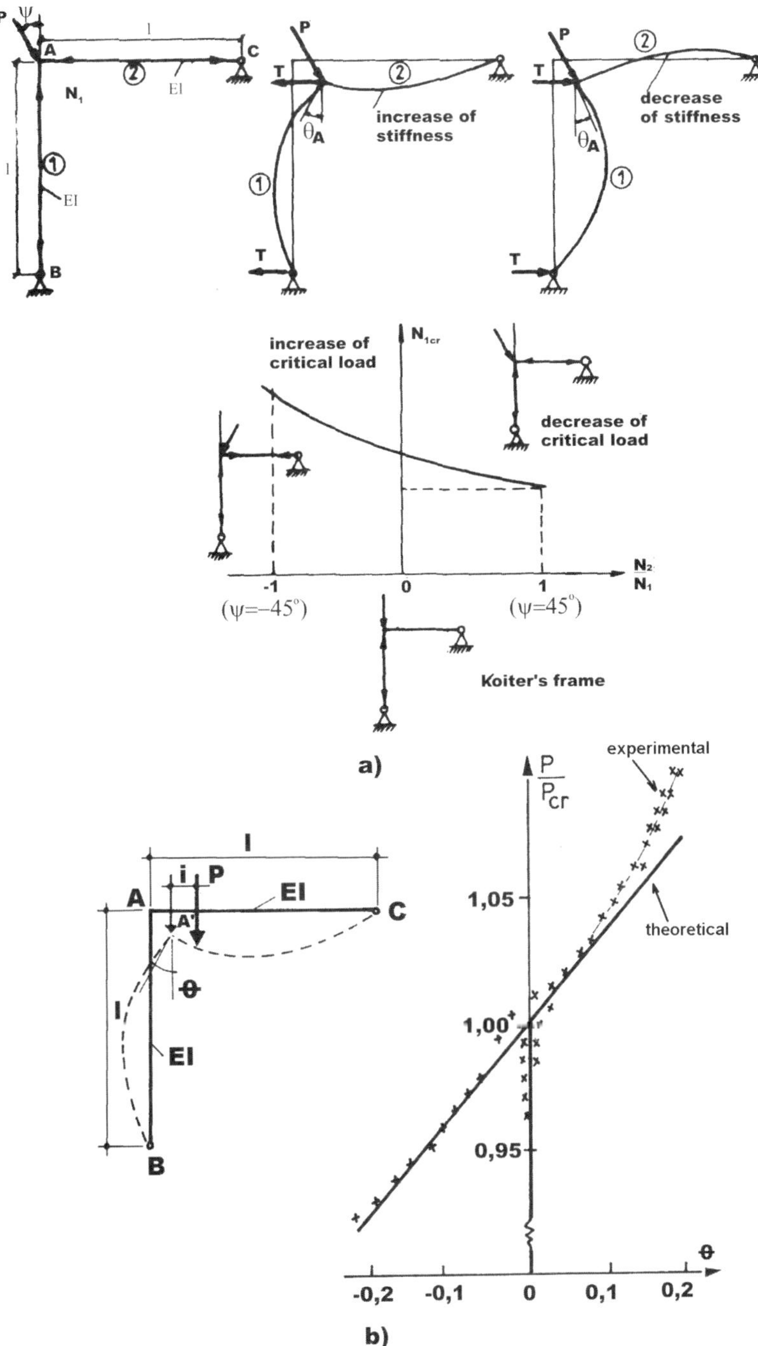

Figure 44. Two-bars frame.

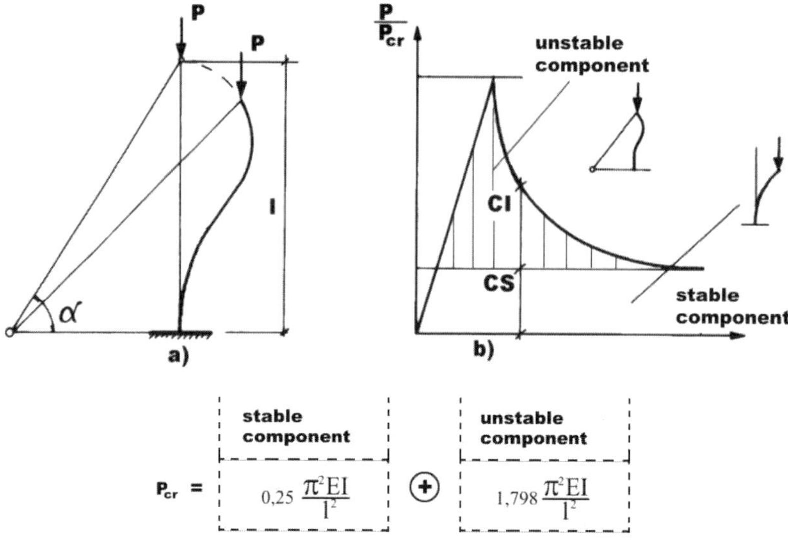

Figure 45. Guyed mast.

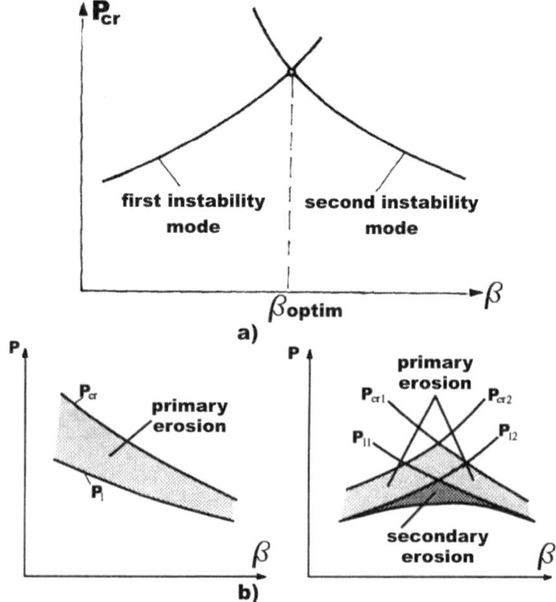

Figure 46. Primary and secondary erosions.

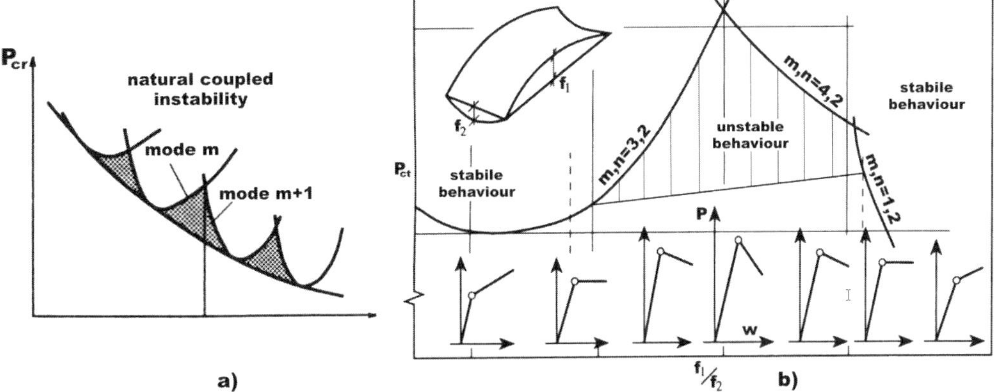

Figure 47. Coupled instability for hypar shell.

imperfections is the erosion of the cusp formed by the intersection corresponding to the two instability modes. Thus, similar to the primary erosion, the imperfections are trying again to cut this cusp. Based on this observation, one must understand that the reducing of critical load due to the coupled instability does not exist without the presence of imperfections. The reason of the increasing of geometrical imperfections is the presence of some supplementary post-critical curves (see Figs. 18 and 19), always more unstable than the uncoupled curves. Examining the cases of coupled instabilities, it can be find that two very different types exist:

- *naturally coupled instability*. This results from garland curves, two forms of instability being possible at the point of the intersection of the curves (Fig. 47a). The post-critical curves can be stable for uncoupled modes, but by coupling they become unstable. This phenomenon results very clear from the critical curves determined for hypar shell (Janko, 1985) in function of different number of half-waves.

- *coupling due to design*. In this case the geometric dimensions of the structure are chosen by designer such that two or more instability modes are possible for the same critical load (Fig. 48a). For this coupling type, the optimization principle plays a very important role, and it is the most important type in design practice. Fig. 48b shows, as example, the lateral bucking of a thin-walled beam, where the two instability modes are the local and lateral buckling (Reis and Roorda, 1977).

As for simple instability, there are coupled instabilities for which the effects of imperfections are not very important, but structures exist with very important reduction in the critical loads. Depending on the erosion size, Gioncu (1994) has the following interaction classification:

- weak interaction, if erosion is under 10%;
- moderate interaction, if erosion is under 30%;
- strong interaction, for erosion up to 50%;

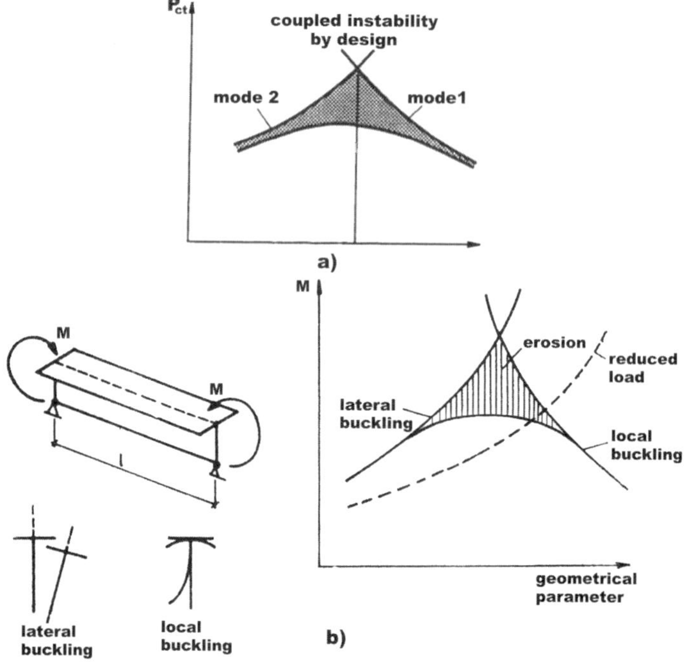

Figure 48. Coupled instability due to design.

- very strong interaction, if erosion is over 50%.

It is very important to know the position of a coupled instability in these interaction classes, because the attitude towards assumed design methods is different. The weak interaction could be neglected, being covert by the safety coefficients used in design. For the moderate interaction, simple methods can be used. In the case of strong or very strong interaction, special methods must be developed.

The main problem is to find the criterion to frame a coupled instability in one of these classes. This framing is function of the ratio of the lengths of the two instability deformations.

a) In some cases the significant bucking modes have wavelengths of about the same order (Fig 49a). The unstable post-critical path shows a weak or moderate interaction and simplified design method can be used. An example is the interaction between *flexural buckling* and *torsional-flexural buckling* (Fig. 49b), having the same buckling lengths.

b) A very different post-critical behaviour is generated in the case of interaction between a *long wavelength (global buckling)* and the *short wavelength (local buckling)* (Fig. 50a). Local buckling is produced by the periodic waves. In this case the coupled post-critical path shows a more marked gradient than in the above case. This coupled instability can be included in the moderate to strong interaction. An example is the laced column, where the overall buckling interacts with the local buckling (Fig. 50b).

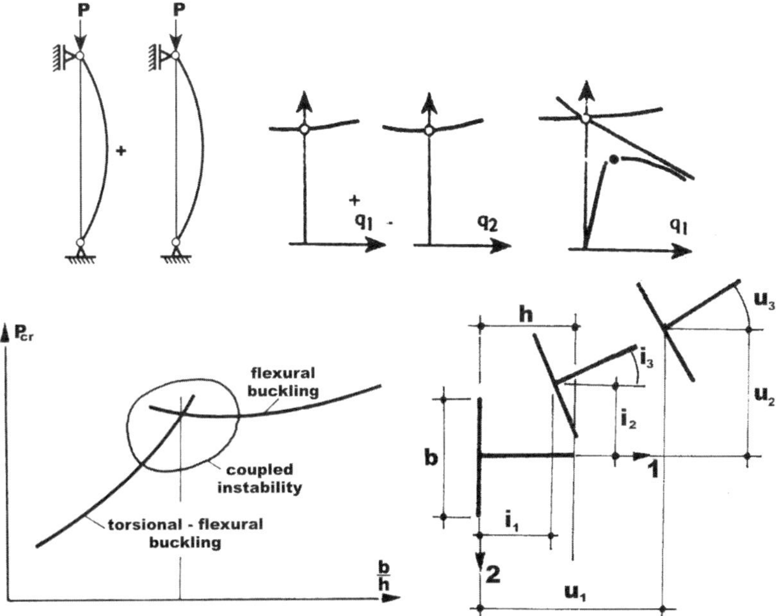

Figure 49. Interaction between flexural and torsional-flexural buckling.

c) If large number of local buckling modes occurs simultaneously under the same local critical load, two types of interaction are produced. The first is due to local multiple local modes and gives rise to an unstable post-critical behaviour. The second interaction of the local buckling modes with the global buckling mode yields to a very unstable post-critical behaviour, with great erosion due to the imperfections (Fig. 51a). Thus, this case of the multiple-buckling modes interaction causes very destabilizing effects. Strong and very strong interaction is the result of this coupled instability. For this type of interaction, very special design methods must be developed. This is the case of compression thin-walled column (Fig. 51b).

5.5 Effect of Plastic Deformations

Three major directions can be distinguished in the approach to the problem of elasto-plastic instability.

a) The first is mainly devoted to the analysis of the *theoretical aspects of the elasto-plastic deformations*. In spite of the important development of the studies, compared with the elastic buckling, plastic buckling is an underdeveloped subject.

b) The second direction, with a market practical character, is materialized by the elaboration of design specifications. The result is the well-known ECCS column buckling curves and they represent the bases of design against buckling in many codes.

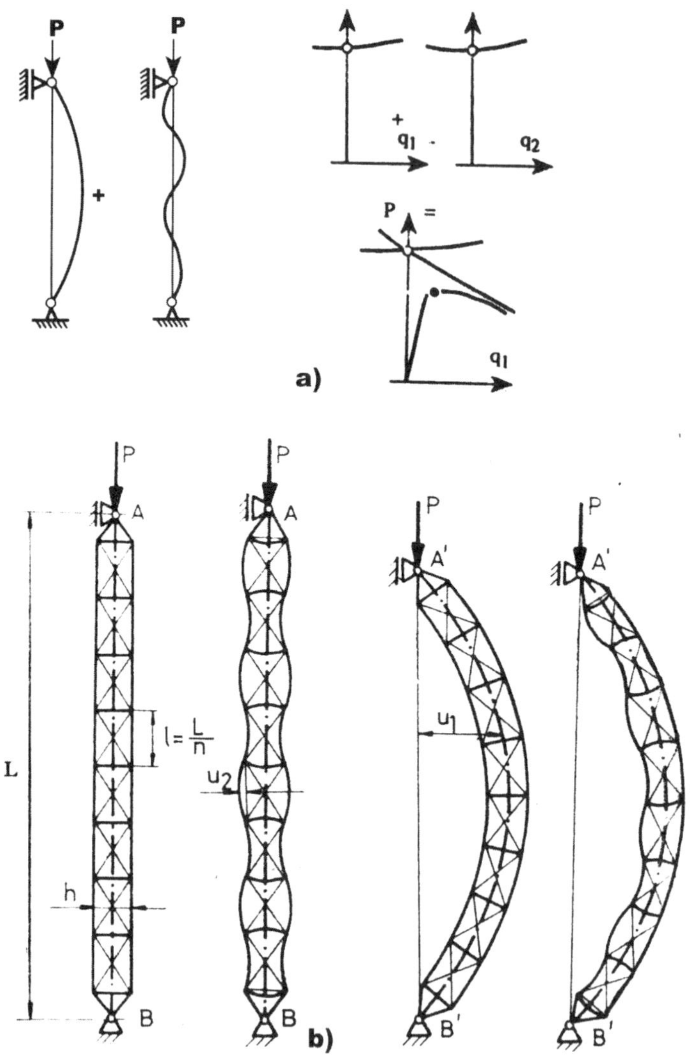

Figure 50. Interaction buckling for laced column.

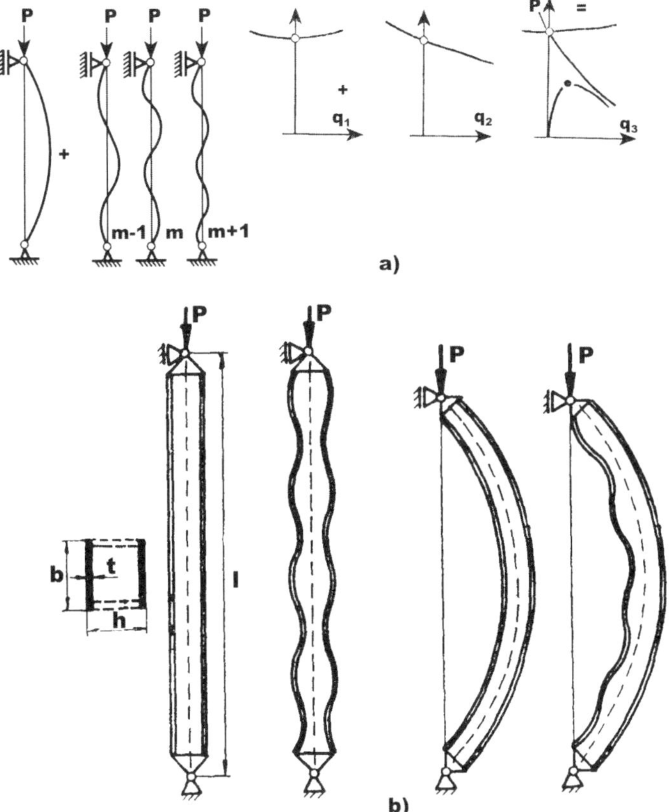

Figure 51. Interaction buckling for thin-walled compression column.

c) The third direction, which is developed in the spirit of the present lecture, transforms the elasto-plastic buckling into a coupled instability.

Thompson and Shorrok (1975) have demonstrated the existence of a bifurcation at the level of a close-packed atomic lattice. The equilibrium and the stability of the unit cell are examined under axial tensile stress. This equilibrium becomes unstable at the bifurcation point, where shearing stress would develop explosively (Fig. 52a). Initial small shearing stress or shearing strain acts as an imperfections and the loss of atomic lattice is produced at a limit point (Fig. 52b). It is easy to understand that the plastic behaviour of a steel member is analogous to that of the unit cell. Regarding the post-critical behaviour of an elastic structure and the elasto-plastic curve, one can see many similarities (Fig. 53a): the same plateau and hardening path and the same effects of imperfections. The similarities are most pronounced if the curves from Fig. 53b are examined. In the first figure one can see the well-known elasto plastic buckling curves, for an ideal and actual member, while in the second, the coupling between two instability modes, with the effects of the

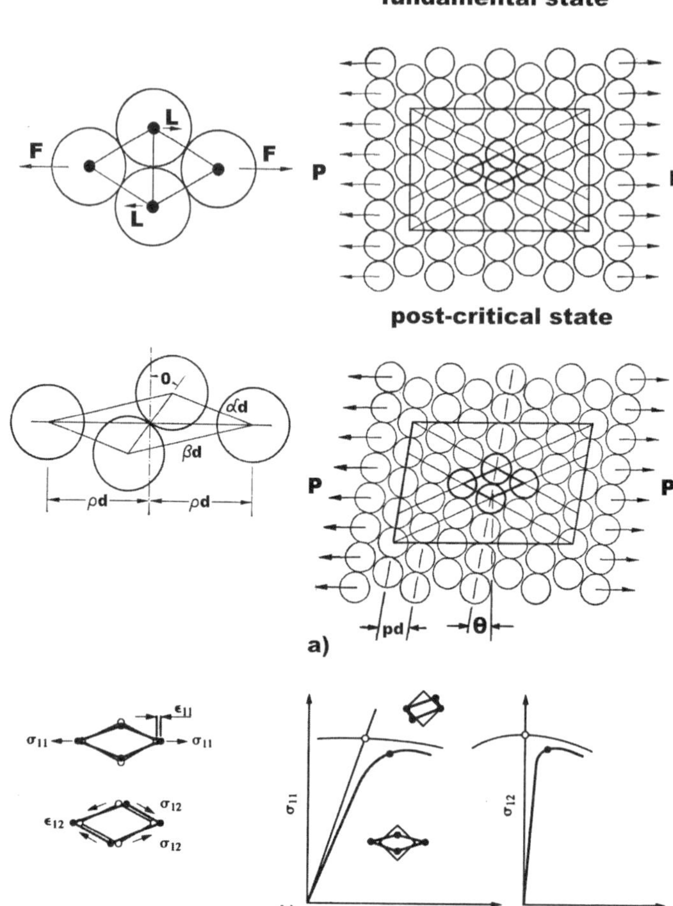

Figure 52. Instability of atomic lattice.

geometrical imperfections. The analogy is almost perfect, the yield stress playing the same role as the local buckling stress. This observations confirm that the elasto-plastic buckling can be considered as a coupled instability between the local bifurcation of the atomic lattice and the elastic buckling of the column.

5.6 Proposed Methodology for Design

Based on these considerations, a simple methodology for structural design is proposed by Gioncu (1985, 2003).

a) *Classes of erosion.* Introducing the ratio between stable component and bifurcation critical load:

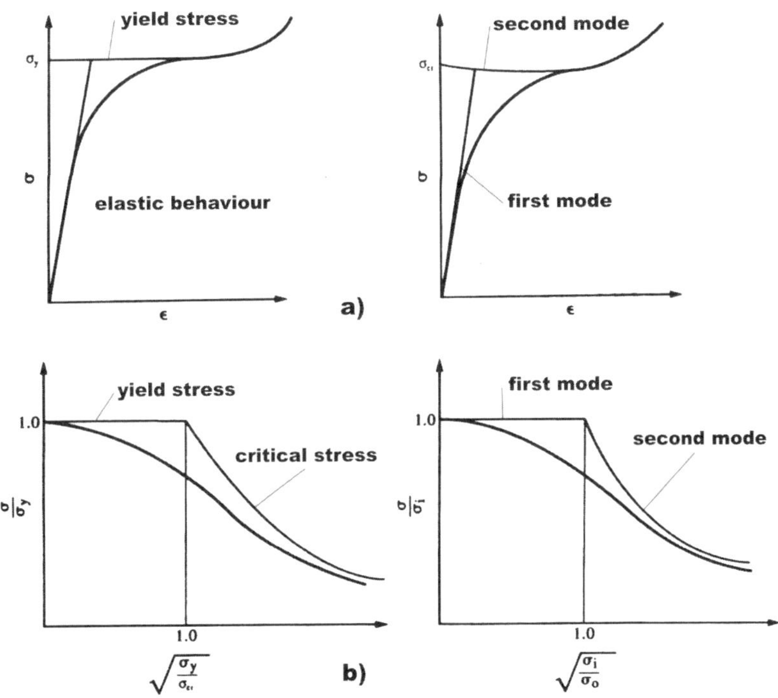

Figure 53. Similarities between plastic buckling and coupled instability.

$$k = (\text{stable component}) / (\text{critical bifurcation load})$$

which takes the form of the well-known knock-down coefficient, a classification of the structures based on this erosion concept is proposed (Fig.54):

- class 1 of erosion for $k = 1.00\ldots0.70$
- class 2 of erosion for $k = 0.69\ldots0.40$
- class 3 of erosion for $k < 0.40$.

This classification is very important for structural designer as it allows the degree of safety to be established. Structures belonging to class 1, in which the erosion is not so pronounced, may be built with greater tolerances than structures of class 2 and special attention must be paid for structures of class 3.

b) *Classes of precision.* Since the unstable component may be totally or partially eroded by imperfections, depending on the size of these imperfections, the introduction of classes of precision is proposed such that structures built with greater attention and accuracy form a superior category. The buckling load should be calculated from the following relation:

$$\text{buckling load} = (\text{stable component}) + r\,(\text{unstable component})$$

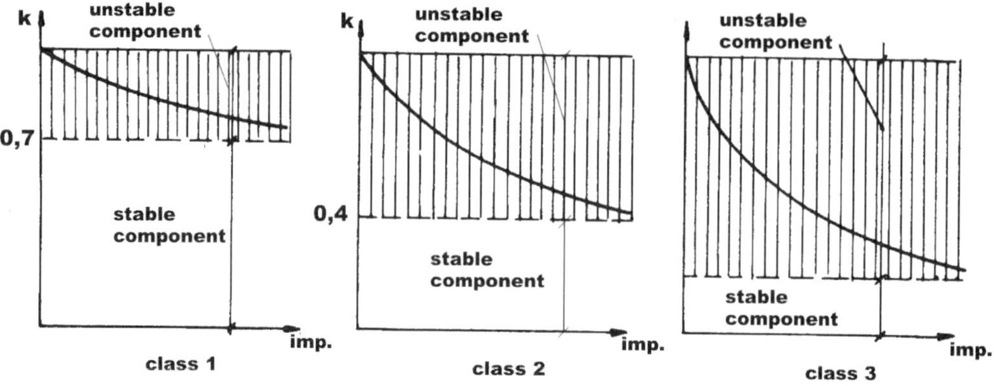

Figure 54. Erosion classes.

r being a reduction coefficient determined in function of control of imperfections (Fig. 55):

- class 1 of precision, r = 0.6
- class 2 of precision, r = 0.3
- class 3 of precision, r = 0

Precision class 1 involves very careful execution and severe control of imperfections. Class 2 involves average execution and class 3 includes poorly built structures, however without permitting grave imperfections or faults.

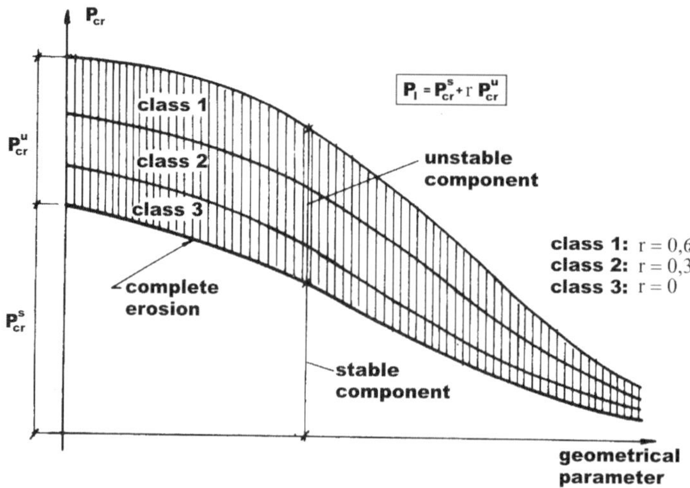

Figure 55. Buckling loads for different classes of precision.

References

Abedi, K. and Parke ,G.A.R. (1996). Progressive collapse of single-layer braced domes. *Journal of Space Structures*, Vol. 11, No.2, 291-306.

Batista, R.C. (1979). *Lower Bound Estimates for Cylindrical Shell Buckling*. Ph. Thesis, University College London.

Croll, J.G.A. (1981). Lower bound of elasto-plastic buckling of cylinders. *Proceedings of Institution of Civil Engineering*, Vol. 71. P.2, 231-261.

Dym, C.L. (1974). *Stability Theory and its Applications to Structural Mechanics*. Leyden, Northoff International Publishing.

El Naschie, M.S. (1975). The initial post-buckling of an extensional ring under external pressure. *International Journal of Mechanical Science*, Vol. 17, 387-388.

El Naschie, M.S. (1977). Der Einfluss von Schubverformung auf die Stabilitat von axial gedruckten Staben in uberkritischen Berich. *Der Stahlbau*, Vol. 46, No..2, 52-54.

Gioncu, V. (1986). Stable and unstable components of the critical load. In J.Szabo, Zs. Gaspar and T.Tarnai, eds. *Post-Buckling of Elastic Structures*, Budapest, Akademiai Kiado, 93-117.

Gioncu, V. (1994). General theory of coupled instabilities. General report. *Thin-Walled Structures*, Vol. 19, No. 2-4, 81-127.

Gioncu, V. (1996). Coupled instabilities under dynamic loading. General report. In J. Rondal, D. Dubina and V. Gioncu, eds. *Coupled Instabilities in Metal Structures*, London, Imperial College Press, 469-481.

Gioncu, V. (2003). *Instabilities and Catastrophes in Structural Engineering*, Bucuresti, Editura Academiei (in Romanian). Manuscript submitted for publication.

Gioncu, V. and Dinculescu, M. (1996). Progressive buckling of reticulated structures. In R.C. Batista, E.M. Batista and M.S. Pfeil, *Stability Problems in Designing, Construction and Rehabilitation of Metal Structures*, SSRC Brazilian Session, Rio de Janeiro, 651-664.

Gioncu, V. and Ivan, M. (1983). *Fundamentals of Structural Stability Analysis* (in Romanian). Timisoara, Editura Facla

Gioncu, V. and Lenza, P. (1993). Propagation of local buckling of reticulated shells. In G.A.R. Parke and C.M. Howard, eds., *Space Structures* 4, London, Thomas Telford, Vol. 1, 147-155.

Haken, H. (1978). *Synergetics. An Introduction*. Berlin, Springer Verlag.

Hunt, G.W. (1986). Hidden (a)symmetry of elastic and plastic bifurcations. *Applied Mechanics Reviews*, Vol. 39, No. 8, 1165-1186.

Hutchinson, L.M. and Budiansky, B. (1966). Dynamic buckling estimates. *AIAA Journal*, Vol. 4, No. 3. 525-530.

Ivan, A. and Gioncu V. (2000). Dynamic propagation of local instability for single-layer reticulated domes In D. Camotim, D, Dubina and J.Rondal, *Coupled Instabilities in Metal Structures*, London, Imperial College Press, 515-522.

Janko, L. (1985). Initialpost-buckling behaviour of shallow saddle-shaped hypar shells supported by shear diaphragms, under uniform load. *Acta Technica*, Vol. 30, Nos .1-4, 239-278.

Koiter, W.T. (1956). Buckling and post-buckling behaviour of a cylindrical panel under axial compression. NLR S 476, Amsterdam, 1-33.

Koiter, W.T. (1967a). *On the Stability of Elastic Structures*. Ph Thesis, Delft, NASA-10

Koiter, W.T. (1967b). Post-buckling of a simple two-bar frame. In *Recent Progress in Applied Mechanics*, Stockholm, Almquist and Wiksell, 337-354.

Kyriakides, S. and Babcock, C.D. (1983). Buckle propagation phenomena in pipelines. In J.M.T. Thompson and G.W. Hunt, *Collapse: The buckling of structures in Theory and Practice*, Cambridge, University Press, 75-91.

Nakamura, T and Uetani K. (1979). The secondary buckling and post-secondary-buckling behaviours of rectangular plates. *International Journal of Mechanical Science*, Vol. 21, 269-286.

Pignataro. M, Rizzi, N. and Luongo, A. (1980). *Stabilita, Bifurcazione e Comportamento Post-critico delle Strutture Elastiche*, Rome, ESA-Editrice.

Prigogine, I. and Stengers, I. (1979). *La Nouvelle Alliace*. Methamorphose de la Science. Paris, Gallimard.

Ravinger. J. (1994). Vibration of imperfect thin-walled panel. *Thin-Walled Structures*, Vol. 19, 1-22, 23-36.

Ravinger, J. and Sokol, M. (1994). Dynamic post-buckling behaviour of slender web subjected to impulse load. *Building Research Journal*, Vol. 42, No. 42, 1-14.

Reis, A. and Roorda, J. (1977). The interaction between lateral-torsional and local plate buckling in thin-walled beams. In *Stability of Steel Structures*, Liege, Preliminary Report, 415-425.

Sheinman, I. and Simitses, G.J. (1982). Axially loaded stiffened and unstiffened cylindrical shells. *Transactions of ASME*, Vol. 20, No. 1-4, 139-149.

Souza, N.A. (1994). Coupled dynamic instability of thin-walled structural systems. *Thin-Walled Structures*, Vol.20, No 1-4, 139-149.

Thom, R. (1972). *Stabilite Structurelle et Morphogenese. Essai d'une Theorie Generale des Modeles*. Paris, Benjamin Publishing House.

Thompson J.M.T. (1965). Discrete bracing points in general theory of elastic stability. *Journal of Mechanical and Physical Solids*, Vol. 15, 311-330.

Thompson, J.M.T. and Hunt , G. W. (1973). *A General Theory of Elastic Stability*. London, John Wiley & Sons.

Thompson, J.M.T., and Shorrock, P.A. (1976). Hyperbolic umbilic catastrophe in crystal fracture. *Nature*, Vol. 260, No. 5552, 598-599.

Modelling Buckling Interaction

Eduardo de Miranda Batista[1,2]

[1] Civil Engineering Program, COPPE/Federal University of Rio de Janeiro, Rio de Janeiro, Brazil

[2] Applied Mechanics and Structures Department, Polytechnic School, Federal University of Rio de Janeiro, Rio de Janeiro, Brazil

Abstract The following six sections deal with the problem of coupled instability. The presentation is mainly addressed to steel structures because of the large application of light construction, which conduct to thin-walled members. In this case two main phenomena are to be considered: torsional behavior and local buckling. The former recommends the knowledge of the Vlasov's theory of beams and the last is to be taken on the basis of the theory of elastic stability. Additionally, interaction between buckling modes may conduct to coupled behavior with important consequences to the member strength. To accomplish a general overview of the buckling interaction of thin-walled members one must be able to deal with the general theory of the elastic stability, as well as to apply numerical solutions for both linear and nonlinear buckling and include experimental analysis results to obtain more profound comprehension of the problem. Finally, one must consider that practical design rules must incorporate the fundaments of the coupled instability theory.

1 Coupled Instability Phenomena

1.1 Introduction

Saving weight is one of the major tasks of structural engineers, leading to solutions with less consuming material. As a consequence, light steel construction is frequently found as an appropriate structural solution, which enables cost saving in the structure as well as in the foundations. The adoption of thin-walled members is a usual choice not only because of engineering aspects but also because of architectural concepts, which take advantage of the most recent design concepts for slender structural members. Light gage steel structural members, on the other hand, bring additional stability problems to be solved, especially those related to local and torsional buckling modes. The buckling modes interaction phenomenon follows the choice of thin-walled steel members to compound the structural solution, which obliges design engineers to be familiar with the basis of structural stability principles, including interaction and coupling behavior. Coupled instabilities deal with at least two buckling modes interaction, as we can find in thin-walled structural members, bar trussed spatial systems, laced columns and shells, among others. In fact, the original Theory of Elastic Stability only deals with linear coupled buckling, as for the case of the flexural-torsional buckling of open cross-section members, for example. The non-linear buckling interaction, which combines at least two linearly uncoupled buckling modes, induces erosion of the limit load, which means loss of ultimate strength of the member. If one disregards the effects of non-linear coupling, a natural solution would be the optimization of the structural solution on the basis of the identity of the critical buckling loads, as originally indicated by Bleich (1952). This design concept was considered a "naïve concept", and so judged inappropriate, after the development of the coupled instability theory. The "naïve" concept of stability design can be presented in the following expression:

$$P_{cr1} = P_{cr2} \tag{1.1}$$

where P_{cri} is associated to two linear uncoupled buckling modes. What seems to be an appropriate solution at first view, can be in fact a dangerous one, since the closer the critical loads the more important the erosion of the limit load, as presented in Figure 1. The more instable the post-critical path the more important the limit load erosion, as can be observed in Figures 1(a) and 1(b). In order to deal with the non-linear behavior of the structural member when buckling interaction is concerned one must take into account the coupling instability theory.

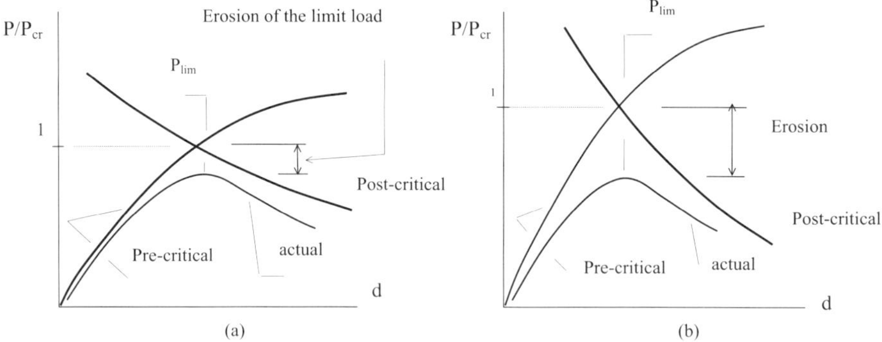

Figure 1. Coupled buckling and the erosion of the limit load.

Several authors applied the coupled instability theory, originally developed by Koiter (1962), as one can find in the fundamental works of Budiansky (1974), Thompson and Hunt (1973), Chilver (1967) and (Van der Neut, 1969), for example. Besides the theoretical developments of the coupled instability, many researches were developed combining experimental analysis with the basis of the non-linear interaction behavior and the obtained results allow incorporating the non-linear interaction behavior in the practical engineering solutions addressed to structural design. Thin-walled structures present the most common buckling interaction, since the local plate or shell buckling usually interacts with the member global modes. This is the case of folded open cross-sections affected by local buckling, in combination with flexural or flexural-torsional buckling of the compressed member, as shown in Figure 2. This type of buckling interaction is found in very commonly used structural members, as girders and the so-called cold-formed steel profiles, for example.

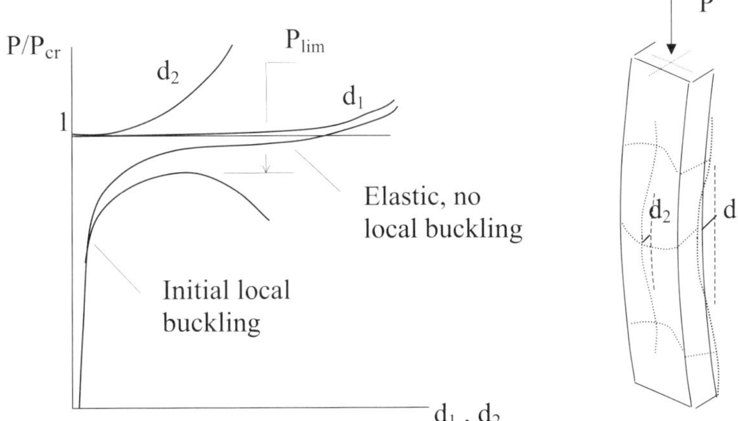

Figure 2. Interaction between local plate and global flexural buckling of a thin-walled member under concentric compression.

As folded steel plates are very commonly used in structural applications, as well as slender cross-sections manufactured by welding process, these are very usual examples of structural members affected by coupled buckling. Figure 3 shows some examples of folded plate thin-walled members that must be currently verified against coupled buckling, and in the same Figure the local plate - cases (a), (b), (e), (f) and (g) - and distortional buckling - cases (c), (d) and (h) - modes are shown.

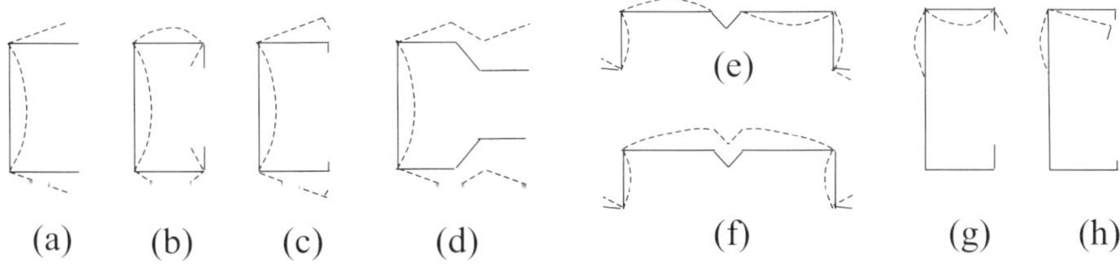

Figure 3. Buckling modes of thin-walled cross-sections.

In order to illustrate a typical example of coupled stability, Figure 4 shows the behavior of a compressed thin-walled member, for which two buckling modes can be in interaction: the local plate mode and the flexural global mode of the member. In this case one may observe:

- The member buckling (d_1) presents a stable post-critical behavior, in a quasi-plateau load-displacement response.
- The local buckling (d_2) usually presents important post-buckling range, with stiffness recovering.

– The coupled post-critical behavior, on the other hand, indicates loss of structural strength, with lower limit load if compared with both fundamental paths (local and global member buckling modes).

This example indicates how important is the consideration of the coupled behavior, since it can dramatically change the post-critical path of the structural member.

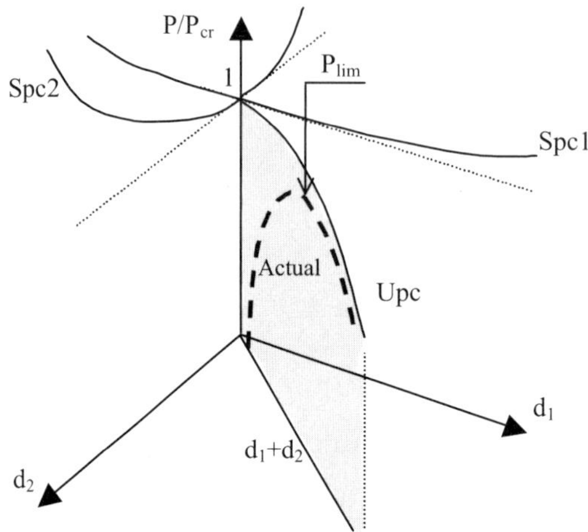

Figure 4. The elastic coupled buckling of a thin-walled member, including the global (d$_1$) and the local plate (d$_2$) buckling. The actual (imperfect) behavior displays unstable post-buckling path, with clear reduction of the limit load.

Finally, it must be pointed that the great majority of the examples and illustrations we will present are placed in the field of elastic behavior, although plasticity has to be considered if the complete structural behavior is aimed, especially for ductile materials as steel. This is so because the main objective in these sections is to present stability principles, which are generally focused in the elastic range. Nevertheless, it should be noted that the inelastic range would be always present at least at the neighborhood of the limit load, when displacements become larger, contributing to the collapse configuration. In fact, for every slender structural member it can be considered that plasticity only appears at the very end of the equilibrium path, with the formation of the collapse mechanism. This is the case of thin plates, for example, where plastic hinge lines, as presented in Figures 5 and 6, form the plastic mechanism. For the case of less slender members, one can expect that the equilibrium path will present inelastic characteristics, which induces additional non-linear effect in the structural behavior.

1.2 Types of coupled instability

Two types of coupled instabilities could be considered: natural and design coupled buckling. This is a useful classification proposed by Gioncu (1994) that serves to distinguish the natural coupled buckling behaviour from the so-called design coupled behaviour. The

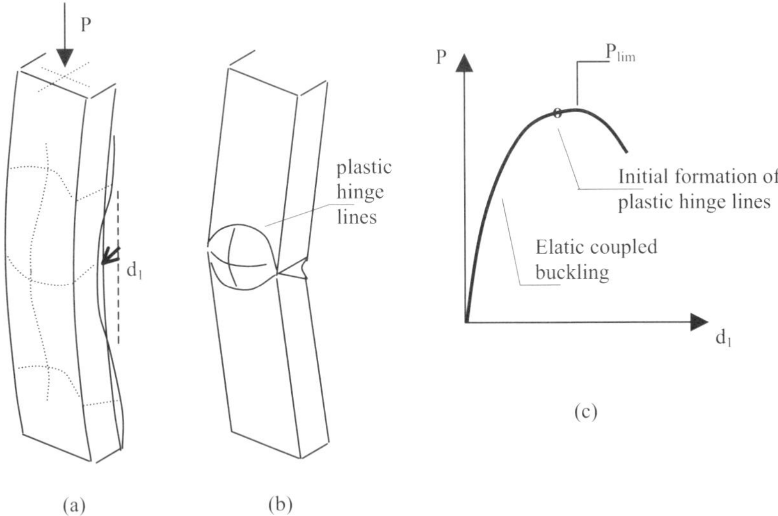

Figure 5. Plastic collapse mechanism of plate in a compressed thin-walled member.

Figure 6. Typical plastic mechanism of plate elements of a steel cold-formed member after compression test.

first arises from the natural coupling between individual modes and cannot be separated by any artificial way. And the second originates in the fact that the geometrical characteristics of the structure or the structural member bring two or more buckling modes very close or even coincident. The former is the case, for example, of the combination of flexural and torsional modes, usually found in open cross-section compressed members (flexural torsional mode); the last is the case of coupled buckling promoted by the fact that the critical buckling loads are close, as the example already presented in Figure 4. A very illustrative example of natural coupling can be obtained from the longitudinally compressed rectangular plate, with four simply supported edges, for which the instability behavior can be represented as shown in Figure 7. In this case, one may observe that the plate presents different buckling modes, depending on its geometrical length-to-width ratio, a/b. The coupled behavior involves the plate modes for one, two, ... and up to n half-wave buckling modes. This is also the case of the local buckling of stub columns of thin-walled members, as shown in Figure 8. In this case, one can observe the analogy between isolated plates and thin-walled folded sections, concerning its stability behavior, with both displaying harmonic buckling modes.

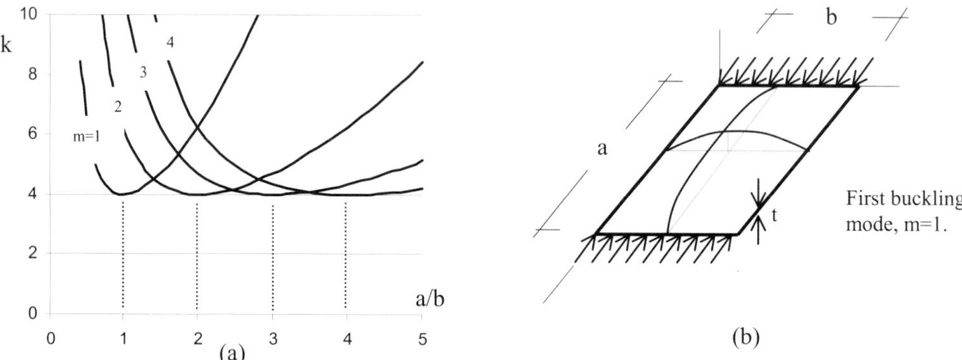

Figure 7. Multiple buckling modes of a compressed simply supported rectangular plate.

In the particular case of the compressed folded member shown in Figure 8, the end conditions were taken as restricted or unrestricted, regarding the transversal in plane displacements of the individual plate elements. In this case, one may also observe in the Figure the effects of restrained conditions, which promotes the reduction of the critical local buckling loading. This is so because of the stress concentration close to the end sections, with the introduction of extra compressive normal stresses σ_y, as shown in the detail of Figure 8(b). This particular observation is important to highlight the fact that not only global member end conditions are important when thin-walled members are concerned but also the plate end conditions. In fact, the behavior of both single plate and thin-walled cold-formed steel members presented in Figures 7 and 8, are based on the first order bifurcation solution. Because of the presence of imperfections, the non-linear load v. displacement behavior of plates has the aspect presented in Figure 9, where the bending and the extensional plate behavior are shown. As can be observed from these figures, one of the major consequences of the unavoidable imperfections is the loss of stiffness of the plate member in the post-buckling range.

1.3   Weak and strong coupling

Depending on the types of the buckling modes that are coupled, and on the ratio between its critical loads, one may experience more or less important coupled behavior. In order to distinguish the differences between weak and strong coupling, it is useful to observe the examples proposed by Gioncu (1994), where the qualitative behavior of three types of compressed structural members are considered:

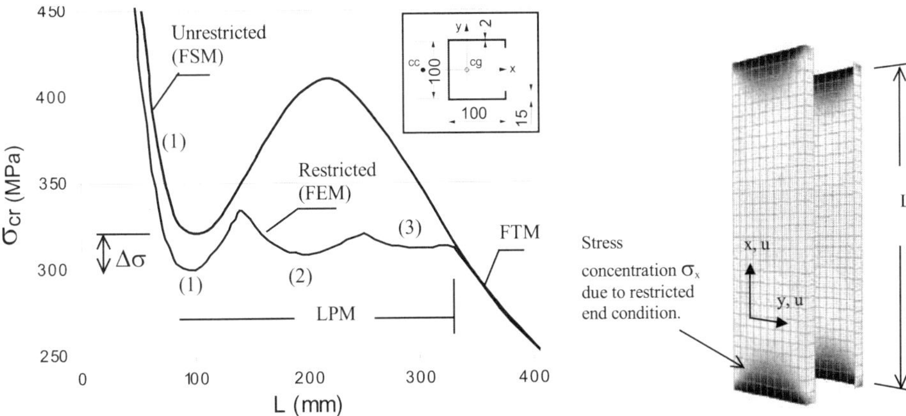

(a) Local buckling critical stresses for restricted and unrestricted end conditions, computed by the FEM and the FSM (plate buckling modes in parenthesis).

(b) Results of the stress distribution σ_y for restricted condition at the end sections, computed by the FEM.

Figure 8. The effects of the end conditions in the local buckling of a stiffened channel member (LPM: local plate mode; FTM: flexural-torsional mode; FSM: finite strip method; FEM: finite element method).

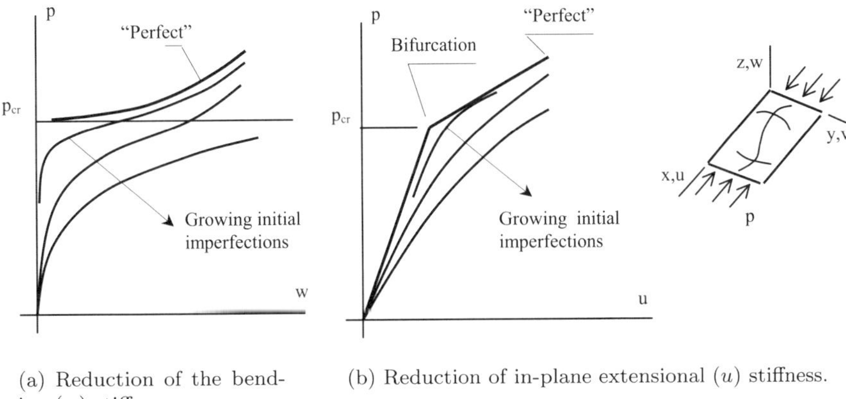

(a) Reduction of the bending (w) stiffness.

(b) Reduction of in-plane extensional (u) stiffness.

Figure 9. The non-linear post-buckling behavior of a rectangular plate element under uniform compression.

(a) an open cross-section member with flexural and flexural torsional coupled buckling (and no local buckling);

(b) a laced column with local bar buckling coupled with global flexural buckling and

(c) a thin-walled member displaying local plate and global flexural coupled buckling.

The first case is shown in Figure 10(a), where the unstable post-critical behavior indicates weak interaction between buckling modes. The second case is in Figure 10(b) and a more pronounced interaction is expected, with moderate-to-strong coupled behavior. Figure 10(c) shows the post-critical behavior for the third case, that of a thin-walled column, for which the interaction effects are much stronger, including the global (single half-wave) and the plate (multiple half-waves) modes. These different behaviors can be explained considering that the weak interaction, case (a), includes two buckling modes with equivalent long wavelengths; and the strong interaction, case (c), includes a long wavelength buckling mode (global flexural) coupled with a multiple short wavelengths (these local buckling modes are of the same type of plate buckling, as shown in Figures 7 and 8, and may include additional local coupling behavior). The intermediate case (b) indicates coupling behavior of the flexural long wavelength with the single buckling mode represented by the local buckling of each individual compressed bar component of the laced column. It is interesting to observe the differences between the two local buckling here presented and the resulting consequences for the post-buckling behavior of the structural members: the one of the strut bar members in Figure 10(b) and the one of the folded plate elements in Figure 10(c). These different buckling interactions induce important consequences for structural engineers, regarding the type of structure or structural member to be designed. It is clear that multiple buckling modes can cause strong interaction, conducting to a very important reduction of the limit load, and especially if these multiple short wavelengths are coupled with a long wavelength buckling mode (flexural or flexural torsional buckling). This is the case of thin-walled structures usually applied in light steel construction, as the cold-formed profiles and the stiffened plates, for which special design methods are expected.

2 Theory of Coupled Instability

2.1 Introduction

The theory of coupled instability is presented, for example, in "A General Theory of Elastic Stability", by Thompson and Hunt (1973). Other references would be the CISM publication "Coupled Instabilities in Metal Structures: Theoretical and design Aspects" (see Rondal (1998)), and "Stability, Bifurcation and Postcritical Behaviour of Elastic Structures" (M. Pignataro et al, 1991). Another important introduction to the theory of coupled stability is the title "Basele Calculului Structurilor la Stabilitate" from Gioncu and Ivan (1983), that unfortunately was not translated from Romanian. We recommend the readers to directly address to the originals to have a clear view of the theoretical principles.

Coupled stability analysis, which includes the buckling modes interaction, may be performed on the basis of the general theory of coupled instabilities, as presented by Thompson and Hunt (1973). At this point, one may perform an energy-based analysis, in order to qualify the post-critical behavior when at least two buckling modes interact, allowing identification of the coupled behavior and the consequent erosion of the limit load. The coupled buckling phenomenon and its consequences in the post-buckling range is only available if the structure is taken with imperfections. In other words, the effects of

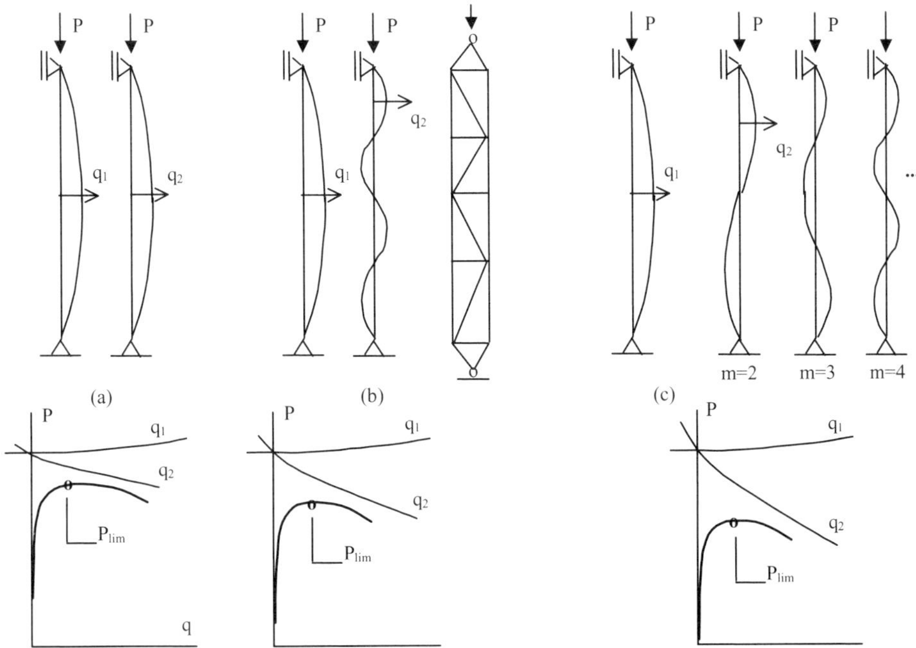

Figure 10. Weak and strong coupled buckling: (a) Flexural and flexural-torsional interaction: weak coupling; (b) Global flexural and local bar buckling modes interaction: moderate coupling; (c) Local plate and global flexural buckling interaction: strong coupling.

the coupled buckling are only to be considered in the non-linear behavior. Appropriate structural stability analysis must incorporate the following steps:

a) Linear analysis, with the identification of the buckling modes and its associated bifurcation critical loads that can be achieved by the theory of elastic stability, with the solution of an eigenvalue problem. Although essential for the comprehension of the phenomenon, the results are addressed to ideal "perfect" structures, what means that the results are only valid for very little displacements of the structure. The obtained results at this step indicate the buckling shape and the critical load when bifurcation-type stability is concerned.

b) Non-linear analysis, to obtain the behavior of the structure in the post-buckling range. The non-linear analysis deals with the real "imperfect" structure, allowing both elastic and inelastic post-critical behavior. Refined numerical methods allow the description of the large displacement behavior of a structure or a single structural member, which must be in close agreement with the buckling modes previously calculated by the linear stability analysis.

Let us consider at this point what an imperfection really means, always addressed to compressed members. The imperfection could be, at first, an initial geometrical imperfection; or an eccentricity of the compressive load; or a local buckling associated to plate buckling, which induces changes in the cross-section and moves the cross-section centroid from its original position. Any of these conditions would change the original "perfect" structure into an "imperfect" one which may now display its non-linear behavior since the beginning of the loading process. Perturbation loadings may also change the original perfect structure into an imperfect one, as transversal loads in framed structures and intermediate transversal load in concentrically compressed members. The application of perturbation loadings is usual in structural analysis of framed structures.

2.2 Phenomenological and mechanical models of coupled instability

The application of simplified models to study coupled buckling phenomena was achieved by several researchers. In these cases simple discrete models may reproduce the essential complex structural behavior. This is the case of the models shown in table 1, originally prepared by Gioncu (1994) and called the phenomenological models. These examples may be analyzed on the basis of the General Theory of Elastic Stability and one might address to the original references to become familiar with the energy formulation applied to solve buckling interaction problems.

Table 1. Phenomenological models for the analysis of coupled buckling (according with Gioncu (1994)).

Author(s)	Year	Model	Type of instability
Koiter[111]	1962		Semi-symmetric
Augusti[1]	1964		Double-symmetric
Chilver[23]	1966		Semi-symmetric Double-symmetric
Thompson Gaspar[207]	1977		Semi-symmetric
Gaspar[41]	1977		Semi-symmetric Double-symmetric
Gioncu-Ivan[61]	1979		Double-symmetric
Luongo-Pignataro[128]	1989		Nearly-symmetry Hidden-asymmetry

2.3   Coupled Instability and Imperfections Sensitivity

Imperfections have a fundamental importance in the coupled instabilities phenomena, since no coupled buckling develops without the presence of some kind of imperfection(s). The first order instability solution that is able to identify the bifurcation problem (critical loads and buckling modes) cannot access the full non-linear behavior of the structure. When coupled buckling modes are concerned, the imperfections tend to reduce the loading path, transforming the ideal bifurcation behavior in the actual interactive buckling, with a maximum limit load that is associated to the structure ultimate strength. Figure 11, for example, shows the result of the coupled behavior of two stable symmetrical buckling modes. The coupled post-critical behavior changes from stable to unstable and the effect of the imperfections is to reduce the limit load the more important are these imperfections. The more pronounced are the imperfections the more pronounced is the limit load reduction.

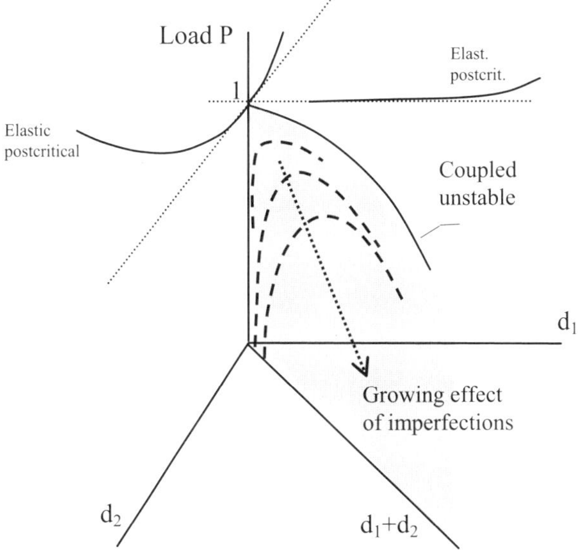

Figure 11. Coupled behavior of two buckling modes and the effect of imperfections.

The so-called imperfections can be considered as those shown in Figure 12.

The geometrical structural imperfections are identified by the inevitable misalignments of the centroidal axis of the member, produced by both the original manufacturing of the profile and the structural assembling. Loading eccentricities are somehow inevitable, caused by the connection system at the ends of the members that are not expected to behave as perfect hinges. The introduction of end bending moments originated by the actual connection system could also be considered in the same category of loading eccentricities.

Local buckling of thin-walled structural members is an additional origin of imperfection, since the change of the cross-section shape introduces compressive loading eccen-

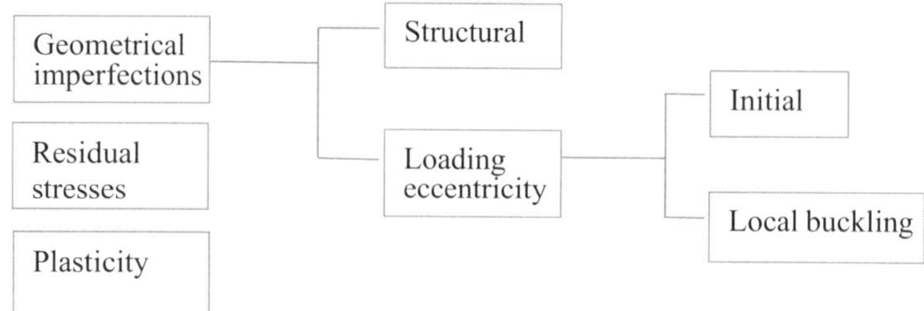

Figure 12. Imperfections to be considered in structural buckling analysis.

tricities along the axis of the member.

Plastic deformations are in fact very usual in steel structures, even for the cases where only elastic behavior is expected. This is so because of the presence of the residual stresses arising from the manufacturing process as well as from the assembling constructional process. Consequently, localized zones of initial plasticity are usually found in structural members, even before the expected initial yielding loading of the structure.

For the particular case of thin-walled members, plasticity is not such an important imperfection factor, since plastic hinges lines tend to develop at the very beginning of inelastic stresses, inducing the typical local plastic mechanism, as presented before in Figure 6.

The above imperfection effects, in combination or not, are in fact the unavoidable origin of the imperfect loading path shown in Figure 11.

Stable and unstable component modes form the coupled structural behavior, the later allows reduction of the limit load. This reduction is also known as the erosion of the limit load, associated to the loss of structural strength. Because of this, when one is dealing with the ultimate strength of structures or members and coupled instability is concerned, the knowledge of the imperfections range is of the up most importance in order to obtain reliable design methods. Also, as the imperfections cannot (obviously) be measured for each individual structure or individual structural member at each structural design process, it should be taken as reference values, on the basis of previous experimental results that must be represented by characteristic values originated from probabilistic procedure. The characterization of the representative imperfections of each type of structural member usually applied in steel construction, for example, has been one of the tasks to be fulfilled by experimental research.

2.4   Measurement of imperfections

Geometrical imperfections and residual stresses have been measured in the case of thin-walled structures and members as for the case of cold-formed profiles, stiffened and sandwich panel plates, and shells. Geometrical initial imperfections can be measured before laboratory (or field) experimental tests and the results can be subsequently applied in numerical calculations, where the actual initial shape of the member can be considered. For this, special apparatus has to be prepared, as the one shown in Figure 13. This is the case of experimental tests of thin-walled cold-formed columns and beam-columns, for which the representative displacements of the expected buckling modes were measured. In this case, as observed in the picture included in Figure 13, the end conditions of the test specimens were fixed against warping and local plate bending, at the same time that spherical hinges assure free global flexural rotations.

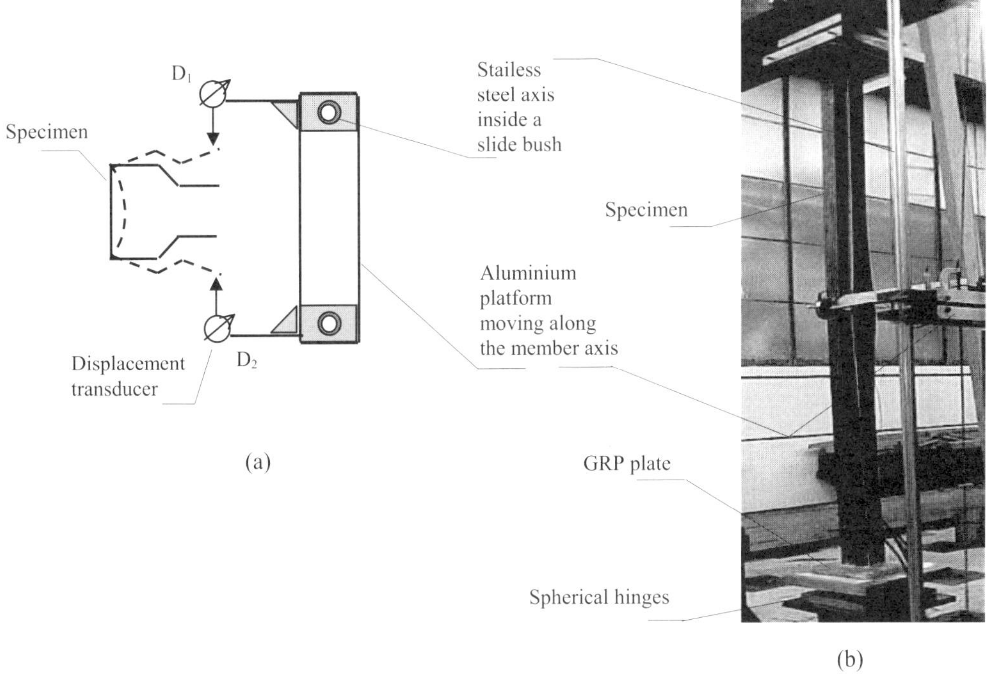

Figure 13. Geometrical imperfections measurements in cold-formed member during compressive tests. (a) Cross-section detail; (b) Photo of the measuring device and the test specimen.

Figure 14(a) shows some results of initial geometrical imperfections, measured for the case of a rack cross-section cold-formed member. In the same figure one can also observe the final deformed shape measured along the member, just before the collapse of the specimen is attained. The final displacements at the collapse are representative of the distortional buckling mode of the section, which is the critical mode in this case.

The experimental results are also useful to observe that the buckling mode shape forms gradually, since the beginning of the loading process, almost independently of the initial shape of the geometrical imperfection, and corresponds to the minimum energy path. In the example shown in Figure 14, the initial shape is a one half-wave that gradually changes to three half-waves, which is in close correspondence with the previously calculated buckling mode. The theoretical buckling mode presented in Figure 14(b) is the result of a first order calculation (eingenvalue problem) with the help of a finite element computational program (shell elements).

For the case of manufacturing residual stresses, many experimental results related to steel profiles were published in the literature. The results are, in general, strongly dependent on the type of manufacturing process: very high residual stresses are found for the case of welded profiles and, on the other hand, low residual stresses are found for the case of cold-formed cross-sections when manufactured with thin steel plates. Figure 15 shows both cases and one may conclude that the residual stresses are very low for the case of thin-walled cold-formed members and, in this case, are not the most important imperfection factor in the stability behavior.

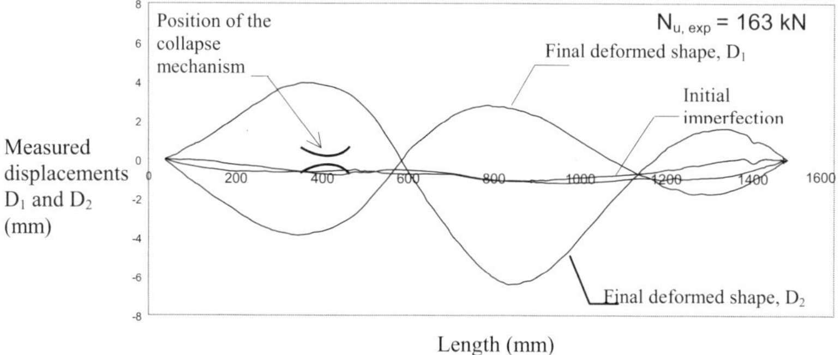

(a) Experimental results of initial geometrical imperfections and distortional buckling mode shape.

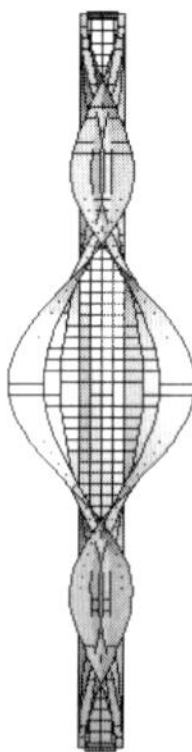

(b) The FEM computed theoretical buckling mode, for fixed warping condition.

Figure 14. Experimental and numerical results of the buckling mode and the initial imperfections.

Finally, it is important to point out that residual stresses experimental techniques were strongly improved with the help of portable rayon-X methods, which allow non-destructive experimental analysis instead of the sectioning strip method (a destructive one) or the controlled hole drilling strain measurements method. Figure 16 shows a portable device for rayon-X diffraction, developed by the Stress measurements laboratory of COPPE/UFRJ (see V. Monin and Gurova (2000)).

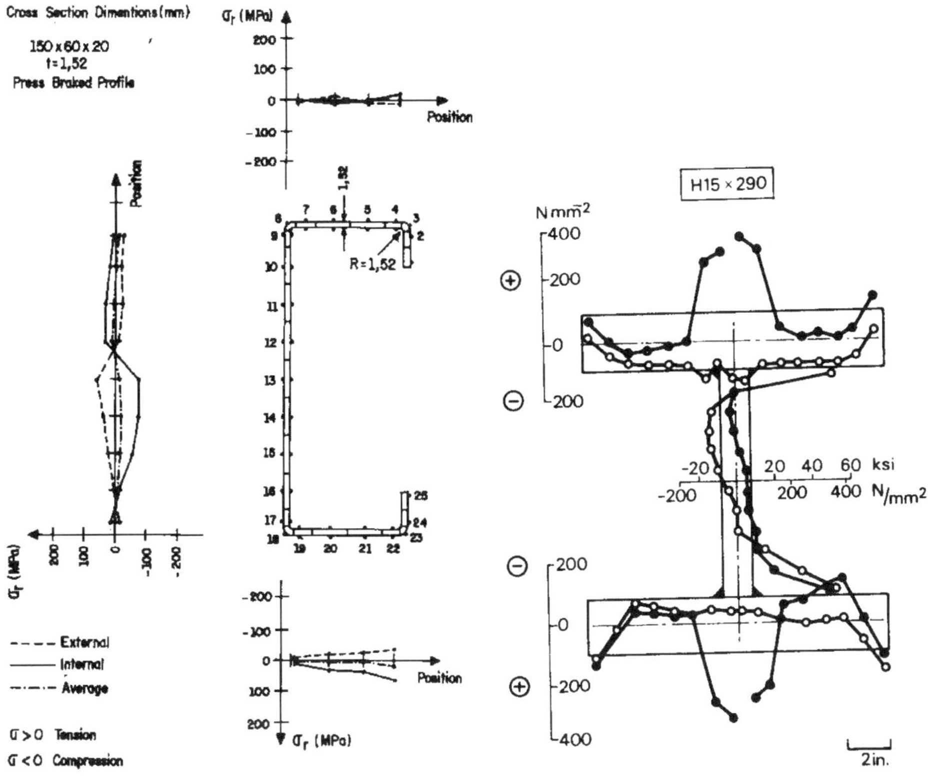

(a) Measured residual stresses in a thin-walled cold-formed channel (Batista and Rodrigues, 1998).

(b) Measured residual stresses in a jumbo welded profile (Alpsten, 1970).

Figure 15. Residual stresses in steel structural members.

3 Local and Distortional Buckling of Thin-Walled Members

3.1 Introduction

Thin-walled members present particular structural behavior, especially because of its torsional behavior and stability problems. In fact, thin-walled structures consist in a particular field of the structural engineering. For thin-walled structural analysis and design, engineers and researchers must be prepared to deal with: (a) the theory of thin-

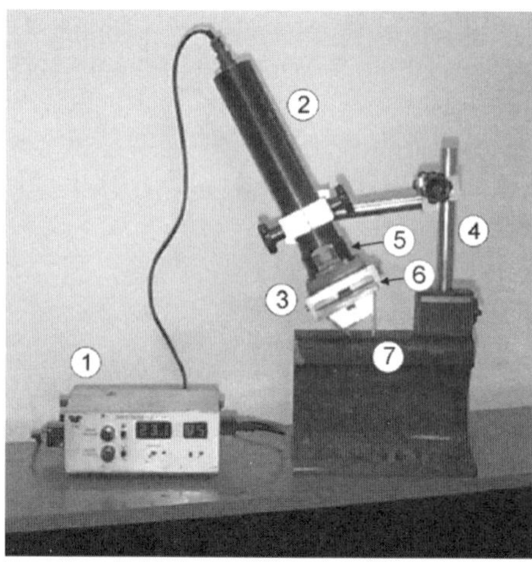

Figure 16. Rayon-X-diffraction portable device performing residual stresses measurements in a rail specimen (COPPE/UFRJ, Stress analysis laboratory).

walled beams; (b) the theory of elastic stability; (c) the theory of coupled instability; (d) the structural non-linear analysis; and (e) experimental analysis.

The theory of thin-walled beams is based on the fundamental developments of Vlasov (1961), and is presented in subsequent publications from Timoshenko and Gere (1961) and Murray (1986), for example. In this matter, the fundamental concepts of non-uniform torsion of beams, the cross-section warping and the flexural torsional buckling mode of concentric and eccentric compressed members must be pointed out. The Vlasov's theory, on the other hand, takes the cross-section of the member with unchanged shape and, as a consequence, no local or distortional cross-section buckling is allowed.

For the theory of elastic stability one may address to the original works of Bleich (1952), Timoshenko and Gere (1961) and Allen and Bulson (1980). The critical buckling modes of bar and plate elements can be found in these references.

Numerical methods for non-linear analysis permit to follow the actual load-displacement path of the structure or structural member, including its post-critical behavior and taking into account the expected imperfections: geometrical, residual stresses and plasticity.

Finally, experimental research may bring deeper knowledge of the actual structural non-linear behavior and may lead to results that help to calibrate the numerical solutions. In addition, the experimental analysis is of fundamental importance to investigate the collapse mechanisms, which is essential to establish consistent theoretical models for ultimate state structural analysis.

3.2 The local buckling of thin-walled sections

Local buckling of thin-walled sections follows the buckling behavior of plates, as its cross-section may be considered as an association of plates in interaction. This is the case of folded plates, as presented in Figure 17, where it is clear that the buckling mode is defined by the interaction between the individual plate elements. Figure 17(a) shows the buckling mode of an unstiffened channel and in this case the local buckling may be initiated by the loss of stability of element b_w or b_f, depending on the cross-section ratio (b_f/b_w). Figure 18(a) shows the results of the first order buckling analysis of the same cross-section of Figure 17(a) under uniform compression, in which the interaction between plates is considered. In the same Figure we have the results of the buckling analysis of the two isolated plate elements, b_w and b_f, showing how different are the results if the actual interaction behavior is taken or not. In these results the local buckling critical load ratio $\sigma_{cr} = P_{cr}/A$ is represented by the buckling coefficient k, as presented in the following equation:

$$\sigma_{cr} = \frac{P_{cr}}{A} = k \frac{\pi^2 E}{12(1 - \nu^2)} \left(\frac{t}{b_w}\right)^2 \tag{3.1}$$

Figure 17 shows other types of folded thin-walled sections, for which the local buckling critical load can be identified by the curves presented in Figure 18(b). All the results presented in Figure 18 were obtained with the help of a finite strip computational program, specially developed to perform first order stability analysis of cold-formed cross-sections.

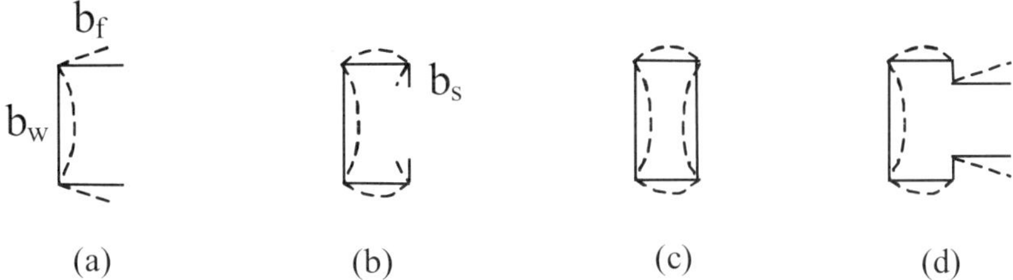

Figure 17. Typical local buckling modes of thin-walled cross-sections: (a) unstiffened channel, (b) stiffened channel, (c) rectangular tube and (d) rack section

As can be observed in Figure 19, where the local buckling of a channel section is shown, the buckling behavior includes many local plate modes until a global mode becomes critical. In the range of the local buckling it can also be observed how the end conditions influence the critical load. The following end conditions are considered: free or fixed warping and free or fixed plate bending. The warping condition is only important when torsional modes are concerned, as in the case of the flexural torsional buckling. One can observe in Figure 19 that fixed plate bending condition at the ends of the concentrically compressed member results in higher local buckling load. For the local buckling, it is observed that several modes are present in the short column range for the case of free end sections, and that the critical load becomes higher if the end conditions are fixed. Also, the local buckling mode tend to displace to the right in the graphic of Figure 19, indicating the effect of the end conditions related to the length of the buckling mode in the column. As the member becomes longer, the end effects tend to become of less importance, with the buckling loads for the two different end conditions becoming closer.

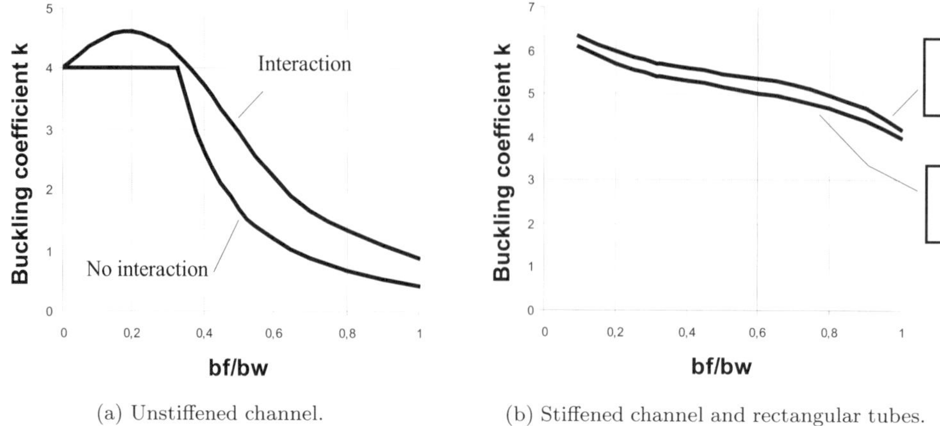

(a) Unstiffened channel.   (b) Stiffened channel and rectangular tubes.

Figure 18. Results from a finite strip method computational program for the local buckling coefficient, k, of thin-walled sections.

These results indicate that the plate bending condition at the ends of a compressed thin-walled member is only important to be considered for the range of short members.

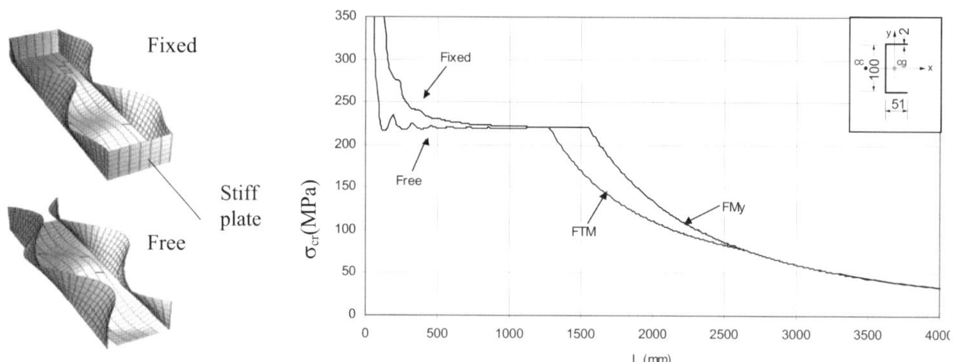

Figure 19. The local plate buckling of unstiffened channel member for fixed and free end conditions of the plate elements: computational results of the buckling load σ_{cr} v. member length L, where $\sigma_{cr} = P_{cr}/A$. (FTM: flexural-torsional buckling; FM$_y$: flexural buckling).

The best way to obtain the critical buckling load for different types of cross-sections is to apply a computational program, based on the FEM, for example. The results from Figure 19 were obtained on the basis of a FEM program, based on shell elements (see Inoue (2003)). The first order buckling analysis of thin-walled folded plates can also be obtained with much less computational effort and in a much easier way with the help of special numerical methods as the finite strip method, for which the buckling mode is previously defined. This type of solution has found many advantageous applications in the field of stability analysis of thin-walled sections, because it is easy to be implemented and allows the preparation of dedicated computational programs. It is very important for researchers and structural engineers to have a complete overview of the possible buckling modes of a typical cross-section, related to its length. If this is possible to be accomplished with the computational tool, one can very easily obtain the complete view of the stability characteristics of the member.

3.3 The distortional buckling of thin-walled sections

The distortional buckling mode is in fact a torsional mode of part of the cross-section, which changes the original cross-section shape, as presented in Figure 20. In this case it is observed that at least one folded corner of the cross-section is laterally displaced, what indicates the essential difference *vis-à-vis* the local plate buckling. The distortional buckling affects open cross-sections with the following geometrical characteristics: large flange-to-web width ratio (b_f/b_w) and low stiffener-to-web width ratio (b_s/b_w). It is also observed that the distortional buckling is the critical mode for an intermediate range of the lengths L of the member, lying between the local and the global buckling, as showed in Figure 21.

Contrary to the case of local and global member buckling, for which there are several direct formulas to help critical load calculation, the distortional buckling does not allow easy direct calculation and, in this case, computational programs, if available, are the best way to identify this buckling mode. For this, the FSM can be very useful to obtain a general overview of the stability problem, with the computational results of the first order buckling showing the variation of the critical buckling loads and its corresponding modes, as presented in Figure 21.

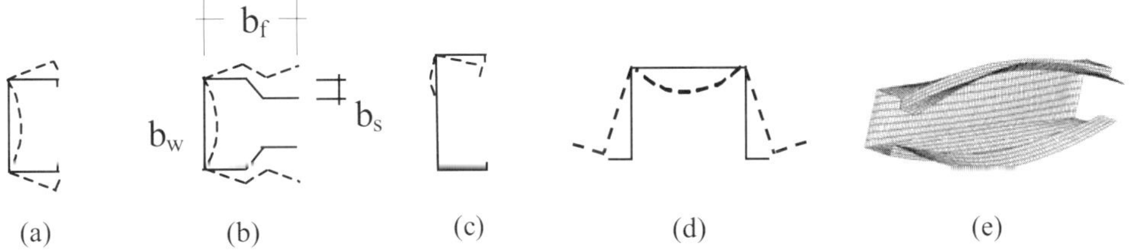

Figure 20. Typical distortional buckling modes of thin-walled cross-sections: (a) stiffened channel under uniform compression, (b) rack section under uniform compression, (c) stiffened channel beam, (d) hat section in bending and (e) first distortional mode, DM(1), of a compressed rack member with free warping condition.

Eccentricity of the compressive loading also promotes important changes in the stability results of the open thin-walled member. In this case, again, computational programs may easily offer very important information on the stability behavior. This is the case of beam-columns, for which the buckling characteristics must be calculated for each

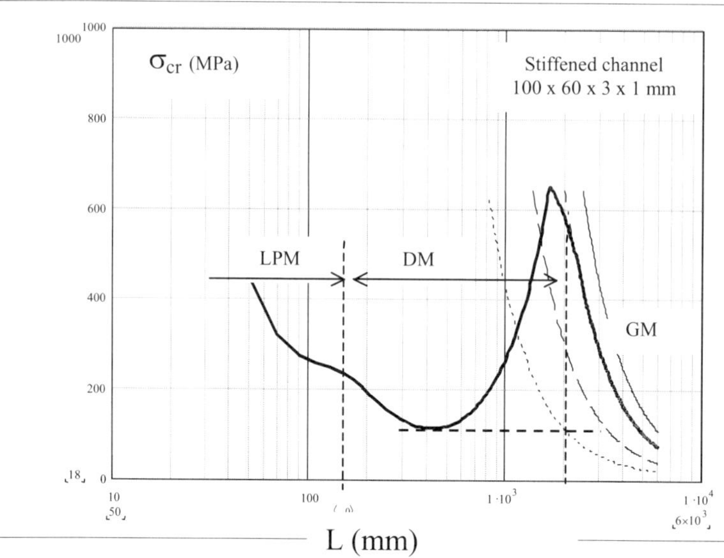

Figure 21. Transition between the local plate (LPM) and the distortional buckling (DM), according to the length L of the member. Computed results with the help of a FSM computational program, for a stiffened channel under uniform compression (according with Nagahama (2003)).

bending-compression combination and no direct equation is available, if the local and/or the distortional buckling are concerned. Figure 22 shows the case of a rack type section under concentric and eccentric compressive load, applied over the symmetric cross-section axis and on the opposite side of the shear center. In this case, the presence of large compressive stresses in the side of the edge stiffeners, b_s, induces lower distortional buckling loads when compared with the concentric compressive loading case.

3.4 Local - distortional buckling interaction

Local plate and distortional buckling interaction can be achieved for certain particular geometrical characteristics of open thin-walled members. This is the case of the results presented in Figure 23, for a concentrically compressed stiffened channel member. Previous numerical calculation indicated the geometrical proportions of the cross-section in order to obtain coupled effects between local and distortional buckling. It can be observed that the coupled mode displays harmonic shapes with three half-wavelengths buckling plate modes combined with one half-wavelength distortional mode, and this is found for member length L multiple of 250 mm. Between two of these coupled modes (for $L \neq 250$, 500, 750 mm), local and distortional buckling are linearly uncoupled. This is confirmed by the results of the computed buckling modes as showed in Figure 24, where the first coupled mode for member length L=250 mm is formed by LPM(3) plus DM(1); the following is an uncoupled LPM(5) for length L=420 mm and the third is the

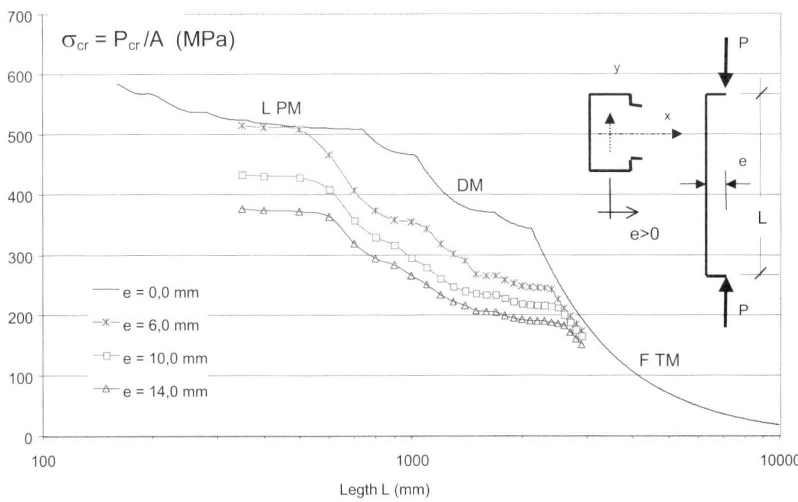

Figure 22. Results of the elastic stability analysis of a rack beam-column with eccentric compressive load. Computed results were obtained with the help of a FEM program, showing the influence of the load eccentricity on the critical buckling load P_{cr}. Compressive load over the symmetric axis of the cross-section (cross-section dimensions are 94 x 55 x 16 x 32 x 2.0 mm).

coupled buckling for member length $L=500$ mm, formed by the combination of LPM(6) plus DM(2).

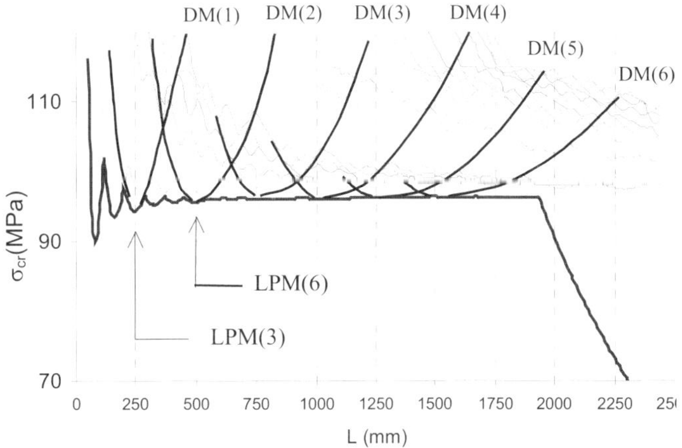

Figure 23. FEM results of the elastic stability analysis of a rack section under uniform compression, with coupled buckling of the LPM and the DM (identical critical buckling loads) for member length $L = n.250$, n integer.

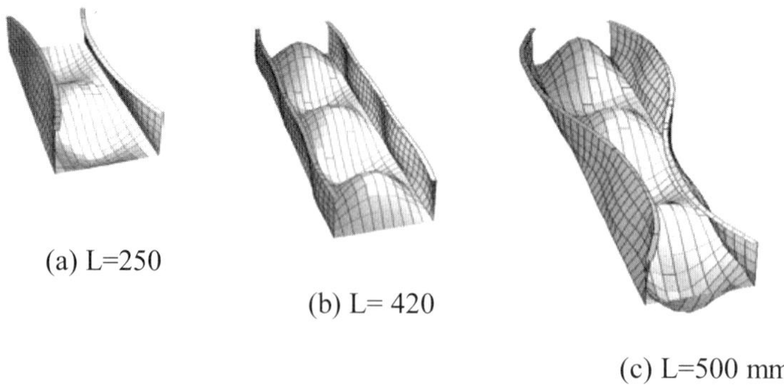

(a) L=250

(b) L= 420

(c) L=500 mm

Figure 24. FEM computational results of the buckling modes related to Figure 23: (a) coupled buckling mode LPM(3)+DM(1), (b) uncoupled bucking LPM(5) and (c) coupled buckling mode LPM(6)+DM(2).

The non-linear analysis, on the other hand, allows the investigation of the local-distortional interaction that should arise along the loading path. The elastic and inelastic non-linear analysis of the post-buckling behavior of thin-walled members permits to clarify the modes interaction. The following results were obtained by Nagahama (2003). This is the case of a concentrically compressed rack member (dimensions: $b_w = 130$ mm; $b_f/bw = 0.5$; $b_s/b_w = 0.2$; $b_a/b_w = 0.328$; $b_w/t = 72$; $L/b_w = 7.04$), for which the local and distortional uncoupled modes are shown in Figure 25, as well as its coupled buckling mode. In this case, the coupled buckling indicates critical loads ratio $\sigma_{LPM}/\sigma_{DM} = 1$. The elastic non-linear behavior was computed with the help of Abaqus (1998) FEM program and the results are presented in Figure 26, for different initial geometrical imperfections combinations, including pure distortional (CD0=1.0) or local buckling mode combinations. The initial imperfections combinations were defined according to the following rule, in which CL0 and CD0 are the local and distortional modes w and v (see Figure 23), respectively, with both related to normalized amplitudes equal to 10% of the plate thickness t:

$$(CL0)^2 + (CD0)^2 = 1 \tag{3.2}$$

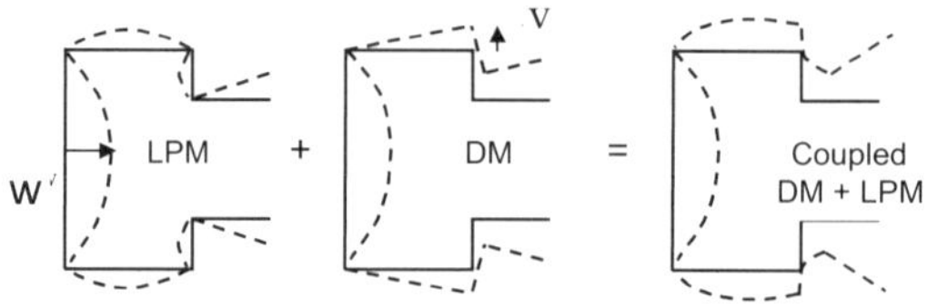

Figure 25. Uncoupled and coupled local plate and distortional buckling modes.

The elastic non-linear post-buckling analysis shows that the distortional mode is dominant in the interaction behavior. Figure 27 confirms this feature, and it can be observed

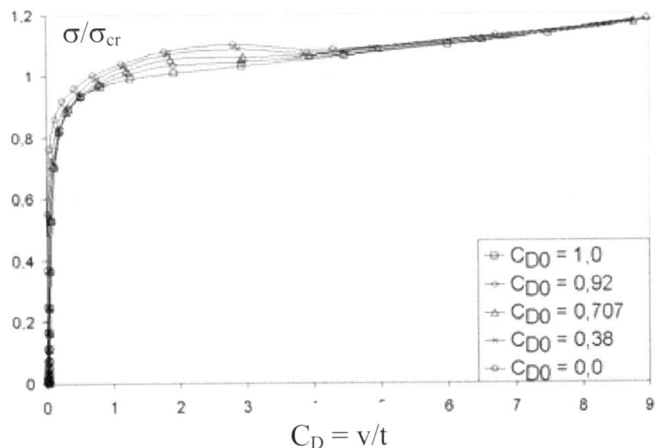

Figure 26. FEM computational results of the elastic non-linear behavior of a rack member under uniform compression and coupled buckling: the load-displacement results (P/A v. v/t), according with the DM for different combinations of the initial geometrical imperfections CD0 and CL0.

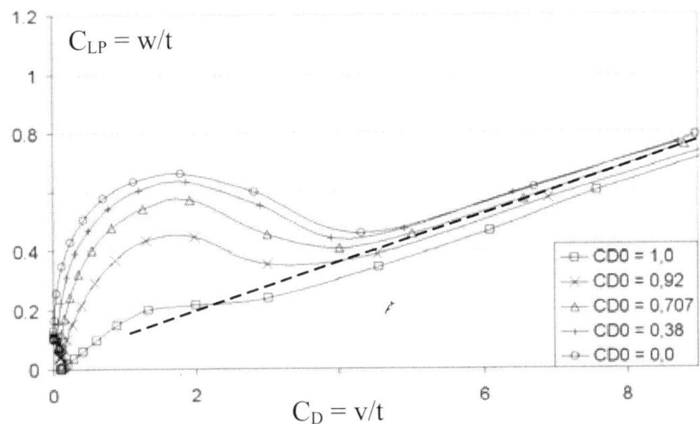

Figure 27. FEM computational results of the elastic non-linear behavior of a rack member under uniform compression and coupled buckling: the DM is dominant in the post-buckling behavior, for any contribution of the LPM in the initial imperfections ($CL0^2 + CD0^2 = 1$).

that even for the case of full local buckling initial imperfection, combined with null distortional initial imperfection, the post-buckling path converges to the distortional mode. In this Figure $C_{LP} = w/t$ and $C_D = v/t$ are the normalized displacements related to the local plate and distortional buckling modes, respectively.

The introduction of inelastic behavior does not change the above observations, as can be followed in Figures 28 and 29. The introduction of plasticity allows the identification of the ultimate load P_u and the initial yielding load P_y. Figure 28 shows how plasticity interferes in the post-buckling range, with the same tendency of the distortional buckling mode to predominate in the interactive process, as can be observed in the large displacements range.

The dashed lines in Figures 27 and 29 indicate the tendency line of the post-buckling behavior, with large distortional displacements CD if compared to the local plate displacements C_{LP}, for both elastic and inelastic material.

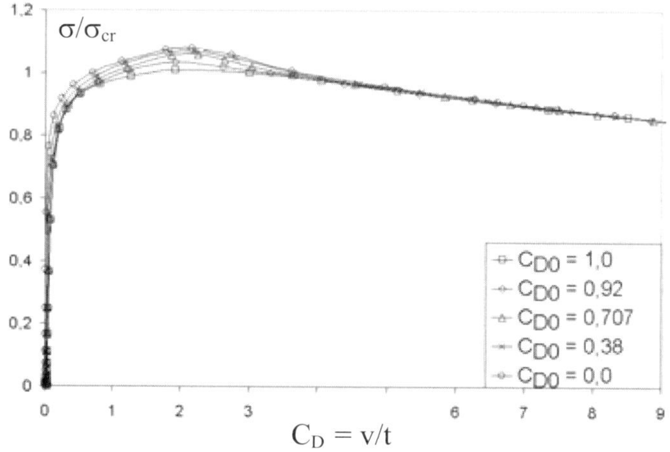

Figure 28. FEM computational results of the elasto-plastic non-linear behavior of a rack member under uniform compression and coupled buckling: the load-displacement results (P/A v. v/t) according to the DM for different combinations of the initial geometrical imperfections CD0 and CL0.

As observed from the above results, the non-linear analysis can give very important information about modes interaction, permitting to highlight the main aspects of the interactive behavior. In the presented case of local-distortional interaction it was shown that:

(a) The distortional buckling has less post-buckling behavior than the local plate mode.

(b) The distortional buckling is dominant in the interaction process with the local buckling.

(c) Depending on the initial imperfections combination, the coupled behavior may never appear before the ultimate load is attained.

(d) The distortional buckling can be dominant even for particular cases where the local buckling is the critical one ($P_{cr} = P_{LM} < P_{DM}$). This is a very important conclusion for design rules, and it indicates how the first order stability analysis is not always enough to identify the correct collapse mechanism of thin-walled members.

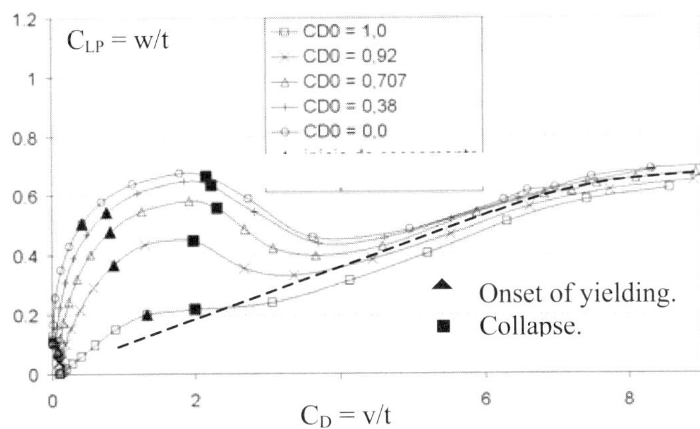

Figure 29. FEM computational results of the elasto-plastic non-linear behavior of a rack member under uniform compression and coupled buckling: the DM is dominant in the inelastic post-buckling behavior, for any contribution of the LPM in the initial imperfections ($CL0^2+CD0^2=1$).

3.5 Experimental Results

Some experimental results will be presented in the following, focusing both local plate and distortional bucking. It is important to clarify that thin-walled stub column tests are not an easy task to carry on, since end conditions of the member are strongly affected by the contact between the plate elements and the stocky templates of the test machine, as can be observed in Figure 30, where two stiffened channel stub-column specimens are shown after compressive test. In fact, stub column test results that are usually applied to analyze the ultimate strength of cold-formed members, have to be carefully considered, as the end conditions bring important influences to the experimental results. This is not the case when long columns are concerned. In both cases of stub and long columns the end conditions are the same, but its influence over the structural behavior is widely different, showing lower importance for the case of long columns and very high importance for stub-column tests. This is one of the reasons why the experimental results of stub columns lie above the theoretical squash load (even applying the actual value of the yielding stress in the calculations).

Figures 31 and 32 show cold-formed stub-columns after test. Figure 31 shows rack cold-formed columns after collapse. This is a case of elastic distortional buckling, followed by the collapse mechanism. Previous numerical computation of this specimen indicated the distortional buckling mode as the critical one, with two half-wavelengths. Also, the experimental test included end conditions with both warping and plate displacements fixed, what was achieved with the help of stiff GFRP (glass fiber reinforced plastic) plates that were manufactured by lamination around and inside the member cross-section.

Figure 32 shows a group of stiffened channel cold-formed stub columns and its collapse mechanism only affected by the local buckling mode. In these cases the elasto-plastic collapse mechanism is in close analogy with typical single plate collapse. Previous numerical calculations with a FSM program showed that the specimens should display three local half wavelengths, what was experimentally confirmed.

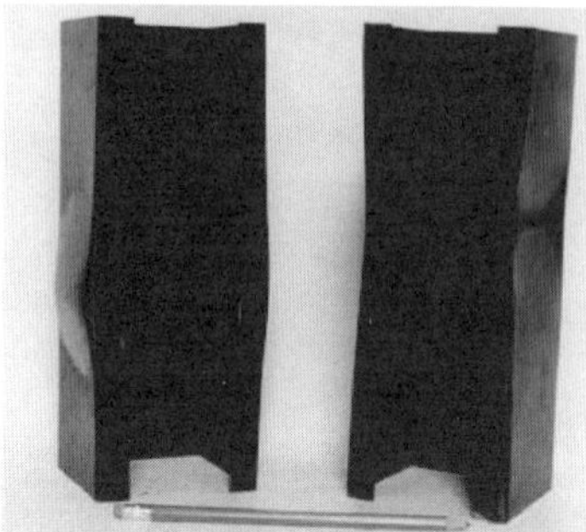

Figure 30. Thin-walled cold-formed stub-columns after compressive test.

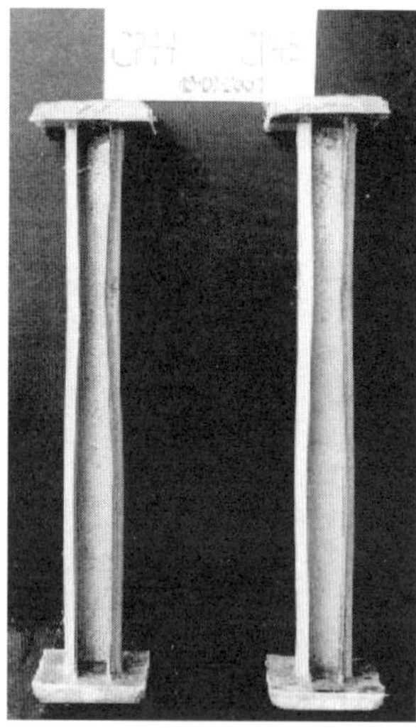

Figure 31. Rack thin-walled cold-formed members after compression test: the distortional buckling mode presents two half-waves and the ends are restrained for warping and plate bending.

4 Global Buckling

4.1 Introduction

The stability of bar members is one of the fundamental problems of structural engineering. In this field it is of fundamental importance to report to the original woks from

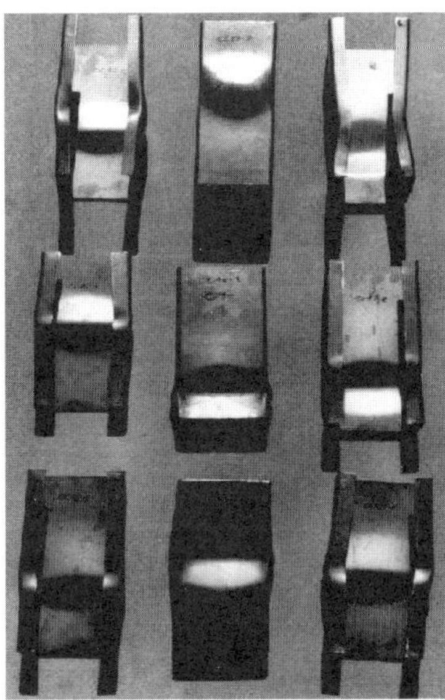

Figure 32. Plastic collapse mechanism of channel cold-formed stub-columns.

Timoshenko and Gere (1961), where the main aspects of the elastic buckling phenomena of bar members are presented. Some of the fundamental first order stability and non-linear behavior of concentrically and eccentrically compressed members will be presented in the following, as well as the lateral buckling of beams. Obviously this will be a short introduction of the problem, and one must address to some of the classical publications on this matter, for what we recommend the above-cited reference, as well as Allen and Bulson (1980) and Reis and Camotim (2001).

4.2   Stable and unstable buckling

The typical behavior of the global stability of bar structural systems can be explored with the help of very useful mechanical models, as those presented by Thompson and Hunt (1973). Note that the advantage of these models is that are discrete models with low number of degrees of freedom, the material can be taken as infinitely rigid and no bending or extensional deformations of the bars are considered. These aspects considerably simplify the mathematical solution, without loosing the main objectives of the stability analysis we are interested to observe. This procedure allows identify and classify equilibrium behavior of different structural systems.

The mechanical models are those presented in Figure 33. Model 1 is of symmetrical-stable post-buckling type; Model 2 is symmetrical-unstable and Model 3 is of unsymmetrical type. The post-critical responses of each model are presented in Figure 34, after the energy-based calculations of the structural systems as presented in the following.

For Model 1 we have the following total energy equation:

$$V = \frac{K\varphi^2}{2} - P\Delta \tag{4.1}$$

where K is the spring constant;

$$\Delta = L(1 - cos\varphi) \tag{4.2}$$

and then

$$V = \frac{K\varphi^2}{2} - PL(1 - cos\varphi) \tag{4.3}$$

Applying the stationary value of the total potential energy principle,$\partial V/\partial\varphi$, we find

$$K\varphi - PLsin\varphi = 0 \tag{4.4}$$

and the load-displacement equilibrium equation is the following, valid for the post-critical behavior of the perfect structural system

$$P = \frac{K\varphi}{Lsin\varphi} \tag{4.5}$$

If we take the equilibrium condition for small displacements, for which we have $\sin\varphi = \varphi$, we obtain the critical buckling load of the Model 1

$$P_{cr} = \frac{K}{L} \tag{4.6}$$

In order to obtain an asymptotic expansion around the buckling load, valid and accurate for small displacements, we can take the power expansion of $\sin\varphi$ that permits to obtain the following equation, taking the terms up to the second degree

$$P = \frac{K}{L} + \frac{1}{6}\frac{K}{L}\varphi^2 \tag{4.7}$$

This equation is applied to obtain the load-displacements $P - \varphi$ of the Model 1 as shown in Figure 34(a), from which we observe that the structural system is of symmetric stable type.

Following the same principles for Model 2 we can have the following results:

$$V = \frac{K\delta^2}{2} - PL(1 - \cos\varphi) \tag{4.8}$$

where

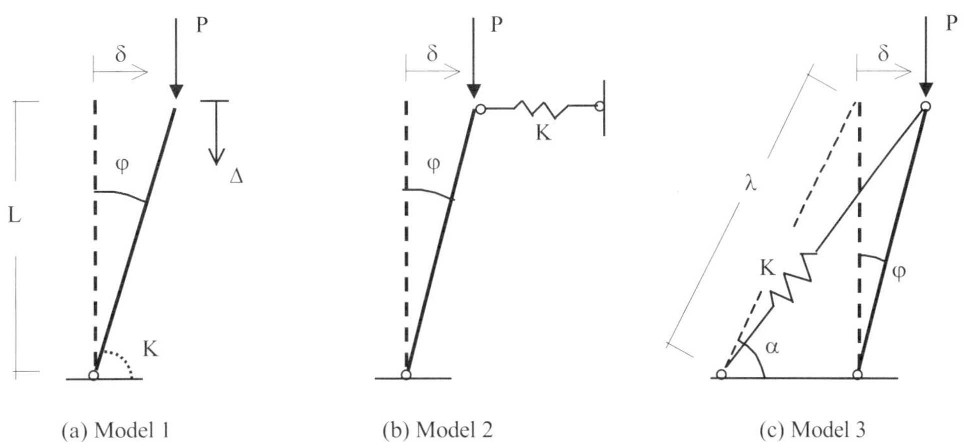

Figure 33. Simple mechanical models applied to the identification of the post-buckling behavior: (a) Model 1; (b) Model 2; (c) Model 3.

$$\delta = L \sin \varphi \tag{4.9}$$

and

$$V = \frac{KL^2 \sin^2 \varphi}{2} - PL(1 - \cos \varphi) \tag{4.10}$$

The expansion in power series up to the fourth order term gives the following equation

$$V = KL^2 \left(\frac{\varphi^2}{2} - \frac{\varphi^4}{6} \right) - PL \left(\varphi^2 + \frac{\varphi^4}{12} \right) \tag{4.11}$$

and the equilibrium solution conducts to the following expression, after applying the total potential energy principle $\partial V / \partial \varphi = 0$

$$P = KL - \frac{1}{2} KL \varphi^2 \tag{4.12}$$

with the critical load defined as

$$P_{cr} = KL \tag{4.13}$$

The post-critical behavior of Model 2 is symmetric unstable, as shown in Figure 34(b). The solution for Model 3 is showed in the following, with the total energy V as follows

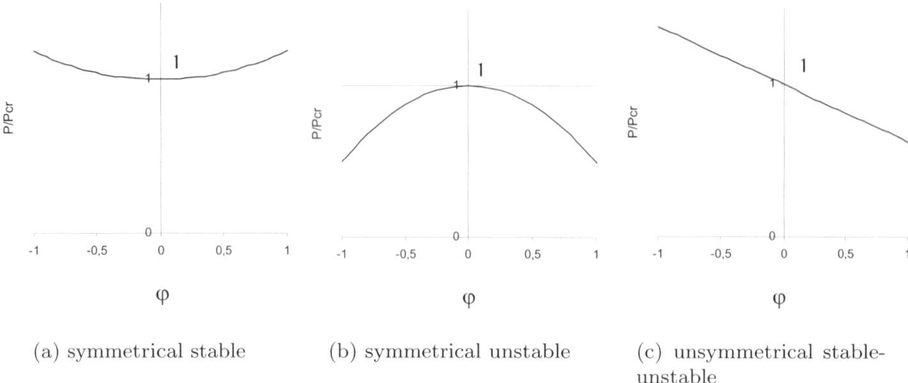

(a) symmetrical stable   (b) symmetrical unstable   (c) unsymmetrical stable-unstable

Figure 34. Post-buckling behavior of the models of Figure 33.

$$V = \frac{K\Delta l^2}{2} - PL(1 - \cos\varphi) \qquad (4.14)$$

After expressing Δl in terms of α and φ and applying $\partial V/\partial\varphi = 0$ we obtain the equilibrium equation

$$P = KL\cos^2\alpha - \frac{3}{8}KL\varphi \qquad (4.15)$$

and the critical load is equal to

$$P_{cr} = KL\cos^2\alpha \qquad (4.16)$$

The structural system is of unsymmetrical-type as presented in Figure 34(c).

The effects of the imperfections for the three structural Models are shown in Figure 35, and it is clear that the imperfections are the origin of the erosion of the limit load when unstable behavior is expected (Figures 35(b) and 35(c)). This can be also observed in Figure 36, where the reduction of the limit load is referred to the initial geometrical imperfections φ_0 of the structural systems for Models 2 and 3. The effects of small imperfections are the most important in the stability behavior of structural systems, since invisible initial deformations may grow in a very sudden way after the onset of buckling (the behavior may be, in this case, close to the theoretical bifurcation problem). Large initial imperfections are very easily observed and usually do not promote sudden changes in the deformed shape of the structure (far from the theoretical bifurcation behavior). This observation leads to the following conclusion: for the case of structural systems with unstable post-critical behavior, it is recommended to promote "important" initial geometrical imperfections, in order to avoid sudden changes in the equilibrium behavior.

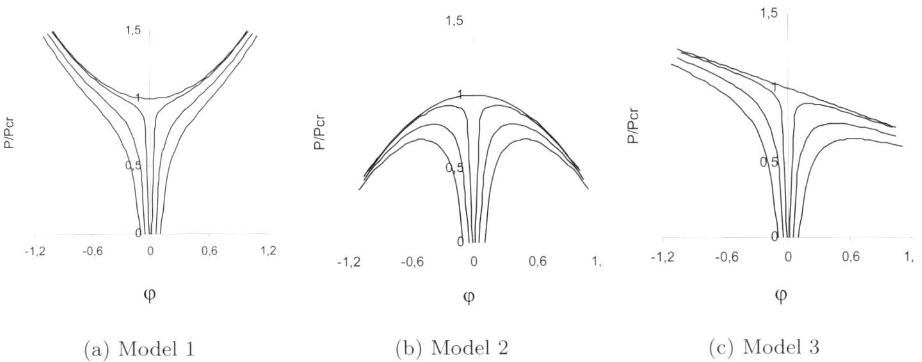

(a) Model 1   (b) Model 2   (c) Model 3

Figure 35. The effects of imperfections for the models of Figure 33.

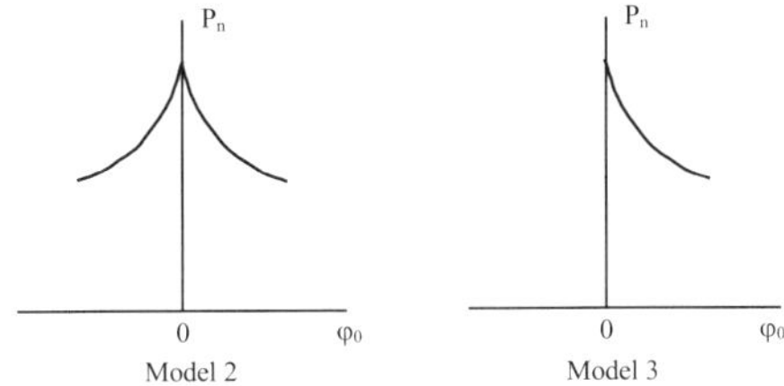

Figure 36. Reduction of the limit load for the models 2 and 3, due to the unstable branch of the post-buckling, related to the models of Figures 34 and 35.

The three cases above presented are related to elastic buckling and only geometrical imperfections are considered in the non-linear paths. The introduction of the inelastic behavior of the material can be taken as an additional non-linear factor and must be considered when actual ultimate limit state is to be found.

4.3   Column flexural buckling

The classical stability problem of columns (Euler) can be stated by the critical buckling load P_E and its buckling mode, where

$$P_{cr} = P_E = \frac{\pi^2 EI}{(kL)^2} \tag{4.17}$$

k is the buckling coefficient related to the end conditions of the column and the buckling mode can be represented by a trigonometric shape. This is the bifurcation problem of the column under flexural buckling.

The non-linear behavior of the column resulting from initial imperfection effects, is in fact the actual behavior of columns and beam-columns and may be represented with the help of the following approximate expression, valid for simply supported columns, where δ_{lin} and δ_{nlin} are the linear and the non-linear lateral displacement of the member

$$\delta_{nlin} = \delta_{lin} \left(\frac{1}{1 - P/P_{cr}} \right) \tag{4.18}$$

The amplification term (at right in the above equation) introduces the bifurcation stability effect in the column behavior and can be applied with error less than 2% if compared with the exact solution, for values of the compressive load less than 0.6 of the Euler critical load. This amplification factor is applied to the case of a simply supported column and the results are showed in Figure 37.

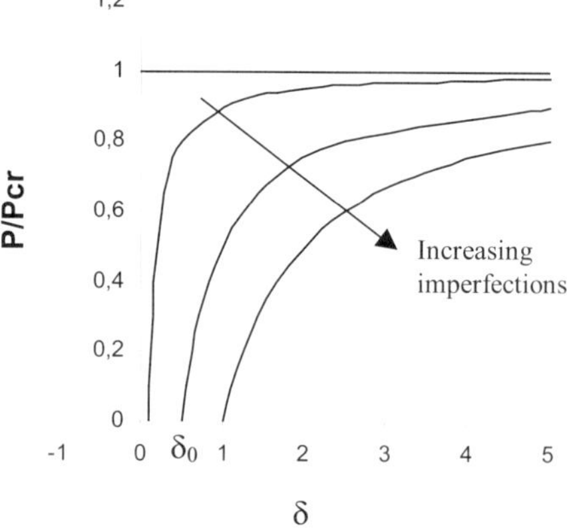

Figure 37. Non-linear amplification effects due to the initial imperfections δ_0 for a column member, after Equation 4.18 and based on the first order stability solution of the column.

The post-buckling behavior of the column is in fact non-linear and can be found by the large displacements elastica equation of the column. By this way one may obtain the solution for the compressive load $v.$ maximum transversal displacement of a simply supported column, as shown in Figure 38 (see Timoshenko and Gere (1961)).

It is observed in Figure 38 that the column is classified as a stable symmetrical equilibrium path, indicating that very slender columns may display post-buckling behavior above the Euler critical load plateau. This is confirmed with the experimental results obtained from a slender steel member that was concentrically tested in laboratory-controlled conditions, as presented in Figure 39. The experimental results indicate how the compressed member displays non-linear load-displacement path, due to the initial imperfections, and how the load value becomes higher than the theoretical Euler buckling load. This large displacement behavior is usually not considered in engineering applications, since structural members are never so slender and only the Euler plateau is to be considered for practical problems.

Open cross section columns, on the other hand, may also display buckling modes associated to torsion, as will be presented in the next item.

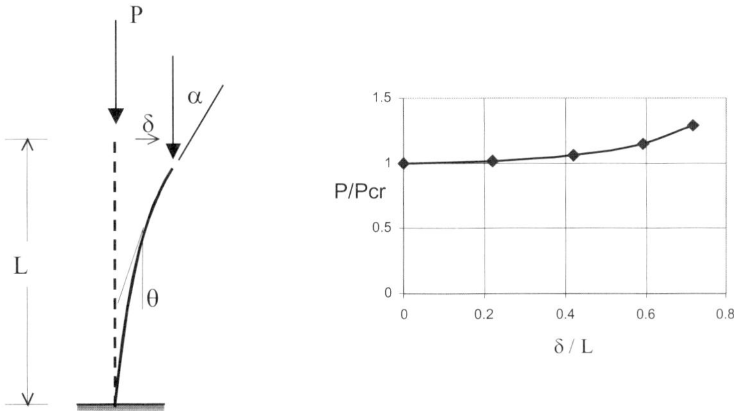

Figure 38. The non-linear post-bifurcation behavior of a cantilever column.

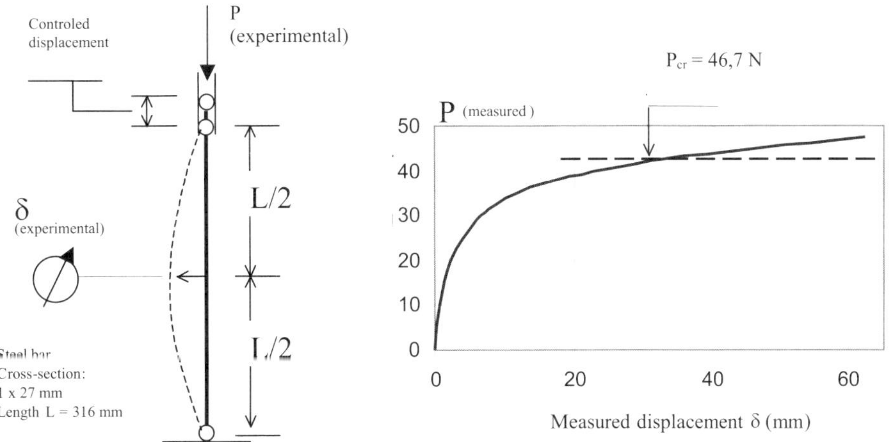

Figure 39. Experimental results of a concentrically compressive test of a very slender steel member, indicating the non-linear post-buckling behavior above the Euler theoretical critical value.

4.4 The flexural-torsional buckling of columns

The bifurcation stability problem of open cross-section members loaded in eccentric condition is the most general problem of beam-columns stability. The problem is defined by the three differential equations system showed in the following, which are related to Figure 40.

$$EI_x \frac{d^4 v}{dz^4} + P \frac{d^2 v}{dz^2} \quad - \quad P(x_0 - e_x) \frac{d^2 \phi}{dz^2} = 0 \tag{4.19}$$

$$EI_y \frac{d^4 u}{dz^4} + P \frac{d^2 v}{dz^2} \quad + \quad P(y_0 - e_y) \frac{d^2 \phi}{dz^2} = 0 \tag{4.20}$$

$$EI_w \frac{d^4 \phi}{dz^4} - \left(GI_t - Pe_y \beta_2 - Pe_x \beta_1 - P \frac{I_0}{A} \right) \frac{d^2 \phi}{dz^2} + P(y_0 - e_y) \frac{d^2 u}{dz^2} - P(x_0 - e_x) \frac{d^2 v}{dz^2} = 0 \tag{4.21}$$

The degrees of freedom are those from bending, u and v, and from torsion, ϕ.

The solution of this problem is obtained with the consideration of sinusoidal buckling modes of the type: $u_i = A_i . \sin(\pi z / L)$, where $i = u$, v, ϕ; and the general solution is obtained after the substitution of these displacement functions into he differential equations, which conducts to the following equilibrium equation

$$[C] u_i = 0 \tag{4.22}$$

for which one may consider the non-trivial solution of the type

$$|C| = 0 \tag{4.23}$$

The above equation gives the third degree polynomial solution presented in the following, the general stability solution of a beam-column, considering free bending and free warping at the end sections of the member.

$$g(P) = (P_y - P)(P_x - P) \left(\frac{I_0}{A}(P_t - P) - P(e_y \beta_2 + e_x \beta_1) \right)$$

$$- (P_y - P)(x_0 - e_x)^2 P^2 - (P_x - P)(y_0 - e_y)^2 P^2 = 0 \tag{4.24}$$

In the above equation P_x, P_y and P_t are the critical loads of the uncoupled buckling modes for bending and torsion, respectively. The torsional critical load P_t is related to the compressive load placed over the shear center of the cross-section.

Depending on the cross-section geometry and the compressive load position, the buckling load may include three or two coupled buckling modes. This can be resumed as shown in Table 2. The solutions addressed to each of the cases appearing in Table 2 are presented in the following, and permits to compute the critical buckling load of each particular case.

For cases 1 and 2 of Table 2 the solution must be obtained by the solution of a third degree Equation 4.24, and the buckling mode will be always the flexural-torsional.

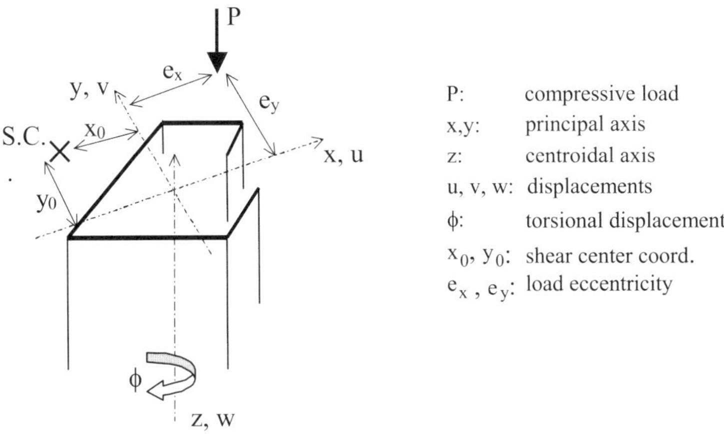

Figure 40. Eccentric compressed member with unsymmetrical open cross-section.

The solution of the Case 3 is a second-degree equation derived from Equation 4.24, which permits the direct solution as follows

$$P_{xt} = \frac{P_x + P_t}{2\left(1 - (x_0/r_0)^2\right)} \left(1 - \sqrt{1 - \frac{4 P_x P_t \left(1 - (x_0/r_0)^2\right)}{(P_x + P_t)^2}}\right) \qquad (4.25)$$

which corresponds to the flexional-torsional buckling mode with the maximum flexural buckling P_x coupled with the torsional mode P_t. In this case, as for any other case of compressive load over the symmetry axis of the cross-section, the minimum flexural buckling P_y becomes an uncoupled mode.

Finally, the case 4 in Table 2 has the following three uncoupled buckling loads

$$P_y = \frac{\pi^2 E I_y}{(k_y L)^2} \qquad (4.26)$$

$$P_x = \frac{\pi^2 E I_x}{(k_x L)^2} \qquad (4.27)$$

$$P_t = \frac{A}{I_0}\left(G I_t + \frac{E I_w}{(k_w L)^2}\right) \qquad (4.28)$$

Equations 4.26 and 4.27 are the Euler flexural buckling loads and Equation 4.28 is related to the pure torsional buckling load of the member.

As commented above, cases 1 to 3 in Table 2 are typical cases of natural coupled buckling that cannot be avoided by any artificial measure. In fact, for a large number of cases currently found in structural engineering, in which the sections are usually singly symmetric and opened, one must deal with the flexural torsional buckling. For the usual cases of concentric compressive load, one may apply the direct second-degree solution, with the direct formula shown in Equation 4.25. The unusual case of compressive load coincident with the shear center, because it is an uncoupled buckling, is directly solved with the three Equations 4.26 to 4.28, the lower value corresponds to the critical buckling mode. This is the case of a great part of columns usually applied in steel construction, with doubly symmetric sections as the H-type profiles.

For all other cases of compressive members with open cross-sections, including asymmetric and singly symmetric sections, the last one with compressive load eccentric to the symmetry axis, the first order stability problem corresponds to a flexural torsional buckling, with interaction between the two flexural modes and the torsional buckling, for which the third degree Equation 4.24 should be solved to obtain the flexural torsional critical load of the member.

4.5   Lateral buckling of beams

Specially for the case of steel structures, many of the beam elements, if not laterally braced, may display lateral buckling, which is a coupled mode that includes the flexural buckling in the plane of minimum inertia (lateral mode) and the torsional mode. Lateral buckling only occurs when the beam is bent in the plane of maximum inertia and the torsion must be considered taking into account the non-uniform torsion of the beam, on the basis of the Vlasov's theory of thin-walled beams. This natural coupled buckling has different particular solutions depending on the type of transversal loading and the beam end conditions. The value of the ratio between Saint Venant and non-uniform torsional stiffness, GJ/EC_w, and the ratio between the principal moments of inertia, I_{max}/I_{min}, are of very high importance in the lateral buckling problem. The higher is the inertia ratio I_{max}/I_{min} the lower is the lateral buckling load. On the other hand, if the moment of inertia ratio is equal to unity (doubly symmetric sections), $I_{max}/I_{min} = 1$, no lateral buckling occurs. Equations 4.29 and 4.30 present two expressions for the critical lateral buckling load, for T cross-section simply supported beam under pure bending or loaded by transversal uniform distributed load. These are the cases shown in Figure 41, for which we have simply supported beams, with free bending and fixed torsion at the ends. Additionally, these beams present free warping at both ends.

$$M_{cr,M} = \frac{\pi}{L}\sqrt{\frac{EI_{min}GJ}{1 - I_{min}/I_{max}}}\sqrt{1 + \frac{\pi^2 EI_w}{GI_t L^2}} \tag{4.29}$$

$$M_{cr,q} = 1.13 M_{cr,M} \tag{4.30}$$

Taking into account that the critical loading of beams is strictly related to the bending moment distribution along the member, usual codes and specifications for steel structural design indicate approximate formulation to obtain the critical bending moment of beams. This may be accomplished on the basis of a flexural coefficient C_b, as presented in Equation 4.31, based on the bending moment distribution in the beam element: M_A and M_C are the bending moments at the quarter and three-quarter points of the laterally unsupported span, M_B is the bending moment at the middle point of the unsupported span, and M_{max} is the maximum moment in the unsupported beam segment.

Table 2. Stability of bar members under eccentric and concentric compressive load.

Case	Cross-section	Loading condition	Stability solution for short member	Stability solution for long member
1	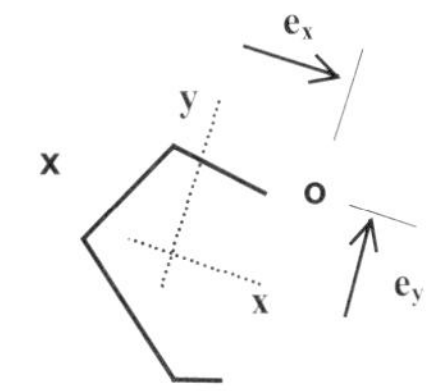	Doubly eccentric: $e_x \neq 0$ $e_y \neq 0$	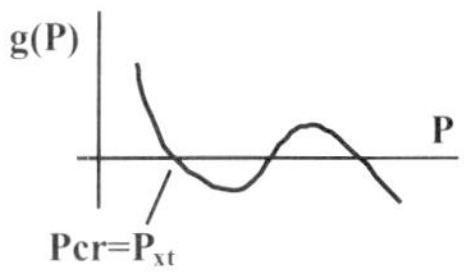 $g(P)$; P ; $Pcr=P_{xt}$	
2	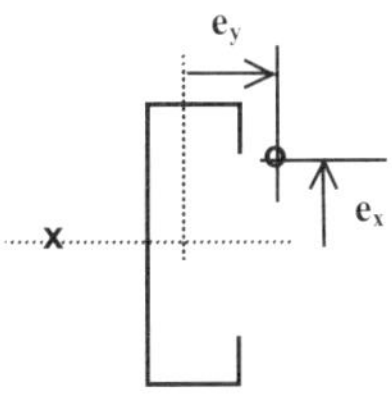	Doubly eccentric: $e_x \neq 0$ $e_y \neq 0$	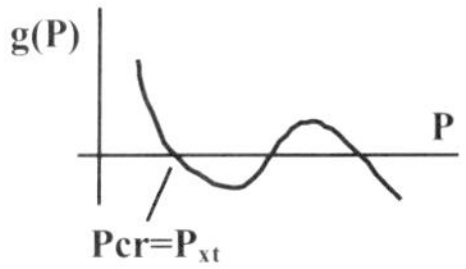 $g(P)$; P ; $Pcr=P_{xt}$	
3	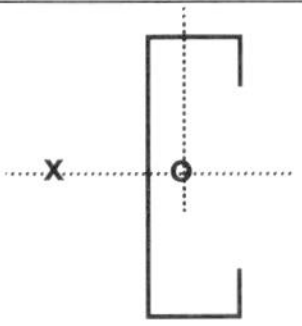	Concentric: $e_x = 0$ $e_y = 0$	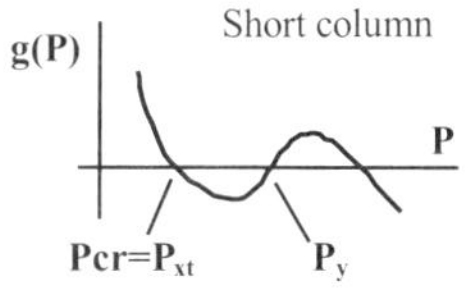 Short column $g(P)$; P ; $Pcr=P_{xt}$; P_y	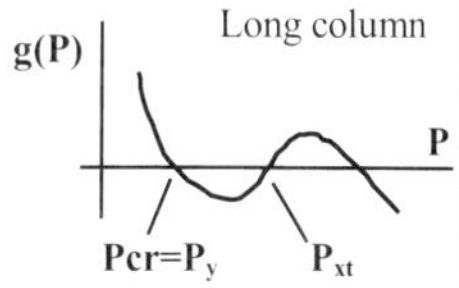 Long column $g(P)$; P ; $Pcr=P_y$; P_{xt}
4	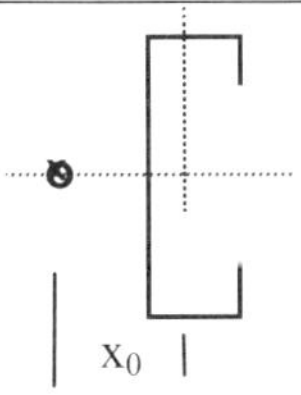	Over the shear center: $e_x = x_0$ $e_y = 0$	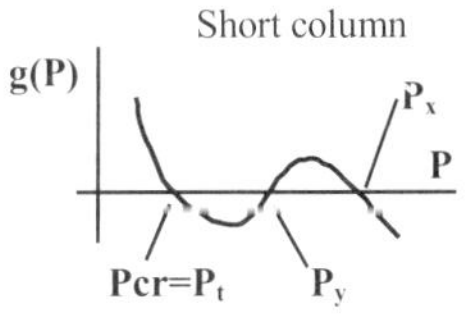 Short column $g(P)$; P_x ; P ; $Pcr=P_t$; P_y	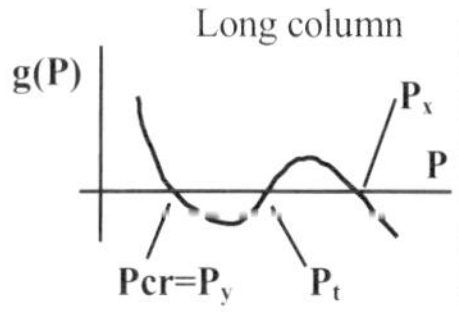 Long column $g(P)$; P_x ; P ; $Pcr=P_y$; P_t

o : compressive load position; x : shear center

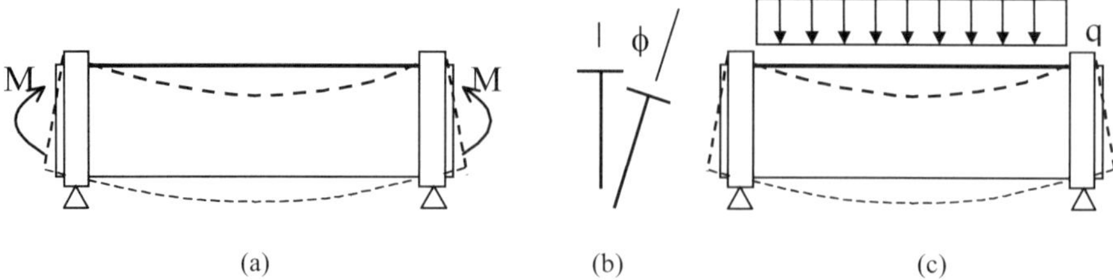

Figure 41. The lateral torsional buckling of slender beams. (a) Beam under uniform bending; (b) The cross-section at the middle span of the beam under lateral torsional buckling; (c)Beam under uniform transversal load.

$$C_b = \frac{12.5 M_{max}}{2.5 M_{max} + 3(M_A + M_C) + 4 M_B} \tag{4.31}$$

The effect of the point of application of the transversal load is also important, since loads placed at the top flange of the beam are more destabilizing than the case of load applied at the bottom flange. This effect is shown in Figure 42, in which it is observed that the position of the transversal load in the cross-section is important for the case of shorter beams (low value of L) and the lower is the uniform-to-non-uniform torsional stiffness ratio (GJ/EI_W).

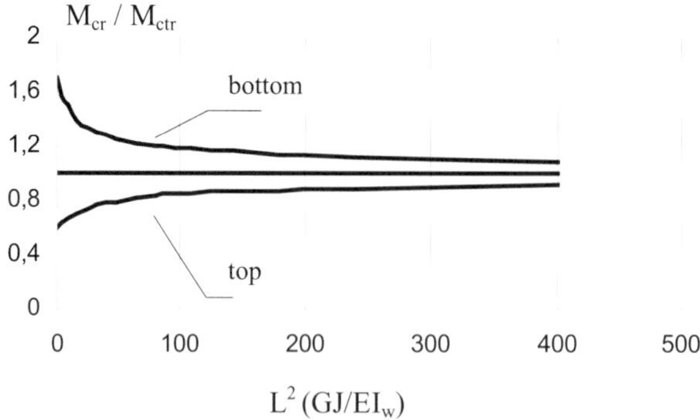

Figure 42. The variation of the critical bending moment ratio, M_{cr}/M_{ctr} , for the case of transversal loading placed at the top flange or the bottom flange of the cross-section (M_{cr} is the critical bending moment, M_{ctr} is the critical bending moment for the transversal load placed at the centroid).

5 Local-Global Buckling Interaction

5.1   Introduction

The interaction between local and global buckling is a very usual phenomenon in metal structures, and it can be found for the case of thin-walled cold-formed sections and girders, for example. Another type of structural member that can be affected by local-global buckling interaction are the GRP (glass reinforced plastic) profiles that are manufactured by pultrusion process. As the GRP is a material that combines high strength with low rigidity (tensile strength $f_u = 500$ MPa; Young modulus $E = 20$ GPa), the pultruded profiles are very sensitive to stability problems. Additionally, the GRP is a brittle material that presents no ductility, what means that the collapse usually develops in a very sudden and dangerous way.

In the field of metal structures, the large applications of light structural solutions (because of architectural and/or costs reasons - usually both), obliges structural design methods to include local plate buckling. This is the case of the cold-formed profiles that are usually manufactured with thin steel plates and sheets. Figure 3 showed some examples of typical steel cold-formed sections and Figure 43 presents some typical GRP pultruded sections.

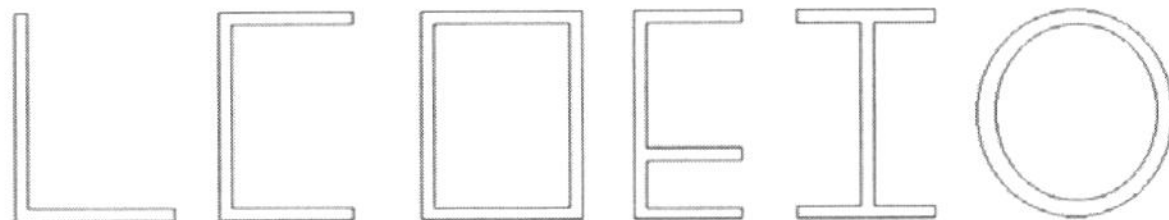

Figure 43. Typical structural GRP pultruded cross-sections.

Due to the stable post-buckling characteristics of the plate element, the interaction between this mode and the global buckling must be considered. Figure 44 illustrates a typical case of local-global buckling interaction, registered during an experimental test of compression of a steel stiffened channel member. It is observed that both local plate and global flexural modes are present and large displacements of the stiffened channel specimen are quite clear. This kind of post-buckling behavior must be considered in the methods of analysis, as well as the structural strength must include the buckling interaction in the design procedures.

As already stated before, the local plate buckling (multiple short wavelengths) interaction with flexural or flexural-torsional mode (long wavelength) may be classified as strong interaction type, conducting to important erosion of the ultimate limit compressive load. Because of its importance, the local-global buckling interaction of thin-walled members is explored in the following. In order to do so, the following points will be commented:

- The structural analysis: how to obtain the buckling modes and its associated critical loads; as well as the complete non-linear behavior of the member, including elastic and inelastic pre and post-buckling ranges.

- The local-global buckling interaction: how to deal with the phenomenon and how to incorporate its effects in practical design applications on the basis of the limit states method.

– The experimental analysis: the main points to be observed during an experimental program to investigate thin-walled members, including buckling interaction.

Figure 44. Short wave local buckling and long wave flexural buckling in interaction, during a compression test of a thin-walled cold-formed stiffened channel.

5.2   Structural analysis

Structural analysis for buckling problems must include calculation methods for first order (bifurcation) buckling and non-linear analysis. The non-linear analysis method must be able to obtain at least the elastic non-linear behavior and, if possible, to include the effects of plasticity. First order computational tools for buckling analysis of thin-walled members are essential for design practice nowadays, and non-linear computational programs are much more applied for research activities.

Linear buckling analysis
In order to identify the buckling modes, a computational tool based on the solution of the eigenvalue problem must be achieved, as expressed by Equation 5.1, where K_0 and K_G are the linear and the geometrical stiffness matrix of the structure, and λ is the load parameter. For this the finite strip method (FSM) or the finite element method (FEM) may be applied.

$$K_0 - \lambda.K_G = 0 \qquad\qquad (5.1)$$

The FSM was applied to solve buckling problems of thin-walled members by Lau and Hancock (1986), and it followed the identification of local and distortional buckling modes for typical rack and stiffened channel cross-sections. The authors indicate the FSM to be directly applied in design applications of thin-walled members.

Later on, Prola and Camotim (2000) developed a computational program based on the FSM to linear and non-linear analysis of thin-walled members. Many cases of thin-walled cold-formed sections were studied, and theoretical results were compared with experimental ones (D. Camotim et al, 2000).

A free share FSM computational program developed by B. Schaffer is available in internet. The author recommends this tool for direct buckling analysis of cold-formed steel sections.

Nagahama (2003) is the author of another version of a FSM-based computational program, which can be applied to isotropic (steel) and orthotropic (GRP) materials. Steel cold-formed sections and GRP pultruded sections may be analyzed, and a friendly computational tool addressed to practical design applications was developed.

In fact, all these FSM computer programs may solve the problem of buckling modes of thin-walled members under any combination of compression plus bending moments, including concentric or eccentric compression. All the results of the buckling analysis obtained with the FSM that will be showed in the following were obtained from the computational program developed by Nagahama (2003). Figure 45 shows the results of the buckling analysis of a stiffened channel cold-formed member under bending, and one may observe the ranges where the local mode (LPM), the distortional mode (DM) or the lateral torsional mode (LTM) are critical. In this case we have free warping end conditions. If we need to analyze end conditions other than free warping, these original formulations of the FSM must be changed. Lau and Hancock (1986) firstly applied Spline functions to solve this problem, what was also achieved by Prola and Camotim (2000). This procedure allows to obtain the buckling behavior for fixed end conditions of the thin-walled members, including, for example, the case of compressed members for which the fixed warping is usually unavoidable, due to connection conditions. Remember that the warping condition is always of great importance when any torsional (or distortional) buckling mode is concerned.

The buckling analysis with the FEM permits to cover any case of thin-walled member and is very useful to calibrate and confirm the results of simplified models, as those from the FSM for example. Figure 46 shows the results of a shell finite element computational program, applied by Inoue (2003) to the case of a concentrically compressed stiffened channel member. This FEM computational program was originally developed to perform linear and non-linear analysis in elastic and inelastic ranges, for single and folded plates. In Figure 46 one may observe the important difference we find if the end conditions are free (Fr-w) or fixed (Fx-w) to warping. The fixed warping implies the displacement of the computed buckling loads, at the same time to higher values and moving to the right of the graphic, if compared with the results of free warping. It is also clear from this Figure that the longer is the member the closer are the results for the two end conditions; what means that the different end conditions may be considered really important for the cases of shorter members. As the member becomes longer and slender the differences between the results obtained with these two end conditions tends to become of less importance.

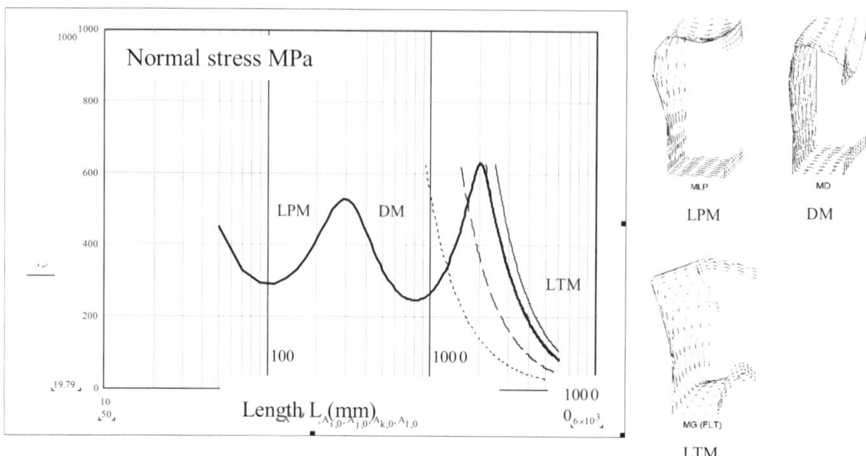

Figure 45. First order buckling analysis of a stiffened channel under bending. Finite strip method computational results (see Nagahama (2003)).

The General Beam Theory, applied for the linear buckling of thin-walled members, bring some advantages when compared to the FS and the FE methods. The GBT, for example, permits the identification of the contributions of each buckling mode in the structural behavior. The GBT was originally developed by Schardt (1983), and it has been applied for steel thin-walled structures by Davis (1998). More recently, Camotim and Silvestre (2002) applied the GBT to derive approximate but very accurate analytical formulas to estimate the distortional buckling loads for columns and beams, with channel and Z section. In a certain way, the GBT and the FSM are analogous methods, regarding the independent consideration of the displacements, related to the cross-section and along the member length. Additionally, the GBT allows the development of analytical solutions to several practical problems.

Finally, it must be pointed that nowadays the linear buckling analysis may be fulfilled on the basis of efficient and practical computational programs, as those based on the FSM, for example, and that the application of numerical solutions may replace the direct and approximate formula methods. It means that one may consider the actual buckling modes in its practical applications. For example, the local buckling of the complete cross-section with interaction between plate elements, avoiding the traditional and simplified procedure of isolated plates. Also, the application of computational analysis permits the structural engineer to have a general view of the buckling problem, conducting to realistic knowledge of the several critical modes interfering in the process.

Non-linear buckling analysis

The post-buckling analysis is to be performed on the basis of non-linear computational methods. For this, numerical methods are available, the FEM being the most general. The non-linear behavior of structural thin-walled members may be obtained with shell

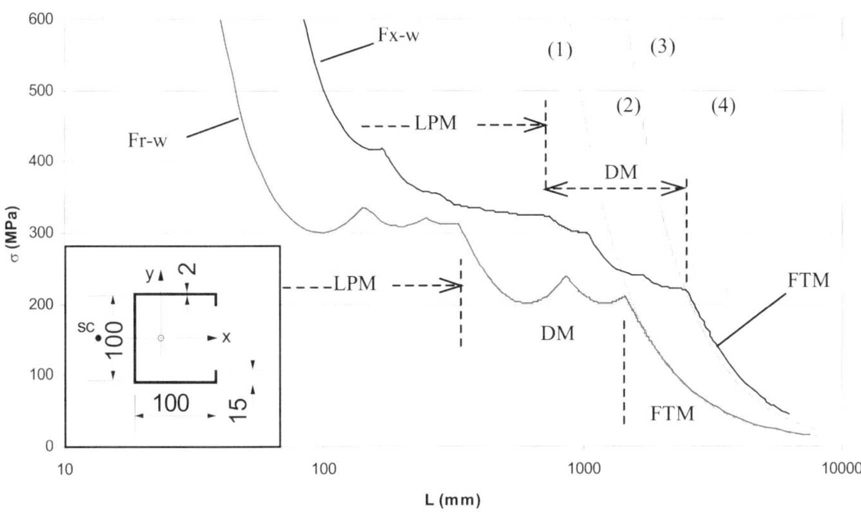

Figure 46. Buckling modes analysis of a stiffened channel under concentrically compression. Finite elements method computational results for free and fixed warping end condition, Fr-w and Fx-w, respectively. The curves indicated as (1) to (4) represents the global buckling and were directly computed by means of Equations (28) to (31): (1) and (2) are the flexural-torsional buckling loads for Fr-w and Fx-w end conditions, respectively; (3) and (4) are the Euler flexural buckling loads.

finite elements and Figure 47 shows the results of a stiffened channel under uniform compression (Inoue, 2003). These results indicate the local plate buckling with large displacements, proportional to the first order-buckling mode. Compressive load P v. out-of-plane displacements of the plate shows the loss of bending stiffness of the plates due to the development of the local buckling. Observing the results of Figure 47, one may also conclude that larger initial geometrical imperfections promote less important changing of the bending stiffness during the loading process, and this is in complete agreement with the following principle: for the post-buckling behavior, the most important imperfections (those that are more deleterious) are not the larger ones but the small ones.

If the FEM formulation includes inelastic material, it is possible to obtain very accurate results for thin-walled members, even if the actual collapse mechanism includes lines of plastic hinges in the plate elements that are of difficult reproduction.

The FEM applied to the non-linear analysis may be performed to follow the local-global interaction for the case of thin-walled members. For this, experimentally measured initial geometrical imperfections may be introduced in the numerical model, as well as initial residual stresses. The previous calibration of the numerical model, with the help of experimental results, permits to apply the FEM for a large number of cases and, in some cases, may replace the experimental analysis, resulting in cost saving. What must be pointed is that experimental, theoretical and numerical results must be carefully combined in order to obtain deeper comprehension of the problem and to verify and

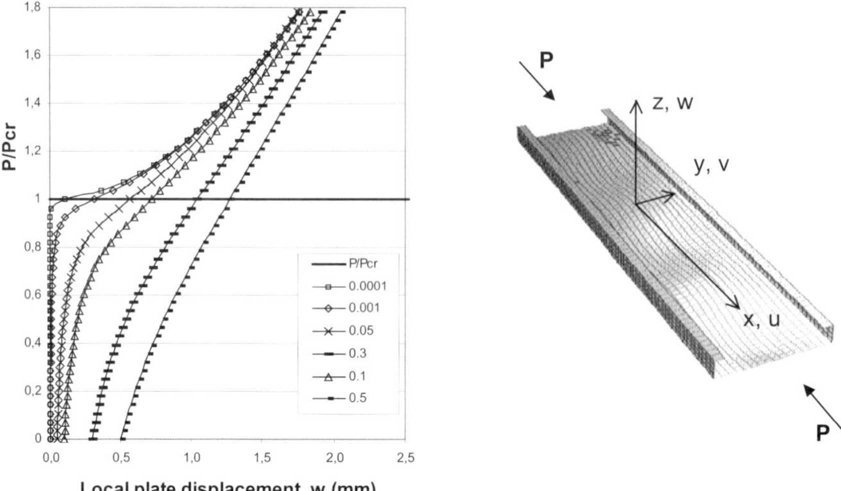

Figure 47. Results of the elastic non-linear analysis of a stiffened channel cold-formed member (600 x 100 x 50 x 1.0 mm, L=2600 mm). Buckling mode is the local plate one and P_{cr} is related to the complete cross-section with interaction between the plate elements. Initial geometrical imperfections follow the buckling mode shape, with maximum amplitude-to-thickness ratio, w_0/t, from 10^{-4} to 0.5.

ameliorate simplified methods of analysis and design.

5.3   The local-global buckling interaction

The interaction between local and global buckling modes is especially addressed to plate buckling with flexural or flexural-torsional global buckling for the case of compressed members, and plate and lateral buckling interaction for the case of beams. Any of these cases should be qualified as strong interaction, as shown in Figure 10, what means that the prescriptions for the design of thin-walled members must incorporate this phenomenon, and the expected consequences are the erosion of the ultimate strength of the member. Buckling modes interaction is always a non-linear problem and numerical methods are essential to access deeper comprehension of the problem. Additionally, experimental results must be available in order to: (a) confirm how the buckling interaction actually develops; (b) obtain the experimental value of the ultimate strength; (c) record the actual geometrical imperfections and residual stresses and (d) observe and qualify the collapse mechanisms.

The results of the local-global interaction can be obtained with the help of the FEM, as the case presented in Figure 48, where we observe the computational results for a thin-walled member under concentric compression (stiffened channel section 200 x 100 x 20 x 1 mm; L=2000 mm).

The FEM program applies shell elements and only elastic results are shown. Also, in

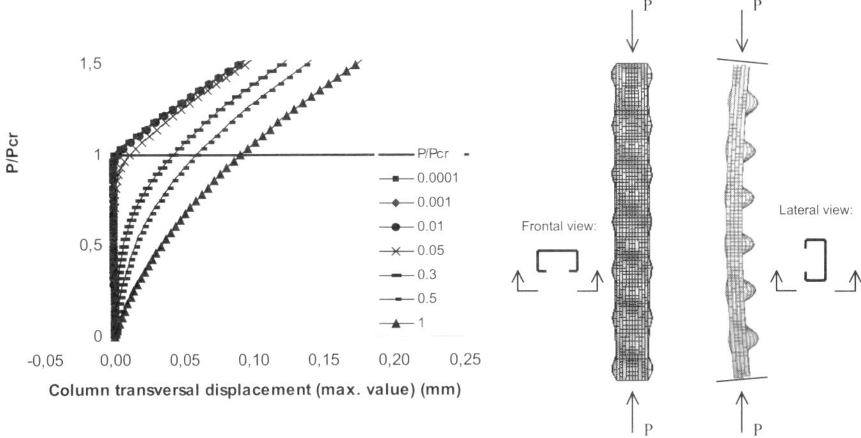

Figure 48. The non-linear results of the interaction between the local plate and the flexural global buckling modes, for a stiffened channel under concentric compression. Large values of the geometrical initial imperfection conduct to soft non-linear path. Small initial imperfections lead to load-displacement results close to the bifurcation-type behavior. FEM computacional results (Inoue, 2003).

this case, initial geometrical imperfections where considered, based on the critical local plate buckling mode - LPM(13), which means that we have 13 local plate half-waves along the member - with maximum amplitude varying from 10^{-4} to 1.0 times the plate thickness t. The results shown in Figure 48 indicate how the local buckling interferes in the global flexural behavior, reducing the structural stiffness of the member. In fact, even for initially perfect structural members, the onset of the local plate buckling induces load eccentricity and global bending effects.

Simplified models other than the FEM may also be applied to obtain the behavior of thin-walled compressed members. The application of direct solution for the non linear behavior of beam-columns - the secant formula for example, that introduces the amplification of the stresses based on a secant amplification factor (see Timoshenko and Gere (1961)) - combined with the approximate effective width method for the local plate mode, may conduct to valuable approximate results of the analysis of thin-walled beam-columns. Computational programs based of the general beam theory or the finite strip method may also result in useful tools for the non-linear analysis of the local-global buckling interaction.

From the point of view of practical structural design, the ultimate strength must be defined including the buckling interaction. The general procedure to do so is commented at the end of the present text, in section 6. For this, both local and global buckling must be conveniently accessed, and the effects of the interaction must be included by means of approximate and calibrated methods.

5.4 Experimental analysis

Experimental results of buckling, post-buckling and buckling interaction may bring very important information. Regarding the experimental analysis, the following main aspects are to be previously considered:

- The precise identification of the objectives of the experimental program.
- How to obtain the main informations by means of experiments.
- The experimental techniques to be applied: preparing the specimens, loading system, sensors and data acquisition and the analysis of the experimental data.

Some of these aspects will be commented in the following.

Specimens for testing

Specimens must be manufactured with the same characteristics of the structural members under investigation, remembering that the manufacturing process always introduces typical imperfections. Actual geometrical dimensions of the specimens must be carefully measured and recorded. One must consider, for example, that the plate thickness is a very important data for torsional behavior of thin-walled open cross-sections and if we are dealing with thin plates, any difference of about 5% in this value may conduct to differences in the torsional stiffness of about 15%.

End conditions for testing

A second factor of up most importance is the end conditions of the specimens. As the subject of investigation is the buckling behavior, end conditions may determine if your experiments are realistic or not and if it is valid to compare the experimental results to the theoretical and numerical results you have. As we are dealing with torsional problems, the warping end conditions must be carefully observed, as well as the plate end conditions of the cross-section. It was already showed in Figure 46 how the end restraints of the member might influence its structural behavior. The structural conditions of both experimental and computational analysis must be in close agreement; if not, the comparisons may not be useful for validation. Figure 49 shows how the apparently simply supported end conditions of a rack cold-formed specimen under concentric compression, displaying the distortional buckling mode, is in fact much more like restrained than free bending of the plate elements. In fact, the contact between the end sections of the member with the stiff template of the test machine results in the actual end conditions with partial restraint of the out-of-plane bending of the plates.

Figure 50 shows another aspect of the problem: the warping condition. One may conclude that the contact of the end sections of the specimen with the templates is sufficient to assure fixed warping. This is true for a great part of the compressive loading path, but it may fail when large torsional displacements develop and some regions of the cross-section may lose contact with the template. Because of this, if fixed warping is to be completely assured, extra measures must be taken, incorporating stiff plates at the ends of the specimen or some other solution as filet welding (what is not recommended because of the introduction of residual stresses).

Figure 51 shows the procedure that has been applied to keep fixed warping condition. This is achieved with the help of a stiff GRP (glass fiber reinforced plastic) plate, molded around the cross section. This procedure conducts to fixed warping and fixed plate condition at the ends of the member.

Global end conditions are also to be carefully observed. Compressive tests of thin-walled cold-formed specimens are usually performed with free bending condition, which is introduced with the aid of special spherical hinges, made of stainless steel and teflon contact elements. Figure 52 shows a typical spherical hinge usually applied in compressive tests of steel members.

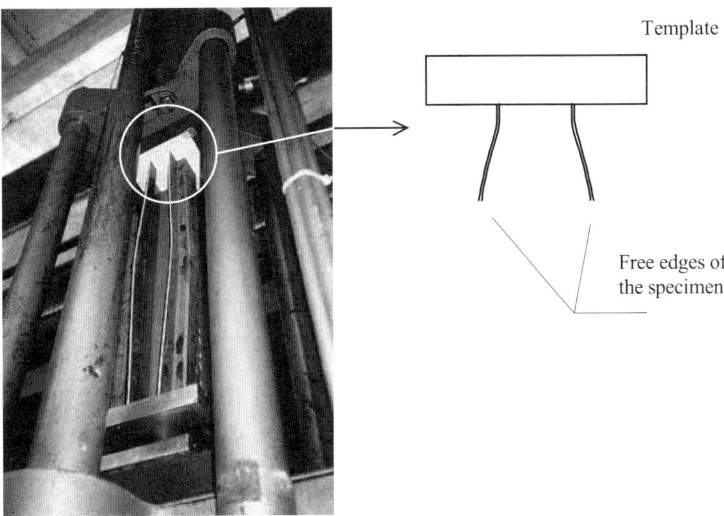

Figure 49. The actual end conditions of the plate elements in a compression test of a thin-walled cold-formed member: the end sections of the specimen, simply fitting the templates, result in restrained out-of-plane bending.

Loading condition

The loading condition is a central aspect of the experimental test. First, one must decide how the position of the compression load will be controlled. Second, how the limit load may be attained, as from this point on the tendency of the structural member is to loose load capacity in unstable behavior.

The position of the resultant compressive load applied by the testing apparatus may be measured by several ways. For example, by measuring the strain/stress distribution around the mid-height cross-section. This is a time and cost expensive technique that, in our point of view, is not the more appropriate. Geometrical procedures may offer very good results, and are simple to be implemented. For this, one may adopt the following steps: (a) draw the actual end cross-sections over sheets of paper and compute its principal axis; (b) place these papers over each template, coinciding the computed principal axis of the section with the centroidal axis of the templates; (c) finally place the specimen in the test machine making the end sections to exactly adjust over its original

Figure 50. Evidence of warping at the end section of a thin-walled cold-formed member, during concentric compressive test. At the presented loading step the member developed the flexural-torsional buckling mode, and the end sections, simply fitting the templates, do not prevent warping.

(a)   (b)

Figure 51. Preparing the end sections of a thin-walled steel member for fixed warping compressive test: (a) before in filling the GRP material; (b) positioning the specimen before testing.

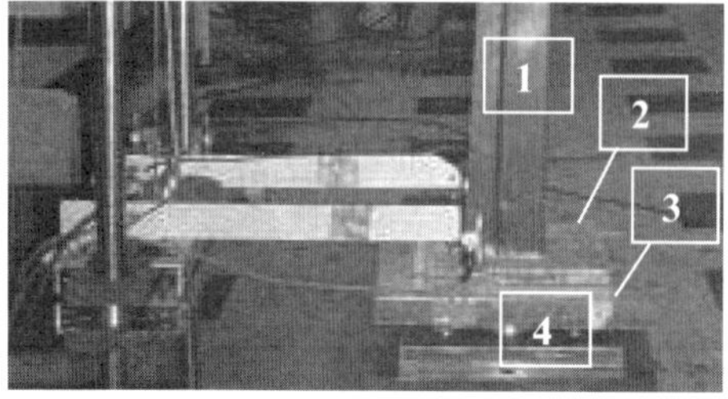

1 : Specimen

2 : GFRP plate

3 : Template

4 : Spherical hinge

Figure 52. End conditions for a compression test of a thin-walled steel specimen.

draws. This procedure results in concentric (or controlled eccentric) compressive loads at the ends of the specimen, what means that the eccentricities along the compressed member will be only those from its initial geometrical imperfections.

The second aspect cited above is related on how the test is performed: load control or displacements control. Traditional test machines are able to induce loading by simply growing the pressure of oil that is generated by a hydraulic system (electrical or mechanical pump system). In this condition, when the limit load is attained, it is not possible to still going with the test in the unloading range and the hydraulic test system, if not disabled, will promote large displacements and deformations in the specimen in an uncontrolled way. To overcome this problem, displacement controlled tests may be performed with the help of electronic controlled systems. In this case servo controlled hydraulic actuators allow controlled displacement tests. This type of structural test system follows a closed-loop control as presented in Figure 53 and include the following elements: (a) the hydraulic supply; (b) the hydraulic actuator; (c) the load, the displacements and the strain transducers; (d) the servo controller that verify the "error" signal; (e) the valve driver that controls the servo valve input signal; (f) the feedback signal from the transducers. This system works with constant high hydraulic pressure and the servo valve is able to command the quick response of the actuator on the basis of the electronic control signal.

It is important to point out that this type of test system is specially recommended for stability analysis, for which the displacements (stroke) control must be achieved in order to follow and record the experimental results up to the limit load, as well as the unstable (unloading) branches of the structural test.

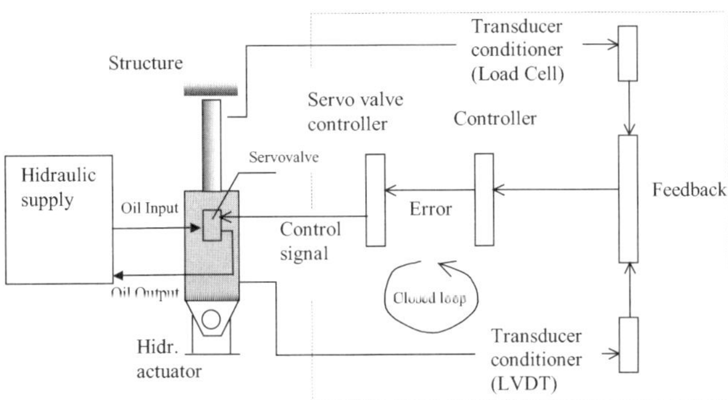

Figure 53. The closed-loop control of the structural test system allows displacement (stroke) controlled tests and the identification of the limit load and the unstable buckling paths.

Data acquisition and analysis

The experimental data acquisition system must combine high frequency collecting the experimental data, software facilities to friendly handle the data acquisition, and the possibility to follow the experimental results of the test in on-line condition. Many types of data-acquisition equipments are available to perform accurate record of experimental results, all of them working together with computational facilities.

As stability experimental tests are usually performed in quasi-static loading condition it is not necessary to use very high frequency acquisition systems, more applicable to dynamic tests, and to perform the acquisition in low-frequency condition or even with the acquisition made in step-by-step strategy. This procedure permits to avoid processing excessive and useless experimental data.

Imperfections measurements

The measurement of initial geometrical imperfections allows structural numerical analysis, based on the actual conditions of the test specimen. For this, it is necessary to have a mechanical apparatus specially addressed to perform these measurements. Figure 54 shows a special arrangement that enables measurements of the geometrical shape of the specimen, which may be performed before (initial imperfections) and during (deformed shape) the compression test. Figure 55 shows an illustrative example of the experimental results of a rack cold-formed member tested under eccentric compressive load, and it can be observed that the initial geometrical imperfections at the free edges of the cross-section presented one half-wave shape. The developed displacements close to the ultimate load correspond to the distortional buckling mode. Figure 56 illustrates the deformed shape and the plastic collapse position of rack cold-formed columns and beam-columns after compressive test.

In the presented example, the displacement sensors (transducer displacements devices that are able to generate analogical electrical signals that are transmitted to the data acquisition system) were conveniently placed in order to take measurements of the distortional buckling mode as shown in Figure 54. Both specimen deformation and the position of the transducers along the specimen's length are simultaneously measured, in a high frequency data acquisition system, permitting to plot the results as shown in Figure 55.

The collapse mechanism

The knowledge of the collapse mechanism is of great importance in structural analysis and theoretical models may be proposed on the basis of experimental observations. Thin-walled members are usually affected by local plate plastic mechanism, with the formation of plastic hinge lines, as shown in Figure 56 (see Pérez (2003)). After the formation of the plastic mechanism, if the member is in simply supported condition (end condition with spherical hinges), the tendency is to concentrate all the deformations in the local plastic collapse region, with the consequent recuperation of the elastic deformations along the member. Typical local buckling-based collapse determines the concentration of the plastic deformations in a short region at the mid-height of the column, which can be associated with the formation of plastic hinges for the case of steel members with compact cross-section. For the case of thin-walled members, the formation of the local plate plastic mechanism usually conducts to almost no plastic range, and the plastic hinge lines are formed for a load step very close to the limit load of the member.

Finally, it is important to point out that experimental observations allow to follow

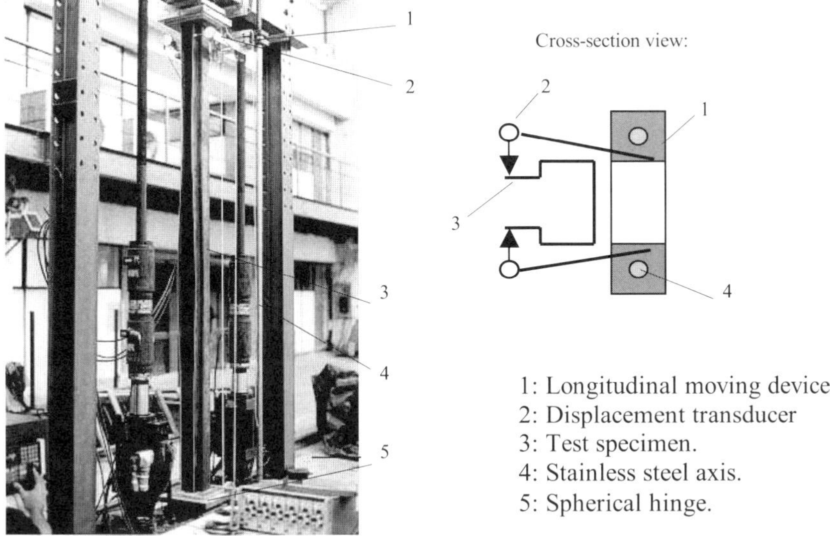

Figure 54. Test arrangement for experimental buckling analysis of thin-walled members. Initial imperfections and buckling shape measurements during the compressive loading test.

the development of the collapse mechanism. As an illustrative example, one may observe the experimental results obtained from cold-formed members with the same rack cross-section, tested under concentric and eccentric compression, and for three different lengths, L. The description of the results is resumed in Table 3 and all specimens failed by distortional buckling. It is interesting to note that the longer are the columns, as well as the larger is the eccentricity of the load (always in the opposite side of the shear center, what results in more pronounced distortional buckling) the more sudden is the collapse. Very sudden collapses were observed for long columns with large eccentricity of compression, which represents a clear tendency as presented in Table 3. Less slender columns, especially if combined with less eccentric load, induce much more ductile collapse, with soft formation of the plastic configuration.

The four collapse mechanisms presented in Figure 56 are related to the rack specimens with total length $L = 1160$ mm, included in Table 3.

6 Design concepts of thin-walled members

The final objective of the theoretical developments on thin-walled structures is to contribute to the definition of accurate methods for structural analysis and design. This last part of our presentation is addressed to structural design concepts, especially for the cases where stability is concerned.

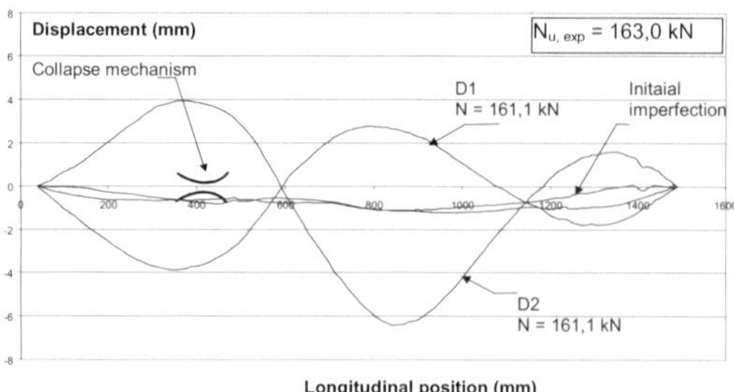

(a) Distortional buckling shape and initial imperfections

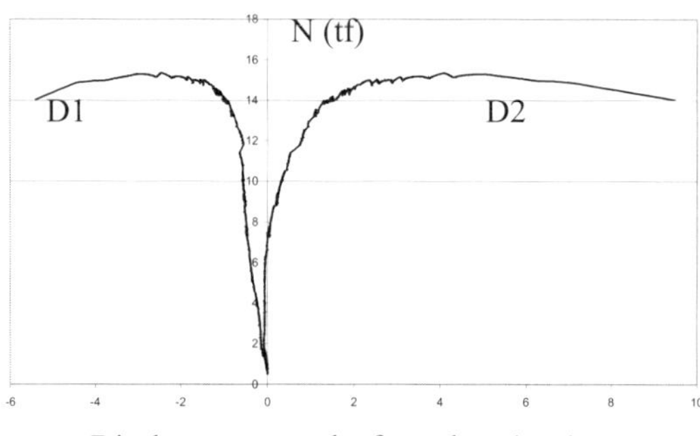

(b) Load-free edge displacements results, including the limit load and the unstable path

Figure 55. Results of experimental measurements.

6.1 Safety principles

The Limit States Method is nowadays found in the great majority of codes and specifications of structural design, independently of which kind of material is considered: steel, aluminum, concrete, fiber-plastic composites, and so on. The main characteristic of this method is its probabilistic concept, which interferes on the side of the load combinations (how the loads must be combined and applied in the structure), as well as on the side of the strength (how the structural strength of typical members must be taken). The strength definition criterion is here our main interest. How to define the characteristic strength of a structural member, as a thin-walled member for example, considering that different sets of experimental results present different probabilistic distributions? This problem is solved on the basis of probabilistic approach. That is the reason why the experimental results must include variations of the imperfection effects, and the results must be processed in order to define the characteristic value of the strength. An assumed level of safety must be firstly defined, by means of the probabilistic value p, that is usually prescribed by each national specification for structural reliability. Based on

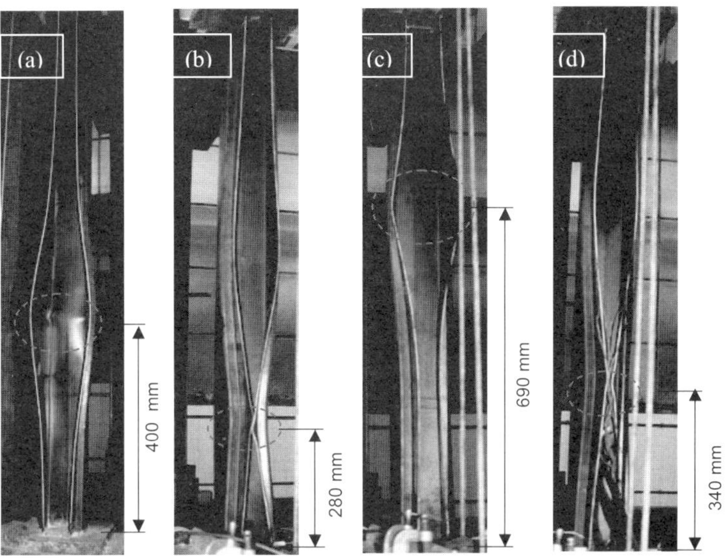

Figure 56. Collapse mechanism of a series of rack cold-formed specimens tested under concentric and eccentric compression (load over the symmetry axis, opposite to the shear center): (a) concentric $e_x = 0$; (b) eccentric $e_x = 6.0$ mm; (c) eccentric $e_x = 10.0$ mm; (d) eccentric $e_x = 14.0$ mm. Nominal cross-section dimensions are 96x57x18x33x2.0 mm; length L is 1160 mm (see Pérez (2003)).

the distribution of the experimental results one may obtain the strength average value R_{av}, the standard deviation value s and the variation coefficient $\delta = s/R_{av}$, which allow to determine the characteristic values R_k

$$R_{k-} = R_{av} - k.s = R_{av}.(1 - k.\delta) \tag{6.1}$$

$$R_{k+} = R_{av} + k.s = R_{av}.(1 + k.\delta) \tag{6.2}$$

Strength values are defined by the lower characteristic value R_{k-}. The probabilistic value p is usually taken between 2% and 5%, and the coefficient k is 2.0 for $p = 2\%$ and 1.64 for $p = 5\%$, for example.

6.2   Strength concept for thin-walled members

The strength of thin-walled members depends on its stability behavior. Because of this, the definition of the strength of the member is based on: (a) the cross-section plastic strength; (b) the reduced strength due to the local buckling effects; (c) the elastic global buckling of the member; (d) the reduced strength curve that might incorporate all the effects of the imperfections. This fundamental idea is presented in Figure 57 and is applied to thin-walled members under compression or bending, for example, allowing the design

188   E. de Miranda Batista

Table 3. Types of collapse mechanism observed during concentric and eccentric compressive tests of cold-formed thin-walled rack members. Load eccentricity is on the opposite side of the shear center and there is no eccentricity out of the symmetric axis of the cross-section. All specimens failed by distortional buckling (according with Pérez (2003)).

Length of the specimen	$e_x = 0.0$ mm	$e_x = 6.0$ mm	$e_x = 10.0$ mm	$e_x = 14.0$ mm
$L_1 = 760$ mm	(x)	(x)	)-(	(-)
$L_2 = 1160$ mm	(x)	)-(	(-)	)-(
$L_3 = 1560$ mm	)-(	)-(	)-(	(-)

e_x : eccentricity of the compressive load, placed over the symmetric axis of the cross-section, opposite to the shear center.

(x) : collapse mechanism with soft opening of the cross-section.

)-(: collapse mechanism with sudden closing of the cross-section.

(-) : collapse mechanism with sudden opening of the cross-section.

Cross-section dimensions: 96 x 57 x 18 x 33 x 2.0 mm.

of columns, beams and beam-columns. This concept is included in the prescriptions for the design of thin-walled members in the national codes addressed to cold-formed structural members.

In all design prescriptions the local-global buckling modes interaction must be considered. The effect of the local plate buckling is usually taken on the basis of the effective width method, which is addressed to transform the non-linear problem of the local plate buckling into a linear formulation that is able, after calibration based on experimental results, to identify the ultimate strength of single plates under uniform or linearly variable compression. The calibrations accomplished by different researchers resulted in several general expressions for the local buckling strength. The next equations are related to the effective width method applied to cold-formed sections, on the basis of the results from Winter (1947), and this is the solution that can be found in several steel design codes.

$$\frac{b_{eff}}{b} = \left(1 - \frac{0.22}{\lambda_p}\right)\frac{1}{\lambda_p} \leq 1.0 \qquad (6.3)$$

$$\lambda_p = \sqrt{\frac{\sigma}{\sigma_{cr}}} \qquad (6.4)$$

where b is the width of the plate element, b_{eff} is the reduced effective width, σ is the reference compression stress in the plate element and σ_{cr} is the critical buckling stress of

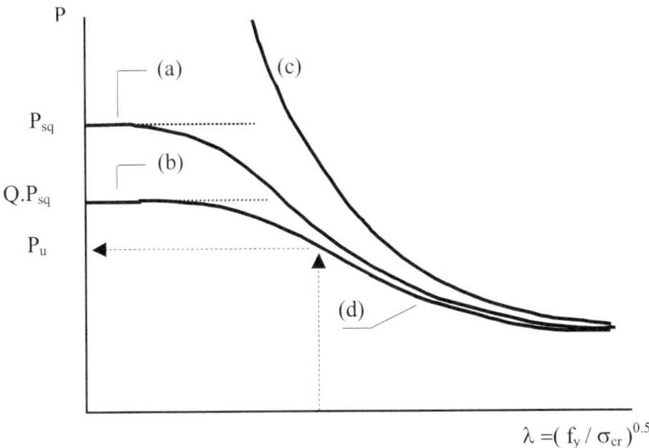

Figure 57. The strength definition of compressed thin-walled structural members: (a) the squash load without local buckling, P_{sq}; (b) the squash load taking into account the local buckling, for which $Q(<1)$ is the reduction coefficient due to the local buckling; (c) the elastic buckling; (d) the ultimate strength curve, P_u.

the element.

The previous equations are to be applied to each single plate element of the cross-section of the cold-formed thin-walled member, permitting to transform the original cross-section A in an effective cross-section, A_{eff}. The latter means that if $A_{eff}/A < 1$ we have a slender cross-section and, in the contrary, we have a compact cross-section. The results of the application of the Winter's formula, Equation 6.3, for the case of the usual cold-formed cross-sections are in good agreement with the experimental results.

An alternative conception, to take into account the local buckling in the ultimate strength of thin-walled members, is the one that directly includes this effect for the complete cross-section. For this, it is firstly necessary to compute the elastic local buckling solution of the section, resulting in the curves of the type shown in Figure 18, for example. As these results incorporate the actual local buckling behavior, it is expected that its application in a convenient strength curve (experimentally calibrated) would conduct to accurate results. That is the case of the experimental results shown in Figure 58, where a collection of experimental results of channel cold-formed stub columns are compared with a proposed strength curve that implicitly includes the effects of the interaction between the plate elements of the cross-section.

This procedure was defined as the effective area method, and it permits the computation of the reduced effective area A_{eff} in one step only, with the help of the results of the linear buckling analysis of the cross-section, and can be resumed by the following equations

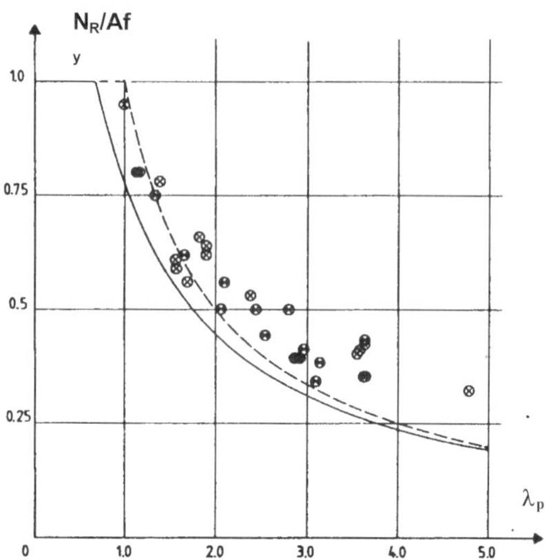

Figure 58. Experimental results of cold-formed thin-walled stub-column tests, compared with the effective area method (Batista, 1989).

$$P_{R,stub} \;=\; A_{eff}f_y \tag{6.5}$$

$$A_{eff} \;=\; \left(1 - \frac{0.22}{\lambda_p}\right)\frac{A}{\lambda_p} \tag{6.6}$$

where $P_{R,stub}$ is the strength of the stub column, and the relative slenderness of the section, $\lambda_p = \sqrt{\sigma/\sigma_{cr}}$, is obtained from the local buckling analysis of the section.

In fact, the obtained results confirmed that the consideration of the actual buckling mode of the complete cross-section, instead of the individual plate elements, conducts to better agreement with the experimental results.

For the strength calculation of long columns, the local buckling effect is to be taken into account by the following expression

$$P_R = \rho A_{eff}f_y = \rho Q A f_y \tag{6.7}$$

where we have two parameters reducing the ultimate strength: $Q = A_{eff}/A$ (< 1 for slender sections and $= 1$ for compact sections) and ρ ($= 1$ for stub columns and < 1 for long columns). The former is related to the local buckling (effective area) and the last is referred to the global buckling of the member (appropriate buckling curve of the profile).

The central idea for consistent design calculation of thin-walled members is that: more accurate results of the elastic buckling analysis conduct to more reliable results of strength. In other words, accurate design means appropriate methods for the elastic buckling analysis.

6.3   The special case of combined compression and bending

The identification of the ultimate strength of thin-walled members, for the case of simultaneous compression and bending, is based on approximate methods with interaction equations of the type shown in Equation 6.8. This type of solution is found in several codes addressed to the design of steel structures. In this case, several non-linear effects are present, including the beam-column behavior, the plates local buckling, the interaction between local and global buckling modes, as well as the inelastic behavior of the material.

$$\frac{P}{P_R} + \frac{C_{mx}M_x}{M_{Rx}(1 - P/P_x)} + \frac{C_{my}M_y}{M_{Ry}(1 - P/P_y)} = 1.0 \qquad (6.8)$$

The above equation is in fact a general expression for the strength of the member and it indicates a linear (conservative) interaction between the three strength ratios. The buckling behavior is included in this formulation in the strength values of P_R and M_R, which correspond to the compressive and the flexural strength of the member. The compressive strength P_R takes into account the flexural, the flexural-torsional, the local and the distortional buckling modes. Variations in this procedure may be applied, including, for example, the load eccentricity effect in the compressive strength P_R, at least to consider the flexural-torsional buckling load with eccentric compression, based on the elastic buckling solutions presented in section 4.

6.4   Direct strength approach

As described before, designing for stability is especially taking the correct critical buckling mode and introducing its critical load in a calibrated strength curve. The elastic behavior of the structural member is considered inside the buckling mode, and the effects of the imperfections and plasticity are to be included in the ultimate strength curve. It is as simple as this, and the better the elastic buckling previsions the better the ultimate strength results (also depending on the experimental strength curve: accurate or inaccurate experimental calibration). So, we must have efficient methods to obtain the elastic buckling characteristics and an experimentally calibrated strength curve (currently known as "buckling curve").

In this direction, we find the recent proposition from Schafer (2000), which should be considered in the next versions of the specifications for the design of cold-formed members. The central idea is the one stated above, for which the actual elastic buckling result is to be taken as accurate as possible, with the help of a numerical computational solution, as the FSM or the FEM. In fact, if one has any other way to find the critical buckling loads for the complete cross-section, the fundaments of the design process remain always the same.

With this concept, the traditional effective width method may be discarded and the local buckling is taken in a much more rational way. This is the same idea we find in the effective area method related above, where the local buckling effect is taken for the complete cross section and the partial treatment of the isolated plate elements is discarded.

The direct strength design approach may take advantage of the following computational methods for buckling analysis: (a) the classical FSM, with free end conditions; (b) a refined FSM (Spline functions based, for example) that allows fixed (warping) end conditions; (c) a GBT based computational program; or (d) the FEM computational results. The following observations may be addressed to the each of these elastic buckling solutions:

- The classical FSM is easy to be implemented and allows friendly computational programs. Not considering end conditions other than free warping may be taken as a lack of precision but, on the other hand, free end condition will never conduct to unsafe results.

- Refined FSM is an upgrade to the classical formulation of the FSM and it would be recommended for more precise design.

- Computational programs based on the general beam theory give a very clear view of the stability problem but, for design purposes a friendly computational program is absolutely necessary.

- As a quite general and accurate method, FEM-based computational programs may be applied to solve the first order elastic buckling. Nevertheless it may be taken as somehow excessive, especially for standard and repetitive design calculations. Also, in this case, it should be considered that a specialized preprocessor for the generation of geometry, end conditions and type of loading will be absolutely necessary.

It must be considered that direct calculation methods for buckling analysis must always be considered, instead of computational programs. This is so because design codes are always obliged to propose direct methods of solution and "go to the following computational solution" is not the kind of recommendation expected by engineers.

The distortional buckling mode is quite useful to illustrate the opposition between direct solutions and numerical-based solutions. In this case the direct and approximate method originally proposed by Lau and Hancock (1987) and included in the Australian code (AS/NZS, 1996) is still possible for practical applications. Nevertheless, it can be applied only for certain types of cross-sections, namely stiffened channels and rack sections, and it is accurate enough for a certain range of geometrical dimensions of the cross-section. Additionally, the Australian direct formulation is not appropriate for the case of eccentric compression of the member. Anyhow, in the absence of computational tools for stability analysis, designers may obviously take advantage of this procedure, and this is the case of the Brazilian code for the design of cold-formed members (ABNT, 2001), in which the Australian method is recommended in one of the annexes of the code.

Bibliography

Abaqus Standart. *Theory Manual v. 5.8.*, 1998.

ABNT. *Design of Steel Cold-Formed Structural Members NBR 14762 (in portuguese).* Associação Brasileira de Normas Técnicas, 2001.

H. G. Allen and P. S. Bulson. *Background to Buckling.* McGraw-Hill, London, 1980.

G. A. Alpsten. Residual Stresses in a Heavy Welded Shape 23H681. Technical Report 337.9, Fritz Engineering Laboratory, Lehigh University, 1970.

AS/NZS. *Cold-Formed Steel Structure AS/NZS 4600*. Australian/New Zealand Standard, 1996.

E. M. Batista. *Etude de la Stabilité des Profils à Parois Minces et Section Ouverte de Types U et C*. PhD thesis, Collection des Publications de la Faculté des Sciences Appliquées no. 119, University of Liège, 1989.

E. M. Batista, H. Inoue, K. Nagahama, F. L. B. Ribeiro and D. Camotim. Stability Analysis of Cold-Formed Members: Finite Element and Finite Strip Method Results. In M. Iványi, editors, *Stability and Ductility of Steel Structures* 25–29, Akadémiai Kiadó, Budapeste, 2002.

E. M. Batista and F. C. Rodrigues. Residual Stress Measurements on Cold-Formed Profiles. *Experimental Techniques*, 16:25–29, 1998.

F. Bleich. *Buckling Strength of Metal Structures*. McGraw-Hill, New York, 1952.

B. Budiansky. Theory of buckling and post-buckling behavior of elastic structures. In Chia-Shum Yih, editor, *Advances in Applied Mechanics*, 14:1–65, Academic Press, New York, 1974.

D. Camotim, E. Batista, L. C. Prola and E. G. Vasquez. Local Post-Buckling Behaviour of Cold-Formed Steel Rack Columns. In D. Dubina D. Camotim and J. Rondal, editors, *Proceedings of the Coupled Instabilities in Metal Structures CIMS'2000* Imperial College, London, 2000.

D. Camotim and N. Silvestre. Application of the Generalized Beam Theory (GBT) to Analyse the Stability Behaviour of Cold-Formed Steel Members (in portuguese). In *Proceedings of the II Congresso Internacional da Construção Metálica IICICOM (full paper in CD-rom*, São Paulo, 2002.

A. H. Chilver. Coupled Modes in Elastic Buckling. *J. Mech. Phys. Solids*, 15:15–28, 1967.

J. M. Davis. Generalized Beam Theory (GBT) for Coupled Instabilities Problems. In J. Rondal, editor, *Coupled Instabilities in Metal Structures: Theoretical and Design Aspects, CISM Course n. 379*, 151–223, Springer-Verlag, Wien, 1998.

V. Gioncu. General Theory of Coupled Instabilities. *Thin-Walled Structures*, 19:81–127, 1994.

V. Gioncu and M. Ivan. *Bazele Calculului Structurilor La Stabilitate*. Editura Facla, Timisoara, 1983.

H. Inoue. FEM Stability Analysis of Thin-Walled Members (in portuguese). Technical Internal Report, Civil Eng. Program, COPPE-Federal University of Rio de Janeiro, 2003.

W. T. Koiter. Theory of Elastic Stability and Post-Buckling Behaviour. Technical Report, Delft University, 1962.

S. C. W. Lau and G. J. Hancock. Buckling of Thin Flat-Walled Structures by a Spline Finite Strip Method. *Thin-Walled Structures*, 4:269–294, 1986.

S. C. W. Lau and G. J. Hancock. Distortional Buckling Formulas for Channel Columns. *Journal of Structural Engineering*, 113(5):1063–1078, 1987.

V. Monin, J. R. Teodosio and T. Gurova. A Portable X-Ray Apparatus for Both Stress Measurements and Phase Analysis under Field Conditions. *Int. Centre for Difraction Data, Advances in X-Ray Analysis*, 43:66–71, 2000.

N. W. Murray. *Introduction to The Theory of Thin-Walled Structures*. Claredon Press, Oxford, 1986.

K. J. Nagahama. *Local Stability Analysis in Open Section Steel and Fiber Glass Reinforced Resin Members (in portuguese)*. PhD thesis, Federal University of Rio de Janeiro, 2003.

K. J. Nagahama, D. Camotim and E. M. Batista, Elastic and Elastic-Plastic Local-Plate/Distortional Mode Interaction in Cold-Formed Lipped Channel Steel Columns In *Proceedings of the 2nd Int. Conference on Structural Stability and Dynamics*, Singapore, 2002.

S. V. S. Pérez. *Experimental Analysis of the Distortional Buckling of Thin-Walled Open Section Members under Eccentric Compression. (in portuguese)*. PhD thesis, Federal University of Rio de Janeiro, 2003.

M. Pignataro, N. Rizzi and A. Luongo. *Stability, Bifurcation and Postcritical Elastic Behaviour of Elastic Structures*. Elsevier, Amsterdam, 1991.

L. C. Prola and D. Camotim. Post-Buckling of Plates and Thin-Walled Sections Using the Finite Strip Method (in portuguese). In *Proceedings of the VI National Congress on applied and Computational Mechanics*, 993–1002, Aveiro, 2000.

A.J.L. Reis and D. Camotim. *Estabilidade Estrutural (in portuguese)*. McGraw-Hill, Lisbon, 2001.

J. Rondal. *Coupled Instabilities in Metal Structures: Theoretical and Design Aspects, CISM Course n. 379*. Springer-Verlag, Wien, 1998.

B. W. Schafer. *Distortional Buckling of Cold-Formed Steel Columns*. Final Technical Report, American Iron and Steel Institute AISI, Washington D. C., 2000.

R. Schardt. *The Generalized Beam Theory*. In *Proceedings of M. R. Horne Conference Instability and Plastic Collapse of Steel Structures*, 469-475, 1983.

J. M. T. Thompson and G. W. Hunt. *A General Theory of Elastic Stability*. J. Wiley and Sons, London, 1973.

S. P. Timoshenko and J. M. Gere. *Theory of Elastic Stability*. McGraw-Hill, New York, 1961.

Van der Neut. The Interaction of Local Buckling and Column Failure of Thin-Walled Compression Members. In *Proceedings of 12th Int. Congr. of Applied Mechanics*, 389–399, Springer-Verlag, Berlin, 1969.

V. Z. Vlasov. *Thin-Walled Elastic Beams*. Israel Program for Scientific Translations, Jerusalem, 1961.

G. Winter. Strength of Thin-Walled Compression Flanges. In *Transactions of the American Society of Civil Engineers ASCE*, vol.112, 1947.

Computational asymptotic post-buckling analysis of slender elastic structures

Raffaele Casciaro

Department of Structure, University of Calabria, Rende, Italy.
e-mail: rcasciaro@labmec.unical.it

Abstract

The lectures provide an introduction to the computational treatment of Koiter's asymptotic strategy for post-buckling analysis of thin elastic structures.

The analysis of slender structures characterized by complex buckling and post-buckling phenomena and by a strong imperfection sensitivity, suffers from a lack of adequate computational tools. Standard algorithms, based on incremental-iterative approaches, are computationally expensive: it is practically impossible to perform the large number of successive runs necessary for the sensitivity analysis, that is, to evaluate the reduction in load due to all possible imperfections. Finite element implementations of Koiter's perturbation method give a convenient alternative for that purpose. The analysis is very fast, of the same order as a linearized stability analysis and new analyses for different imperfections only require a fraction of the first analysis time.

The main objective of the present course is to show that finite element implementations of Koiter's method can be both accurate and reliable.

1 Preliminary notions

This section intends to give a (very) short introduction to the Koiter asymptotic approach to the analysis of slender structures. We start considering the Liapunov instability theory applied to conservative systems, then examine and criticize the Euler method for critical equilibrium and give a rapid insight into bifurcation theory. For deeper insights see Koiter (1978) and Budiansky (1974).

1.1 Stability of the elastic equilibrium

Liapunov's instability Having defined a dynamic, consider two solutions:

- the *reference trajectory* $u^*[t]$, $\dot{u}^*[t]$ following from initial conditions $u^*[0] = 0$ and $\dot{u}^*[0] = 0$;
- the *perturbed trajectory* $u[t]$, $\dot{u}[t]$ following from initial conditions $u[0] = u_0$ and $\dot{u}[0] = \dot{u}_0$.

The reference trajectory is called *stable according to Liapunov* if for any (small) $\epsilon > 0$ there exists $\delta > 0$ such that, for any initial disturbance

$$||u_0|| + ||\dot{u}_0|| < \delta \tag{1.1}$$

we have:

$$||u[t] - u^*[t]|| + ||\dot{u}[t] - \dot{u}^*[t]_0|| < \epsilon \quad \forall t > 0 \tag{1.2}$$

It is worth noting that:

1. *stability is a dynamic concept;*
2. *it depends, in general, on the state variables $u[t]$ chosen to describe the configuration of the system and the assumed metric;*
3. *with this definition, we can have stability or instability: there are no intermediate conditions.*

Conservative systems If we consider equilibrium configurations, the reference trajectory becomes:

$$u[t] = 0, \qquad \dot{u}[t] = 0 \tag{1.3}$$

then, stability implies that, for a *small initial perturbation*, we have

$$||u[t]|| + ||\dot{u}[t]|| < \epsilon \tag{1.4}$$

Note that, also in this case, stability is a dynamic concept.

If the system is characterized by the presence of a *Total Potential Energy* (T.P.E.) we can estimate its stability through static considerations. In the following we will indicate with

$$E[t] = \Pi[u[t]] + T[\dot{u}[t]] \tag{1.5}$$

the total energy of the system, the sum of the T.P.E. $\Pi[u[t]]$ and of the *Kinetic Energy* $T[\dot{u}] \geq 0$. We also make the following assumptions:

1. The system is characterized by a (though small) viscous dissipation so that the function $E[t]$ monotonically decreases along dynamic ($\dot{u}[t] \neq 0$) trajectories.
2. The total potential energy $\Pi[u]$ is an analytical function of u.

The first hypothesis appears "natural" for mechanical systems and will assume a relevant role in the following. The second one, is not so obvious, however it is necessary if aiming to use an asymptotic approach.

Some comments about the analyticity For a three–dimensional continuum, the total potential energy is usually written (Green–Lagrange):

$$\Pi[u] := \frac{1}{2} \int_B E^{ijhk} \varepsilon_{ij} \varepsilon_{hk} dV - \int_B q^i u_i dV - \int_{\partial B} f^i u_i dA \tag{1.6}$$

where B represents the domain occupied by the body in the reference configuration, ∂B is its boundary, u_i is the displacement from this configuration and

$$\varepsilon_{ij} := \varepsilon^0_{ij} + \frac{1}{2}(u_i|_j + u_j|_i) + \frac{1}{2} u^k|_i u_k|_j \tag{1.7}$$

is the strain tensor. The analyticity of $\Pi[u]$ implies the analyticity of the strain energy:

$$\Phi[u] := \frac{1}{2} \int_B E^{ijhk} \varepsilon_{ij} \varepsilon_{hk} dV \tag{1.8}$$

But this is not analytical. In fact we have the following results:

- Tonelli (1929): *The functional $\Phi[u]$ is not analytical in u in the* energy *norm:*

$$\|u\|^2 := \int_B E^{ijhk}(u_i|_j + u_j|_i)(u_h|_k + u_k|_h)dV \qquad (1.9)$$

- Martini (1979): *Indicating with ∂u the gradient of u, the functional:*

$$\int_B F[\partial u]dV \qquad (1.10)$$

is not Fréchet differentiable more than once (for any suitable metric) if $F[\partial u]$ is more than quadratic in ∂u. Therefore it is not analytical in u.

The previous obstacles can be removed in one of the following ways:

1. by including, in the strain energy, a contribution (however small) related to the second derivatives of the displacement;
2. by removing from the admissible displacement space the part providing loss of analyticity;
3. by relating to discrete systems, e.g. obtained by finite element modelling;

Therefore, we will assume in the following that $\Pi[u]$ is analytical. The following known result will also be used:

Lemma 1.1. *Assuming $\Pi[u]$ is analytical in u, and letting $u = 0$ be an equilibrium configuration defined by the condition:*

$$\Pi'[0]\delta u = 0 \qquad (1.11)$$

there exists a neighborhood of the origin $\|u\| < \bar{\epsilon}$ (with $\bar{\epsilon} > 0$) where the following holds:

1. *There are no other equilibrium states with $\Pi[u] \neq \Pi[0]$.*
2. *Let $\gamma[\rho]$ and $\Gamma[\rho]$ be the scalar functions defined by the conditions:*

$$\begin{aligned}
\gamma[\rho] &= min\{\Pi[u], \|u\| = \rho\} \\
\Gamma[\rho] &= max\{\Pi[u], \|u\| = \rho\}
\end{aligned} \qquad (1.12)$$

both functions $\gamma[\rho]$ and $\Gamma[\rho]$ are monotonic.

Stability theorems We report here the two basic theorems on stability strictly following the form given in Koiter (1978).

Theorem 1.2 (Stability Theorem). *If the Total Potential Energy is positive definite in $u = 0$, i.e. $\gamma[\rho]$ is increasing monotonically, the equilibrium configuration $u = 0$ is stable.*

Proof. Let ϵ be such that $0 < \epsilon < \bar{\epsilon}$ and $\delta > 0$ be defined by the condition $\Gamma[\delta] = \gamma[\epsilon]$. For every trajectory starting from $\|u_0\| < \delta$, $\dot{u}_0 = 0$ we have:

$$E[t] \equiv \Pi[u[t]] + T[\dot{u}[t]] \leq E[0] \equiv \Pi[u_0] < \Gamma[\delta] = \gamma[\epsilon] \qquad (1.13)$$

and then, since $T[\dot{u}[t]]$ is positive definite,

$$\Pi[u[t]] < \gamma[\epsilon] \;, \quad \forall t > 0 \tag{1.14}$$

That implies $||u[t]|| < \epsilon$, for any $t > 0$. In fact, if assuming, for some $t > 0$, $||u|| \geq \epsilon$ we obtain the absurd

$$\Pi[u] \geq \gamma[\rho \geq \epsilon] \geq \gamma[\epsilon] \tag{1.15}$$

For dynamic perturbation ($\dot{u}_0 \neq 0$), the proof can be easily extended by making

$$\Gamma[\delta] + T[\dot{u}_0] = \gamma[\epsilon] \tag{1.16}$$

$\square$

Theorem 1.3 (Instability Theorem). If the total potential energy is not positive defined in $u = 0$, in the sense that $\gamma[\rho]$ is decreasing monotonically, the equilibrium configuration $u = 0$ is unstable.

Proof. Let $\delta > 0$, however small, so an initial disturbance u_0 exists i.e.

$$\Pi[u_0] = \gamma[\rho_0] < 0 \;, \quad \rho_0 \equiv ||u_0|| < \delta \tag{1.17}$$

Since

$$\Pi[u[t]] = E[t] - T[t] \leq E[0] \equiv \Pi[u_0] < 0 \tag{1.18}$$

any points of the trajectory starting from u_0 cannot be in static equilibrium. The trajectory is then characterized by a total potential energy

$$E[t] = \Pi[t] + T[t] \tag{1.19}$$

decreasing in a monotonic way. As a consequence, the function $\Pi[t] \leq E[t]$ is not bounded from below.

In these conditions, for every chosen $\epsilon < \bar{\epsilon}$, we will surely obtain $||u[t]|| > \varepsilon$ for some t ¿ 0. In fact, if we negate this occurrence, we would have

$$\Pi[u[t]] \geq \gamma[\rho] > \gamma[\epsilon] > \gamma[\bar{\epsilon}] \tag{1.20}$$

that is absurd, $\Pi[u[t]]$ being unbounded from below. $\square$

1.2   The critical equilibrium method

This section intends to make some links to traditional concepts and procedures. We discuss about linearized stability and the real meaning of the Euler method of critical equilibrium.

Taylor series expansion of the T.P.E.   Since $\Pi[u]$ is analytical, it can be expanded in Taylor series from the equilibrium configuration $u = 0$. Denoting Frechèt differentiations with primes and using the compact notation suggested by Budiansky (1974), we obtain:

$$\Pi[u] = \Pi_0 + \Pi_0'u + \frac{1}{2}\Pi''u^2 + \frac{1}{6}\Pi'''u^3 + \cdots \tag{1.21}$$

where $\Pi_0 = 0$ (for the hypothesis) and $\Pi_0'u = 0$ (from the equilibrium). Representing $\Pi[u]$

$$\Pi[u] \equiv \Phi[u] - (p_0 + p)u \tag{1.22}$$

as a sum of the *strain energy* contribution $\Phi[u]$ and the *load work* contribution $(p_0 + p)u$, the latter being bilinear in the load $(p_0 + p)$ and in the displacement u, the expansion provides:

$$\Pi[u] = \frac{1}{2}\Phi_0''u^2 + \frac{1}{6}\Phi_0'''u^3 + \cdots - pu \tag{1.23}$$

p_0 being the load in equilibrium with $u = 0$.

Linearized solution   If we assume u to be opportunely small, the high order terms $\frac{1}{6}\Phi_0'''u^3 + \cdots$ could be neglected with respect to the second order one $\Phi_0''u^2$.
Then the equilibrium configuration following from a small increase p of the load can be obtained by the *linearized condition*:

$$\Pi[u] \approx \frac{1}{2}\Phi_0''u^2 - pu = stat. \tag{1.24}$$

which obviously provides:

$$\Phi_0''u\delta u = p\delta u \tag{1.25}$$

We call this equation *linearized equation for small displacements*. It is worth considering the following:

1. The linearized solution corresponds to the assumptions:
   a) The equilibrium is imposed in the *deformed* configuration u.
   b) The geometrical quantities used in the analysis are evaluated (in an approximate way) by using the hypothesis of small displacements.
2. It is possible to show that the relative error on the solution is bounded by the worst ratio

$$max\left\{\frac{\frac{1}{6}\Phi_0'''u^3 + \cdots}{\frac{1}{2}\Phi_0''u^2}\right\} \tag{1.26}$$

between the neglected terms $(\frac{1}{6}\Phi_0'''u^3 + \cdots)$ and the terms $(\frac{1}{2}\Phi_0''u^2)$ kept in the equation.

Critical equilibrium   The presence, in Taylor's expansion of the T.P.E., of a term of second order $\frac{1}{2}\Phi_0''u^2 > 0$, $\forall u$ assures the stability. Conversely, if for some u we have $\frac{1}{2}\Phi_0''u^2 < 0$, that implies instability. Therefore, the critical situation

$$\exists v: \quad \frac{1}{2}\Phi_0''u^2 = \begin{cases} 0 & \text{if } u = v \\ > 0 & \text{if } u \neq v \end{cases} \tag{1.27}$$

provides a heuristic 'critical' condition for loss of stability.

We can rewrite the previous condition in the form:

$$\min\left\{\frac{1}{2}\Phi''u^2\right\} = 0 \quad , \quad \forall u \; : \; ||u|| = 1 \tag{1.28}$$

i.e., using the Lagrange multiplier techniques and expressing the norm in the form $||u||^2 := Mu^2$:

$$\frac{1}{2}\Phi''u^2 - \gamma(Mu^2 - 1) = \text{stationary} \tag{1.29}$$

By denoting by v its solution, the condition provides:

$$\Phi_0''v\delta u - \gamma Mv\delta u = 0 \; , \; \forall \delta u \tag{1.30}$$

By choosing $\delta u = v$ we obtain $\gamma = \Phi''v^2 = 0$. Then the critical condition implies:

$$\Phi_0''v\delta u = 0 \tag{1.31}$$

Apart from the presence of the external load $p\delta u$, the previous condition looks identical to the linearized equation seen before, relating small increments of displacements u to small increments of load p. This could heuristically suggest the following interpretation: *Loss of stability occurs when the structure reaches a critical equilibrium configuration u_0 from which small displacements $v \neq 0$ are possible, according to the linearized equilibrium equation, in presence of a null variation of the load ($p = 0$).*

Euler's method of critical equilibrium   Letting $p[\lambda]$ be a load path and $u[\lambda]$ a known reference solution, the search of the critical equilibrium condition is usually performed in the following way:

1. Starting from the configuration $u[\lambda]$, consider a near deformed configuration $u[\lambda] + v$.

2. Write the equilibrium conditions referring to the deformed configuration but within the hypothesis of *small* deviation.

3. Verify if this deviation satisfies the equilibrium in the absence of load variations.

We know that in simple contexts, such as proportional loading, the critical condition results in an eigenvalue equation providing both the *critical mode* v_c and the *critical load parameter* λ_c.

Euler's beam problem is a simple but good example for the implementation of the method. The equilibrium conditions are directly obtained from an inspection of the problem:

$$M + \lambda w = 0 \quad , \quad M := EJ\chi \tag{1.32}$$

and, according to small displacements kinematics ($\chi \approx w,_{xx}$), can be written as

$$EJw,_{xx} + \lambda w = 0 \; , \; w[0] = 0 \; , \; w[\ell] = 0 \tag{1.33}$$

which provides two solutions:

$$w = 0 \; \forall \lambda \quad \text{and} \quad w[x] = \bar{w}\sin\left(\frac{\pi x}{\ell}\right) \; , \; \lambda = \frac{\pi^2 EJ}{\ell^2} \; , \; \forall \bar{w} \neq 0 \tag{1.34}$$

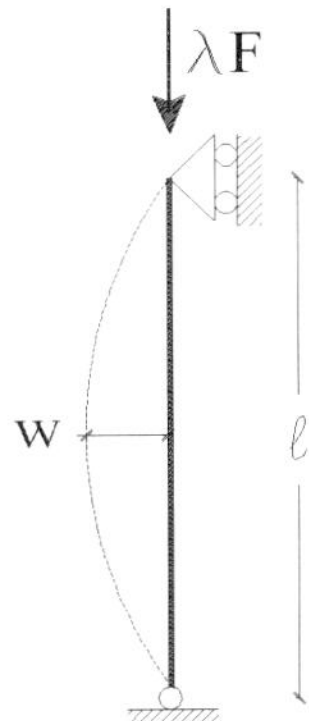

Figure 1. Euler's beam problem

We can also refer, and this will be convenient in the general case, to the strain energy which, by the same kinematic hypotheses, can be written

$$\Phi := \frac{1}{2} \int_0^\ell \left\{ EA\varepsilon^2 + EJ\chi^2 \right\} dx; \quad \text{where} \quad \begin{cases} \varepsilon := u,_x + \frac{1}{2}w,_x^2 & , \quad N := EA\varepsilon \\ \chi := w,_{xx} & , \quad M := EJ\chi \end{cases} \tag{1.35}$$

We obtain

$$\Phi'' \dot{u}\delta u := \int_0^\ell \left\{ EA\dot{u},_x\,\delta u,_x + EJ\dot{w},_{xx}\,\delta w,_{xx} + N\dot{w},_x\,\delta w,_x \right\} dx = 0 \tag{1.36}$$

that provides the solution:

$$\dot{N} := EA\dot{u},_x = 0; \quad \text{and} \quad EJw,_{xx} + Nw = 0 \quad , \quad w[0] = 0\,,\ w[\ell] = 0\,,\ N = -F \tag{1.37}$$

coinciding with that obtained by the direct equilibrium approach.

Implementation in large structures Real structures are usually to complex too be solved by hand–driven procedures. However, the approach already discussed can be used if combined with computational tools such as that provided by the so–called *finite element method*.

The analysis is usually performed in two steps:

1. A linear solution for a proportional loading $\mathbf{p}[\lambda] = \lambda\hat{\mathbf{p}}$

$$\mathbf{u}[\lambda] = \lambda\hat{\mathbf{u}} \quad , \quad \boldsymbol{\sigma}[\lambda] = \lambda\hat{\boldsymbol{\sigma}} \tag{1.38}$$

is known in advance or computed through a linear analysis:

$$\mathbf{K}_0\hat{\mathbf{u}} = \hat{\mathbf{p}} \tag{1.39}$$

where the *initial stiffness matrix* $\mathbf{K}_0$ is defined by the energy equivalence:

$$\delta\mathbf{u}^T\mathbf{K}_0\hat{\mathbf{u}} \equiv \Phi_0''\hat{u}\delta u \tag{1.40}$$

2. Displacements $\mathbf{u}[\lambda]$ are considered so small that the geometry of the structure could be *frozen* to the initial unloaded configuration and the stress/load relationship could be considered to be linear. By this assumption, the *tangent stiffness matrix* $\mathbf{K}_t[\lambda]$ defined by the energy equivalence

$$\delta\mathbf{u}^T\mathbf{K}_t[\lambda]\hat{\mathbf{u}} \equiv \Phi''[\lambda]\hat{u}\delta u \tag{1.41}$$

can be written in the form

$$\mathbf{K}_t[\lambda] := \mathbf{K}_0 + \lambda\mathbf{K}_g \tag{1.42}$$

where $\mathbf{K}_0$ refers to the initial unstressed configuration and coincides with the usual linear stiffness matrix and the *geometric matrix* $\mathbf{K}_g$ accounts for the presence of stresses $\boldsymbol{\sigma}[\lambda]$ varying linearly with λ. The Euler critical condition then reduces to a standard eigenvalue problem

$$\mathbf{K}_0\mathbf{v} + \lambda\mathbf{K}_g\mathbf{v} = 0 \tag{1.43}$$

which is easily solved by standard numerical algorithms (e.g., by the inverse power method).

Some remarks about the Euler method The Euler method has been justified according the following (heuristic) assumptions:
1. The condition $\Phi''v_c\delta u = 0$ corresponds to the loss of the stability.
2. At critical load, a displacement increase $\dot{v} \neq 0$ occurs for a null increase of the load.

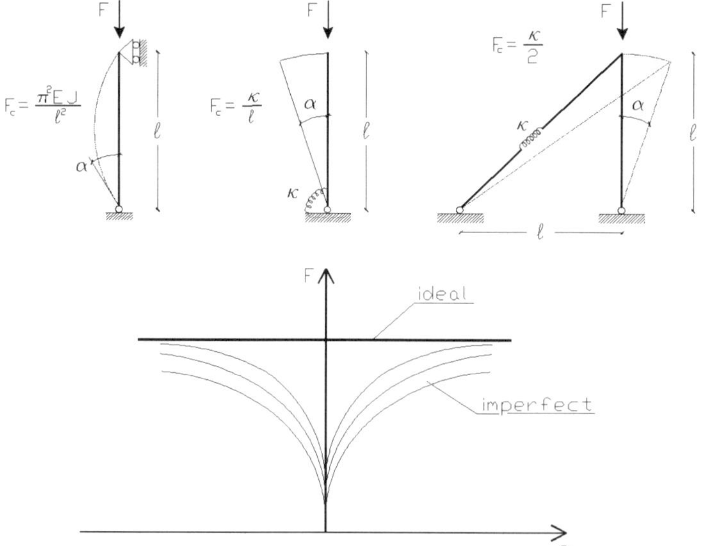

Figure 2. Equilibrium paths recovered by Euler's method

According to these assumptions, the critical equilibrium condition could be considered as a crisis condition for the structure, the critical load as representative of its carrying capacity and all the structures shown in fig.2 would be characterized by the same qualitative load–displacement path. Imperfections, if present, would provide only a rounding of the equilibrium path but not a variation of the carrying capacity of the structure.

Consider, however, that the assumptions made derive from viewing the critical condition as a linearized equilibrium condition, associating a *small* increase in the displacement with a null increase in the load. But the linearization requires the neglected terms $(\frac{1}{6}\Phi'''u^3 + \cdots)$ be considered small with respect to the second order terms $(\frac{1}{2}\Phi''u^2)$, we maintain in the equation. This is not possible in the case $\Phi''v^2 = 0$ we are actually considering. Therefore we can't really assert that the deviation, if possible, will be characterized by a null increase in the load.

A better understanding of the real meaning of the critical equilibrium method requires some insights into the bifurcation theory, without introducing a–priori linearizations of the problem. Before a more general discussion we will examine here two simple test cases:

Example 1   We start considering the cantilever beam of fig.3. From the equilibrium we have

$$F\ell\sin(\alpha) = k\alpha \tag{1.44}$$

That provides two solutions:

$$\alpha = 0 \, , \, \forall F \quad \text{and} \quad F = \frac{k\alpha}{\ell\sin(\alpha)} \, , \, \alpha \neq 0 \tag{1.45}$$

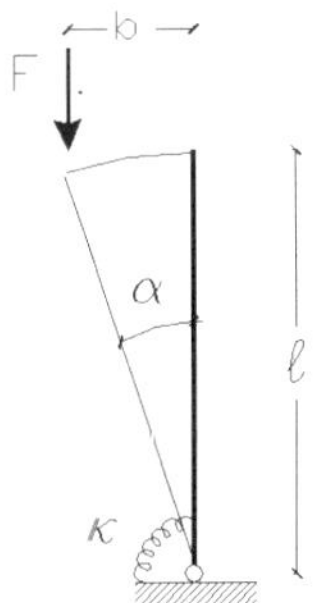

Figure 3. Test 1

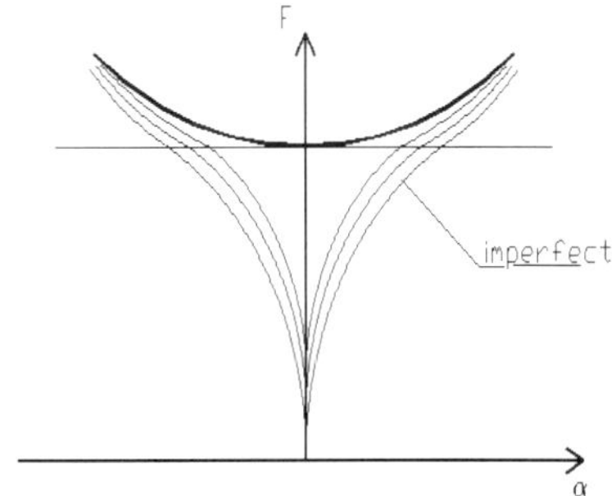

Figure 4. Test 1 - equilibrium paths

Expanding the latter in Taylor series we obtain

$$F \approx \frac{k}{l}(1 + \frac{1}{6}\alpha^2) \tag{1.46}$$

that provides a negligible increase of the load for small values of α. Furthermore note that the critical value $F_c = k/\ell$, while not representing a real limit for the structure, is of technical relevance.

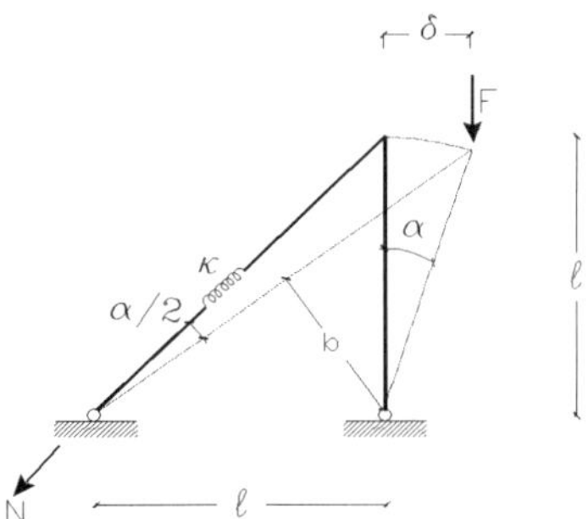

Figure 5. Test 2

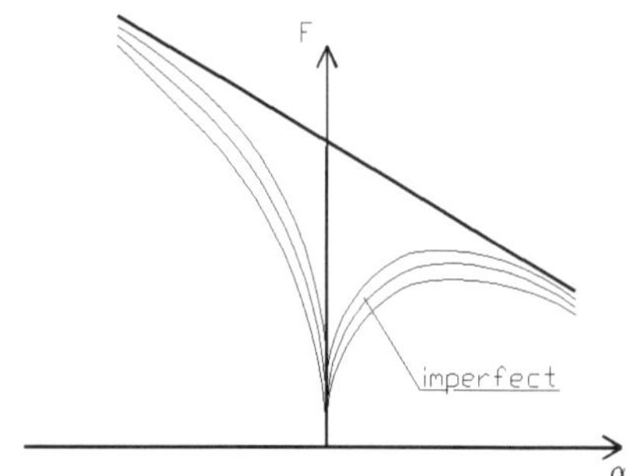

Figure 6. Test 2 - equilibrium paths

Example 2 Let's consider now the simple arc of fig.5
From the equilibrium we have:

$$F\ell\sin(\alpha) = N\ell\sin(\frac{\pi}{4} - \frac{\alpha}{2}) \qquad (1.47)$$

where, by the spring constitutive equation

$$N = k\frac{\Delta\ell}{\ell\sqrt{2}} = \frac{k}{\sqrt{2}}cos(\frac{\pi}{4} - \frac{\alpha}{2}) \qquad (1.48)$$

By these, expanding in Taylor series and simplifying, we obtain

$$F(\alpha - \frac{1}{6}\alpha^3 + \cdots)l = \frac{k}{2}(\alpha - \frac{3}{4}\alpha^2 + \cdots) \qquad (1.49)$$

which gives the 2 solutions:

$$\alpha = 0 \,, \ \forall F \quad \text{and} \quad F = \frac{k}{2}(1 - \frac{3}{4}\alpha + \cdots) \,, \ \alpha \neq 0 \qquad (1.50)$$

The latter shows an unaspected descending behaviour, where a decrease in the load is produced for positive values of α. This is relevant from a technical point of view, because the structure exploits *imperfection sensitivity,* that is, in the presence of imperfections the carrying capacity of the structure can be noticeably inferior to the critical value $F = k/2$.

1.3 Bifurcation theory

Equilibrium path Let a structure be subjected to a proportional load:

$$p[\lambda] = \lambda\hat{p} \qquad (1.51)$$

The equilibrium condition, expressed as a stationarity condition of T.P.E.

$$\Pi[u; \lambda] \equiv \Phi[u] - \lambda\hat{p}u = stat. \qquad (1.52)$$

or in the equivalent form of virtual work equation:

$$\Phi'[u]\delta u - \lambda \hat{p}\delta u = 0 \ , \quad \forall \delta u \tag{1.53}$$

defines a curve eventually composed of several branches (sometimes intersecting).

For a better understanding of this equation we have to consider that:

1. u is an element of the kinematically compatible displacement set $\mathcal{U}$; δu is an element of the space $\mathcal{T}$ tangent to $\mathcal{U}$ in u.

2. In the presence of nonlinear constraints (external or internal) $\mathcal{U}$ is a nonlinear manifold and then $\mathcal{T}$ depends on the considered configuration u.

In the following we will suppose $\mathcal{U}$ to be *linear* and then $\mathcal{T}$ to be *constant*. It is worth mentioning that this assumption is actually needed when we want to operate through finite element discretization, for practical reasons. However it prevents the use of internal or external nonlinear constraints (such as axial incompressibility, ..). A simple example of this will be shown in section 2.3.

Asymptotic representation   The local representation of the equilibrium curve starting from the point (u_0, λ_0) can expressed by the asymptotic expansion:

$$\begin{cases} u & := u[\xi] = u_0 + \dot{u}_0\xi + \frac{1}{2}\ddot{u}_0\xi^2 + \cdots \\ \lambda & := \lambda[\xi] = \lambda_0 + \dot{\lambda}_0\xi + \frac{1}{2}\ddot{\lambda}_0\xi^2 + \cdots \end{cases} \tag{1.54}$$

where the quantities $(\dot{u}_0, \ddot{u}_0, \cdots, \dot{\lambda}_0, \ddot{\lambda}_0, \cdots)$ are obtained through the perturbation equations:

$$\Phi_0''\dot{u}_0\delta u - \dot{\lambda}_0\hat{p}\delta u = 0$$

$$\Phi_0''\ddot{u}_0\delta u + \Phi_0'''\dot{u}_0^2\delta u - \ddot{\lambda}_0\hat{p}\delta u = 0 \tag{1.55}$$

$$\Phi_0''\dddot{u}_0\delta u + 3\Phi_0'''\dot{u}_0\ddot{u}_0\delta u + \Phi_0''''\dot{u}_0^3\delta u - \dddot{\lambda}_0\hat{p}\delta u = 0$$

Note that the use of the asymptotic expansions requires $u[\xi]$ and $\lambda[\xi]$ to be (possibly smooth) analytical functions in the region of interest. However, the asymptotic representation is possible, with a suitable choice of the parameter ξ, either in the regular zones of the curve or in correspondence to critical zones (near to *limit* or *bifurcation* points).

Limit and bifurcation points   Points (u_c, λ_c), defined on the equilibrium curve by the condition:

$$\Phi_c''\dot{v}_c\delta u = 0, \quad \forall \delta u \in \mathcal{T} \tag{1.56}$$

are called *critical*. In this case condition

$$\Phi_c''\dot{u}_c\delta u - \dot{\lambda}_c\hat{p}\delta u = 0, \quad \forall \delta u \in \mathcal{T} \tag{1.57}$$

can define a non–unique solution $(\dot{u}, \dot{\lambda})$. We have two cases:

1. If $\hat{p}\dot{v}_c \neq 0$, we obtain a unique tangent:

$$\dot{u}_c = \dot{v}_c \ , \quad \dot{\lambda}_c = 0 \tag{1.58}$$

corresponding to a regular *limit point*.

2. If $\hat{p}\dot{v}_c = 0$, the tangent is defined by the combination:

$$\dot{\lambda}_c \hat{u}_c + \beta \dot{v}_c \tag{1.59}$$

where $\dot{\lambda}_c \hat{u}_c$ represents a particular solution and the scalar quantity β can be obtained from the high–order perturbation equation:

$$\Phi_0'' \ddot{u}_0 \delta u + \Phi_0''' \dot{u}_0^2 \delta u - \ddot{\lambda}_0 \hat{p} \delta u = 0, \quad \forall \delta u \in \mathcal{T} \tag{1.60}$$

Making $\delta u = \dot{v}_c$, we obtain the second order equation in β:

$$\dot{\lambda}_c^2 \Phi_c''' \hat{u}^2 \dot{v}_c + 2\dot{\lambda}_c \beta \Phi_c''' \hat{u}_c \dot{v}_c^2 + \beta^2 \Phi_c''' \dot{v}_c^3 = 0 \tag{1.61}$$

If we choose $\hat{u}_c$ so that $\Phi_c''' \hat{u}^2 \dot{v}_c = 0$, the equation provides 2 solutions:

$$\left\{ \begin{array}{rcl} \dot{u} &=& \dot{\lambda}_c \hat{u}_c \\ \dot{\lambda} &=& \dot{\lambda}_c \end{array} \right\}, \qquad \left\{ \begin{array}{rcl} \dot{u} &=& \dot{\lambda}_c \hat{u}_c + \dot{v}_c \\ \dot{\lambda} &=& -\frac{1}{2} \Phi_c''' \dot{v}_c^3 / \Phi_c''' \hat{u}_c \dot{v}_c^2 \end{array} \right\} \tag{1.62}$$

In this case we have a *bifurcation point*.

Fundamental and bifurcated paths Consider a bifurcation point belonging to two paths and let the first one, which we call *fundamental path*, be defined in explicit form as:

$$u^f[\lambda] = u_c + (\lambda - \lambda_c)\hat{u}_c + \frac{1}{2}(\lambda - \lambda_c)^2 \hat{\hat{u}} + \cdots \tag{1.63}$$

The second path, which we call *bifurcated path*, can be obtained by means of a perturbation technique. We obtain:

$$\left\{ \begin{array}{rcl} u[\lambda] &=& u^f[\lambda] + \xi \dot{v}_c + \frac{1}{2}\xi^2 \ddot{v}_c + \cdots \\ \lambda &=& \lambda_c + \xi \dot{\lambda}_c + \frac{1}{2}\xi^2 \ddot{\lambda}_c + \cdots \end{array} \right. \tag{1.64}$$

where the quantities are defined through the relations:

$$\Phi_c'' \dot{v}_c \delta u = 0 , \quad \forall \delta u \in \mathcal{T}$$
$$\Phi_c'' \ddot{v}_c \delta u + \Phi_c''' \dot{v}_c^2 \delta u = 0 , \quad \forall \delta u \in \mathcal{T} \tag{1.65}$$

$$\cdots$$

$$\dot{\lambda}_c = -\frac{1}{2} \frac{\Phi_c''' \dot{v}_c^3}{\Phi_c''' \hat{u}_c \dot{v}_c^2} \tag{1.66}$$

$$\ddot{\lambda}_c = -\frac{1}{3} \frac{\Phi_c'''' (\dot{v}_c^4 + 3\dot{\lambda}_c \hat{u}_c \dot{v}_c^3 + 3\dot{\lambda}_c^2 \hat{u}_c^2 \dot{v}_c^2)}{\Phi_c''' \hat{u}_c \dot{v}_c^2} - \frac{\Phi_c''' (\dot{v}_c^2 \ddot{v}_c + \dot{\lambda}_c \hat{u}_c \dot{v}_c \ddot{v}_c + \dot{\lambda}_c^2 \hat{\hat{u}}_c \dot{v}_c^2)}{\Phi_c''' \hat{u}_c \dot{v}_c^2} \tag{1.67}$$

$$\cdots$$

It is worth noting that, because of the singularity of $\Phi_c''(.)(.)$ in direction $\dot{v}_c$, the component of $\ddot{v}_c$ parallel to $\dot{v}_c$ is not uniquely determined. Different values of that component correspond to different (but equivalent in asymptotic sense) choices for the arbitrary expansion parameter ξ.

2 A first insight into asymptotic post-buckling analysis

This section presents a computational–oriented approach to the asymptotic analysis of slender elastic structures that has been initially proposed and investigated in Casciaro, Aristodemo (1977) and Brezzi et al. (1986). For a further insight, the reader could also refer to Casciaro et al. (1991), Casciaro et al. (1992), Casciaro et al. (1998).

2.1 The simple linear algorithm

Basic elements of the analysis The bifurcation theory can be a powerful tool to determine the essential features of the nonlinear behaviour of an elastic slender structure defined by the condition:

$$\Phi[u] - \lambda \hat{p}u = stat. \tag{2.1}$$

However, the following basic elements are required:

1. *The fundamental equilibrium path*, expressed by the analytical function $u^f[\lambda]$, solution of the problem:

$$\Phi'[u^f[\lambda]]\delta u - \lambda \hat{p}\delta u = 0 , \quad \forall \delta u \in \mathcal{T} \tag{2.2}$$

2. *The bifurcation point* $u_b = u^f[\lambda_b]$ defined by the condition:

$$\Phi''_b \dot{v}_b \delta u = 0 , \quad \forall \delta u \in \mathcal{T} \tag{2.3}$$

Since $u^f[\lambda]$ is analytical, (u_b, λ_b) corresponds to a bifurcation point.

3. *The bifurcated path* defined by the expansion:

$$\begin{aligned}
\lambda &= \lambda_b + \xi \dot{\lambda}_b + \cdots \\
u[\xi] &= u^f[\lambda] + \xi \dot{v}_b + \cdots
\end{aligned} \tag{2.4}$$

where

$$\dot{\lambda}_b = -\frac{1}{2}\frac{\Phi'''_b \dot{v}_b^3}{\Phi'''_b \hat{u}_b \dot{v}_b^2} \tag{2.5}$$

We will examine each of these in detail.

How to define the fundamental path A convenient way of describing the fundamental path within an automatic solution process, is to define it as Taylor extrapolation from the initial given equilibrium point $(u_0, \lambda = 0)$, that is, using superscript "^" for denoting derivatives with respect to λ,

$$u^f[\lambda] = u_0 + \lambda \hat{u}_0 + \cdots \tag{2.6}$$

$\hat{u}_0$ being provided by the 1st-order asymptotic equation

$$\Phi''_0 \hat{u}_0 \delta u = \hat{p}\delta u , \quad \forall \delta u \in \mathcal{T} \tag{2.7}$$

Operating in this way, the path will be characterized by the out of balance

$$\tilde{p}[\lambda]\delta u \equiv (\Phi'[u^f[\lambda]] - \lambda \hat{p})\delta u \tag{2.8}$$

in general non–null. Assuming a linear extrapolation, it can be evaluated by the expression

$$\tilde{p}[\lambda]\delta u = \frac{1}{2}\lambda^2\Phi_b'''\hat{u}^2\delta u + \cdots \tag{2.9}$$

which is obtained from the 2nd–order perturbation equations. We will take into account this imbalance by an a–posteriori treatment as an *implicit imperfection*.

Note that, from a computational point of view, the analysis requires the solution of the linear–elastic problem:

$$\mathbf{K}_0\hat{\mathbf{u}} = \hat{\mathbf{p}} \tag{2.10}$$

$\mathbf{K}_0$ being the initial stiffness matrix of the structure and $\mathbf{u}$ and $\mathbf{p}$ the nodal displacement and load vectors.

How to search for the bifurcation point   Along the path $u^f[\lambda]$ we have to search for the first bifurcation point, characterized by the condition:

$$\Phi_b''\dot{v}_b\delta u = 0 , \quad \forall\delta u \in \mathcal{T} , \; \dot{v} \neq 0 \tag{2.11}$$

that is, in matrix form

$$\mathbf{K}_t[\lambda]\dot{\mathbf{v}} = 0 , \; \dot{\mathbf{v}} \neq 0 \tag{2.12}$$

By referring to the Taylor expansion:

$$\Phi''[u^f[\lambda]] = \Phi_0'' + \lambda\Phi_0'''\hat{u}_0 + \cdots \tag{2.13}$$

and discarding high–order terms, the bifurcation condition can be reduced to a linear eigenvalue problem expressed in matrix form as

$$(\mathbf{K}_0 + \lambda\mathbf{K}_1)\dot{\mathbf{v}} = 0 \tag{2.14}$$

This linear eigenvalue equation can be solved by means of standard methods, like the standard inverse power method (an even more efficient and general method will be discussed in section 2.3).

Note that this phase has the same computational cost as an ordinary eigenvalue analysis. If we use the inverse power method, the assembly and decomposition of the initial matrix $\mathbf{K}_0$, which represents the more onerous part of the whole process, has already been performed in the previous step. Furthermore note that, until now, the whole analysis does not present any differences with respect to an ordinary linearized stability analysis.

How to determine the bifurcated path   Starting from the known bifurcation point, the evaluation of the bifurcated path

$$\lambda = \lambda_b + \xi\dot{\lambda}_b + \cdots$$
$$u[\xi] = u^f[\lambda] + \xi\dot{v}_b + \cdots \tag{2.15}$$

requires, if we consider (as the fundamental path) a linear extrapolation, only the evaluation of $\dot\lambda_b$ from the equation:

$$\dot\lambda_b = -\frac{1}{2}\frac{\Phi_b'''\dot v_b^3}{\Phi_b'''\hat u_b\dot v_b^2} \tag{2.16}$$

It is an explicit formula for expressing the ratio between two integrals of known functions which can be obtained (at least) by means of Gauss numerical integration. Then, the computational cost required by this phase, representing the only difference with respect to a standard application of the Euler method for linearized stability analysis, is quite negligible.

Implicit imperfection and imperfect path We now have to take into account the presence of the imbalance $\tilde p[\lambda]$ along the fundamental path. Consider that:

1. The imbalance $\tilde p[\lambda]$ can be considered as an *imperfection* in the modeling with respect to the *idealized perfect scheme* we have analyzed.
2. The behaviour of the *real imperfect structure*, in any case, remains strictly related to that of the *perfect scheme*.

Therefore, an approximate evaluation of the equilibrium path of the real structure can be obtained by assuming a displacement field in the form

$$u[\lambda,\xi] = u^f[\lambda] + \xi\dot v_b \tag{2.17}$$

provided by the perfect scheme and by determining the relation $\lambda \div \xi$ from a Galerkin solution of the equilibrium problem:

$$(\Phi'[u[\lambda,\xi] - \lambda\hat p)\dot v_b = 0 \tag{2.18}$$

After some manipulations we obtain:

$$(\lambda - \lambda_b)\xi\Phi_b'''\hat u_b\dot v_b^2 + \frac{1}{2}\xi^2\Phi_b'''\dot v_b^3 + \frac{1}{2}\lambda^2\Phi_b'''\hat u_b^2\dot v_b = 0 \tag{2.19}$$

that is:

$$\lambda + \lambda^2\mu/\xi = \lambda_b + \xi\dot\lambda_b \tag{2.20}$$

having introduced the *imperfection factor*

$$\mu = -\frac{1}{2}\frac{\Phi_b'''\hat u_b^2\dot v_b}{\Phi_b'''\hat u_b\dot v_b^2} \tag{2.21}$$

Remarks on the linear algorithm The perturbation linear algorithm does not require particular manipulations, from both the theoretical and computational points of view, with respect to an ordinary analysis based on linearized stability.

The new features (calculation of $\dot\lambda_b$ and μ) correspond to the solution of explicit formulas requiring the calculation of integrals of known functions.

Despite that, the provided solution is extremely accurate. A theoretical analysis performed in Brezzi et al. (1986) provided the following results:

1. the error due to the projection in the *bifurcation manifold* $\lambda \div \xi$ is:

$$\begin{aligned} \approx \xi^3 &\quad : \text{in the description of the path} \\ \approx \xi^5 &\quad : \text{in the evaluation of the limit load} \end{aligned}$$

2. the error deriving from the use of an incomplete expansion, of order k ($k = 2$ in our case), of the equilibrium equation is

$$\begin{aligned} \approx \xi^{k+1} &\quad : \text{in the description of the path} \\ \approx \xi^{2k+2} &\quad : \text{in the evaluation of the limit load} \end{aligned}$$

Great accuracy is then expected ($err \approx \xi^5$) for the limit load value. In practical application, the relative error is unlikely to exceed some $1/1000$ (see for instance Casciaro et al. (1992)).

Additional imperfections   When analyzing a structure, it is difficult to characterize its geometry and loads exactly. In fact they are affected by a random distribution of small imperfections.

The presence of these, while preserving the general behaviour of the structure behaviour, changes some aspects of its response and often causes a reduction of the carrying capacity (imperfection sensitivity).

An evaluation of the imperfection sensitivity requires the analysis be fully repeated for any possible imperfection combinations, so for a relevant number of times. Such a necessity renders the approaches based on incremental–iterative analysis unsuitable for the purpose, because of their severe computational demand.

The asymptotic approach provides a natural way for performing an imperfection sensitivity analysis, because the presence of imperfections can be taken into account in a post–processing phase where only the final (scalar) equation in λ and ξ is considered and then, in practice, without any computational extra–cost.

Essentially we can consider two possible sources of imperfections: small geometrical differences in the real structure from the ideal theoretical scheme which is modeled in the analysis (*geometrical imperfections*), and small disturbances in the applied loads with respect to that theoretically foreseen in the analysis (*load imperfections*). Both can be easily treated by introducing an additional extra–term in the scalar equation (2.19):

1. The presence of a load imperfection $q = \lambda \tilde{q}$ can be treated by the additional term

$$\lambda \mu_q \quad , \quad \mu_q := -\frac{\tilde{q}\dot{v}_b}{\Phi_b''' \hat{u}\dot{v}^2} \tag{2.22}$$

2. The presence of an initial geometrical imperfection $\tilde{u}$ can be treated by the additional term

$$\lambda \mu_u \quad , \quad \mu_u := \frac{\Phi''' \hat{u}\tilde{u}\dot{v}}{\Phi''' \hat{u}\dot{v}^2} \tag{2.23}$$

Then, considering all imperfection types (implicit and additional) we obtain the following final formulas:

$$\begin{cases} \lambda + \lambda(\mu_q + \mu_u)/\xi + \lambda^2 \mu/\xi = \lambda_b + (\xi + \frac{4}{3}\mu_u)\dot{\lambda}_b \\ u = \tilde{u} + u^f[\lambda] + \xi \dot{v}_b \end{cases} \tag{2.24}$$

2.2 Implementation to plane frames

For a better insight into the linear asymptotic algorithm, it is convenient to discuss some relevant aspects of its computer implementation. Plane frames, provide a simple context particularly suitable for that purpose.

Technical beam theory A rational theory for one–dimensional modelling of beam elements is quite difficult (we will consider this later). For our present aim we can refer to the so called simplified technical beam theory which is defined by the strain energy expression

$$\Phi := \frac{1}{2} \int_0^\ell \{EA\varepsilon^2 + EJ\chi^2\}dx; \tag{2.25}$$

ε and χ being the axial and flexural components of strain, defined by

$$\varepsilon = \varepsilon_0 + u_{,x} + \frac{1}{2}w_{,x}^2$$
$$\chi = w_{,xx} \tag{2.26}$$

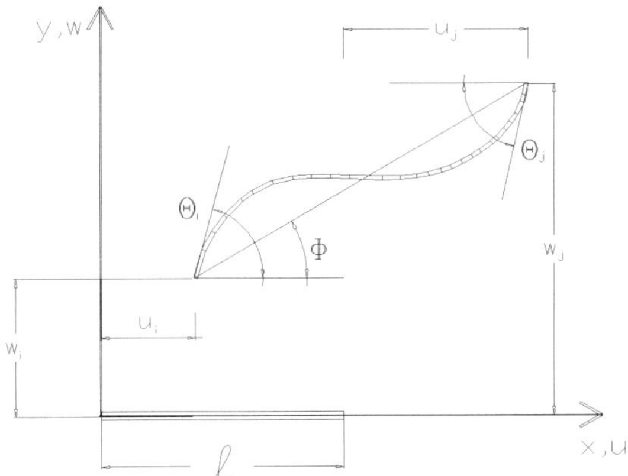

Figure 7. Plane beam

By combining the two equations, $\Phi[u]$ and expanding in Taylor series, we obtain:

$$\Phi[u] = \Phi_0 + \int_0^\ell N_0 u_{,x}\,dx + \frac{1}{2}\int_0^\ell \{EAu_{,x}^2 + EJw_{,xx}^2 + N_0 w_{,x}^2\}dx$$
$$+ \frac{1}{6}\int_0^\ell 3EAu_{,x}\,w_{,x}^2\,dx + \frac{1}{24}\int_0^\ell EAw_{,x}^4\,dx \tag{2.27}$$

where $N_0 := EA\varepsilon_0$ is the initial axial force.

Variations of the strain energy The previous assumptions provide the following energy variations:

- *First variation:*
  (needed for equilibrium equation)

$$\Phi' \delta u := \int_0^L N_0 \delta u,_x \, dx \qquad (2.28)$$

- *Second variation:*
  (needed for evaluating the fundamental path and the bifurcation point)

$$\Phi'' \dot{u} \delta u := \int_0^L \{ EA\dot{u},_x \, \delta u,_x + EJ\dot{w},_{xx} \, \delta w,_x \, x + N_0 \dot{w},_x \, \delta w,_x \} dx \qquad (2.29)$$

- *Third variations:*
  (needed for calculating $\dot{\lambda}_b$ and μ_b)

$$\Phi_0''' \dot{u}^3 := 3 \int_0^L EA\dot{u},_x \, \dot{w},_x^2 \, dx$$

$$\Phi_0''' \hat{u}\dot{u}^2 := \int_0^L EA\hat{u},_x \, \dot{w},_x^2 \, dx + 2 \int_0^L EA\dot{u},_x \, \dot{w},_x \, \hat{w},_x \, dx \qquad (2.30)$$

It is worth mentioning that the fundamental path is usually characterized by very small rotations (that is, $\hat{w},_x \approx 0$). So, we can assume

$$\Phi_0''' \hat{u}\dot{u}^2 := \int_0^L EA\hat{u},_x \, \dot{w},_x^2 \, dx \qquad (2.31)$$

This approximation could be useful within an at–hand solution process.

Some at-hand examples Asymptotic post–buckling analysis is substantially quite simple, not more difficult than ordinary linearized stability analysis, To emphasize this, it will be usefull to discuss here some simple test–problems suited to an easy at-hand solution.

We start by considering the simple arc of fig.8. The fundamental path is easily obtained directly from the equilibrium:

$$\alpha_1 = \alpha_2 = 0 \,, \quad N_1 = -F \,, \quad N_2 = 0 \qquad (2.32)$$

then we have

$$\hat{\alpha}_1 = \hat{\alpha}_2 = 0 \,, \quad \hat{N}_1 = -1 \,, \qquad \hat{N}_2 = 0 \qquad (2.33)$$

By simple geometric considerations, the critical mode will be defined by the rotations

$$\dot{\alpha}_1 = 1 \,, \quad \dot{\alpha}_2 = 1/2 \,, \qquad (2.34)$$

and the related axial strengths

$$\dot{N}_2 = k\dot{\alpha}_1 \,, \quad \dot{N}_1 = -k\dot{\alpha}_1/2 \qquad (2.35)$$

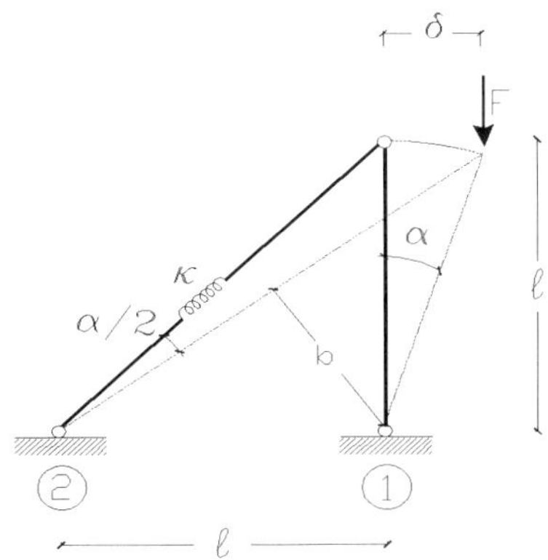
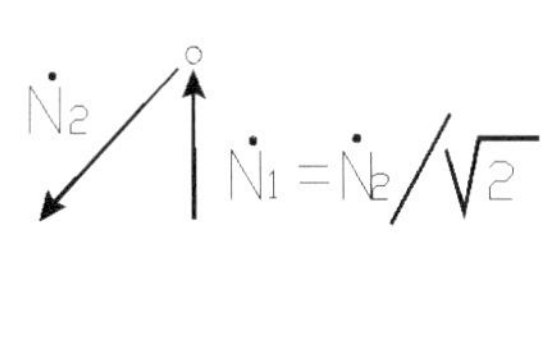

Figure 8. Example 1 : simple arc   **Figure 9.** Vertical equilibrium

the first being derived from the spring constitutive equation, the second by relaxing the internal constraint of axial inextensibility of the vertical bar. Imposing critical condition we obtain

$$F_b\dot\delta = \dot N b \Rightarrow F_b\dot\alpha_1 l = k\frac{\dot\alpha_1 l}{l\sqrt2}\frac{l}{\sqrt2} \implies F_b = k/2 \tag{2.36}$$

It is worth mentioning that, up to now, we have performed no more than a standard linearized stability analysis. Post–buckling analysis further requires the evaluation of the *post–critical slope* $\dot\lambda_b$. For doing so, we only need to compute the two scalar terms appearing in the numerator and denominator of its formula (2.16). From eq.2.30, we obtain

$$\Phi_b'''\hat u\dot v^2 = \int_0^l \hat N_1\dot\alpha_1^2 dx = -l$$

$$\Phi_b'''\dot v^3 = 3\int_0^l \dot N_1\dot\alpha_1^2 dx + \int_0^{l\sqrt2} \dot N_2\dot\alpha_2^2 dx = -\frac34 kl \tag{2.37}$$

and thus

$$\dot F = -\frac12\frac{1-\frac34 kl}{-l} = -\frac38 k \quad\Rightarrow\quad F = F_b + \dot F_b\xi = \frac k2(1-\frac34\alpha) \tag{2.38}$$

which implies

$$F = F_b + \dot F_b\xi = \frac k2\left(1-\frac34\alpha\right) \tag{2.39}$$

exactly recovering equation (1.49).

The second example refers to the simple two–bar frame in fig.10.

Critical analysis is quite easy. Its relevant results are reported in fig.10. we have from (2.30) ,

$$\Phi_b'''\hat u\dot v^2 = \int_0^\ell \hat N\dot w_{,x}^2\, dx \quad , \quad \Phi_b'''\dot v^3 = 3\int_0^\ell \dot N\dot w_{,x}^2\, dx \tag{2.40}$$

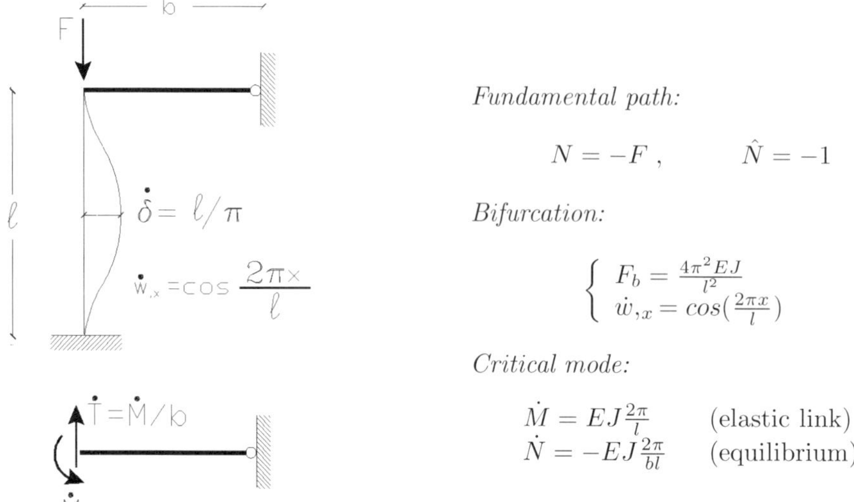

Fundamental path:

$$N = -F \,, \qquad \hat{N} = -1$$

Bifurcation:

$$\begin{cases} F_b = \dfrac{4\pi^2 EJ}{l^2} \\ \dot{w}_{,x} = \cos(\frac{2\pi x}{l}) \end{cases}$$

Critical mode:

$$\dot{M} = EJ\frac{2\pi}{l} \qquad \text{(elastic link)}$$
$$\dot{N} = -EJ\frac{2\pi}{bl} \qquad \text{(equilibrium)}$$

Figure 10. Example 2: simple frame

and the post–critical slope is easily recovered as

$$\dot{F}_b = -\frac{1}{2}\frac{\Phi_b'''\dot{v}^3}{\Phi_b''\hat{u}\dot{v}^2} = -\frac{1}{2}\frac{3\dot{N}}{\hat{N}} = -\frac{3}{2}EJ\frac{2\pi}{bl} = -\frac{3}{4}\frac{F_b l}{\pi b} \tag{2.41}$$

We then obtain the final path equation

$$F = F_b + \dot{F}_b\xi = F_b\left(1 - \frac{3}{4}\frac{\delta}{b}\right) \tag{2.42}$$

Finite element modelling The finite element method is based on three basic ideas:

1. The structure is considered as an assemblage of interconnected elements.

2. Two equivalent but different descriptions are used for both the static and kinematic quantities:
   - an *external* one chosen with the aim of making the connection laws between the elements and the expression of the external work as simple as possible;
   - an *internal* one chosen with the aim of making the expression of the element strain energy as simple as possible.

3. The strain energy and the external work of the overall structure is obtained by summation of all the element contributions.

   When referring to a framed structure, we can see it as an assemblage of beam elements and choose as external kinematical description the $[n \times 1]$ vectors $\mathbf{u}$ and $\mathbf{p}$ collecting displacements u_x, u_y and rotation φ, and nodal forces f_x, f_y and nodal couple m, respectively, for all the n nodes of the structure:

$$\mathbf{u} := \left\{ \begin{array}{c} u_{1x} \\ u_{1y} \\ \varphi_1 \\ \cdots \\ \varphi_n \end{array} \right\} \qquad \mathbf{p} := \left\{ \begin{array}{c} f_{1x} \\ f_{1y} \\ m_1 \\ \cdots \\ m_n \end{array} \right\} \tag{2.43}$$

The external load pu is so expressed by the scalar product

$$pu := \mathbf{p}^T \mathbf{u} \tag{2.44}$$

while the inter-element connection laws are identically satisfied by direct identification of the displacements of the element end–sections with those of the corresponding nodes.

Internal description   Apart from inessential rigid body motions, the element displacements can be viewed as the combination of the 4 natural modes shown in figure: rigid rotation φ_r, flexural symmetric φ_s, flexural hemisymmetric φ_e and extensional ε.

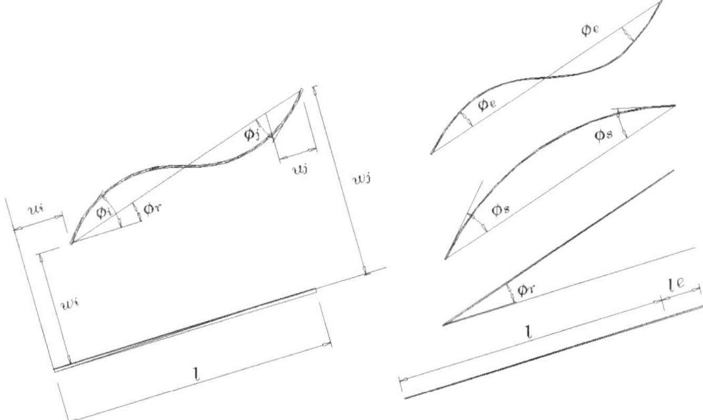

Figure 11. Natural modes for the beam element

When collecting the element natural modes in a local vector $\mathbf{u}_e$, we obtain:

$$\mathbf{u}_e := \left\{ \begin{array}{c} \varphi_r \\ \varphi_s \\ \varphi_e \\ \varepsilon \end{array} \right\} = \left[\begin{array}{cccccc} -c & -s & 0 & c & s & 0 \\ 0 & 0 & \frac{1}{2} & 0 & 0 & -\frac{1}{2} \\ -s & c & \frac{1}{2} & s & -c & \frac{1}{2} \\ s & -c & 0 & -s & c & 0 \end{array} \right] \bar{\mathbf{u}} = \mathbf{A}_e \bar{\mathbf{u}} \tag{2.45}$$

where vector $\bar{\mathbf{u}}$ collects the nodal element displacements in the global external description:

$$\bar{\mathbf{u}} := \{ u_{ix} , u_{iy} , \varphi_i , u_{jx} , u_{jy} , \varphi_j \}^T \tag{2.46}$$

i and j being the element end–nodes, and

$$c := \frac{\cos \alpha_j}{\ell_n} \quad , \quad s := \frac{\sin \alpha_j}{\ell_n} \tag{2.47}$$

express the initial beam orientation.

Finite element discretization   Assuming standard shape functions ($u[x]$ linear, $w[x]$ cubic) for expressing the internal element displacement as a function of their nodal values, the element contributions to the relevant energy variations are obtained as:

- Second energy variation:

$$\Phi_e'' \dot{u} \delta u = \mathbf{u}_e^T \mathbf{K}_e \mathbf{u}_e = \mathbf{u}_e^T [\mathbf{E}_e + N_0 \mathbf{G}_e] \mathbf{u}_e \tag{2.48}$$

where the element matrices $\mathbf{E}_e$ and $\mathbf{G}_e$ are given by:

$$\mathbf{E}_e = \begin{bmatrix} 0 & 0 & 0 & 0 \\ 0 & 4EJ/\ell & 0 & 0 \\ 0 & 0 & 12EJ/\ell & 0 \\ 0 & 0 & 0 & EA\ell \end{bmatrix} , \quad \mathbf{G}_e = \begin{bmatrix} \ell & 0 & 0 & 0 \\ 0 & \ell/3 & 0 & 0 \\ 0 & 0 & \ell/5 & 0 \\ 0 & 0 & 0 & 0 \end{bmatrix} \tag{2.49}$$

- Third variations:

$$\Phi_e''' \dot{u}^3 = 3\dot{N} \dot{\mathbf{u}}_e^T \mathbf{G}_e \dot{\mathbf{u}}_e$$
$$\Phi_e''' \hat{u} \dot{v}^2 = \hat{N} \dot{\mathbf{u}}_e^T \mathbf{G}_e \dot{\mathbf{u}}_e + 2\dot{N} \dot{\mathbf{u}}_e^T \mathbf{G}_e \hat{\mathbf{u}}_e \tag{2.50}$$

We have, obviously,

$$\Phi''' \dot{u}^3 := \sum_e \Phi_e''' \dot{u}^3 \quad , \quad \Phi''' \hat{u} \dot{v}^2 := \sum_e \Phi_e''' \hat{u} \dot{v}^2$$

$$\Phi'' \dot{u} \delta u := \sum_e \Phi_e'' \dot{u} \delta u = \dot{\mathbf{u}}^T \left[\sum_e \mathbf{A}_e^T \mathbf{K}_e \mathbf{A}_e \right] \delta \mathbf{u} \tag{2.51}$$

The latter, through the energy equivalence

$$\Phi'' \dot{u} \delta u \equiv \dot{\mathbf{u}}^T \mathbf{K} \delta \mathbf{u} \tag{2.52}$$

defines the assemblage lows for the stiffness matrix $\mathbf{K}$:

$$\mathbf{K} = \sum_e \mathbf{A}_e^T \mathbf{K}_e \mathbf{A}_e \tag{2.53}$$

which can be used for obtaining both fundamental mode $\hat{u}$ and buckling mode $\dot{v}$ as a solution of the matrix equations:

$$\mathbf{K}[0]\hat{\mathbf{u}} = \mathbf{p} \quad , \quad \mathbf{K}[\lambda]\dot{\mathbf{v}} = \mathbf{0} \tag{2.54}$$

Using exact shape functions   The use of the standard shape functions (linear–cubic) makes the analysis particularly simple. However these functions correspond to an exact solution for $u[x]$ and $w[x]$ only if $N = 0$ and gives a certain discretization error for $N \neq 0$.

The error can be reduced by refining the discretization, i.e. by subdividing each beam in a sufficient number of elements of smaller length, but it is much better to use the so called *exact shape functions* corresponding to the solutions of the *2nd* order equation:

$$\int_0^L \{EA\ddot{u}_{,x}\,\delta u_{,x} + EJ\ddot{w}_{,xx}\,\delta w_{,x}\,x + N\ddot{w}_{,x}\,\delta w_{,x}\}dx = 0 \tag{2.55}$$

with the appropriate boundary conditions. These are well known from classical books (e.g. see Timoshenko, Gere (....) or Livesley (1975)).

Using exact shape functions and making:

$$\sigma := \frac{N\ell^2}{EJ} \; , \quad \alpha := \sqrt{|\sigma|} \tag{2.56}$$

we obtain, after some algebraic manipulations, the following expression for element tangent stiffness matrix $\mathbf{K}_e$:

$$\mathbf{K}_e = \begin{bmatrix} N\ell & 0 & 0 & 0 \\ 0 & k_s EJ/\ell & 0 & 0 \\ 0 & 0 & k_e EJ/\ell & 0 \\ 0 & 0 & 0 & EA\ell \end{bmatrix} \tag{2.57}$$

where

$$k_s = \begin{cases} 2\alpha/\tan(\alpha/2) & \text{if } N < 0 \\ 4 + \sigma/3 & \text{if } N \approx 0 \\ 2\alpha/\tanh(\alpha/2) & \text{if } N > 0 \end{cases} \; , \quad k_e = \begin{cases} 4\sigma/(4 - k_s) & \text{if } N \neq 0 \\ 12 + \sigma/5 & \text{if } N \approx 0 \end{cases} \tag{2.58}$$

The contribution of the element to the *geometrical matrix* $\mathbf{K}_g$, which is required for computing third energy variations, becomes

$$\mathbf{G}_e = \begin{bmatrix} L & 0 & 0 & 0 \\ 0 & g_s L & 0 & 0 \\ 0 & 0 & g_e L & 0 \\ 0 & 0 & 0 & 0 \end{bmatrix} \tag{2.59}$$

where:

$$\begin{cases} \text{if } N < 0 \quad g_s = \dfrac{\alpha - \sin(\alpha)}{2\alpha \sin^2(\alpha/2)} \; , \quad g_e = \dfrac{-4 + \alpha^2 + 4\cos(\alpha) + \alpha\sin(\alpha)}{2(\alpha\cos(\alpha/2) - 2\sin(\alpha/2))^2} \\[3ex] \text{if } N > 0 \quad g_s = \dfrac{\alpha - \sinh(\alpha)}{2\alpha \sinh^2(\alpha/2)} \; , \quad g_e = \dfrac{-4 + \alpha^2 - 4\cosh(\alpha) + \alpha\sinh(\alpha)}{2(\alpha\cosh(\alpha/2) - 2\sinh(\alpha/2))^2} \\[3ex] \text{if } N \approx 0 \quad \; g_s = 1/3 - \sigma/30 \; , \quad g_e = 1/5 - \sigma/70 \end{cases} \tag{2.60}$$

It is worth noting that the use of exact shape functions makes the bifurcation search a nonlinear eigenvalue problem. We will see, however, an efficient iteration scheme can be used for that purpose. Therefore, a nonlinear eigenvalue search does not introduce additional difficulties with respect to a standard linearized analysis, while providing more accurate results.

2.3 Solution of the nonlinear eigenvalue problem

We will discuss here in some detail the solution of (2.11) which is required for determining the bifurcation point along the fundamental path $u^f[\lambda]$. A fast and robust numerical iterative algorithm will also be presented, suitable to be used within FEM implementations of the asymptotic analysis.

Once the structure has been modelled by FEM, the solution of bifurcation equation reduces to the form

$$\mathbf{K}_t[\lambda]\mathbf{v} = \mathbf{0}\,, \quad \mathbf{v} \neq \mathbf{0} \tag{2.61}$$

where $\mathbf{K}_t[\lambda]$ is a symmetric $[n \times n]$ matrix, differentiable function of the scalar parameter λ and such as $\mathbf{K}_t[0] > 0$. We are actually only interested in its principal eigen–solution $\{\mathbf{v}_c, \lambda_c\}$, characterized by the minimum positive value of λ.

Linearized solution   Matrix $\mathbf{K}_t[\lambda]$ generally is a nonlinear function of the parameter λ, so (2.61) correspond to a nonlinear eigenvalue problem and, in general, can not be solved by standard procedures.

For a series of reasons, while accepting some approximations in the obtained solution, the problem is usually linearized by assuming

$$\mathbf{K}_t[\lambda] \approx \mathbf{K}_0 + \lambda\mathbf{K}_1 \tag{2.62}$$

such that the equation could be written in the form

$$\mathbf{K}_0\mathbf{v} = -\lambda\mathbf{K}_1\mathbf{v} \tag{2.63}$$

suitable to be solved by the well known inverse power method iteration scheme:

$$\begin{cases} \mathbf{v}_{j+1} := -\,\mathbf{K}_0^{-1}\mathbf{K}_1\mathbf{v}_j \\ \lambda_{j+1} := \|\mathbf{v}_j\|/\|\mathbf{v}_{j+1}\| \end{cases} \tag{2.64}$$

The inverse power scheme is robust and computationally efficient, expecially if considering that the process generally follows a linear analysis phase which makes a virtual $\mathbf{K}_0^{-1}$ obtained through a Choleski decomposition of matrix $\mathbf{K}_0$ already available. It is rapidly convergent to the dominant eigen–solution characterized by the smallest absolute value of λ. Then it provides the sought eigen–solution $\{\mathbf{v}_c, \lambda_c\}$ if there are not negative eigenvalues λ_k less than λ_c in absolute values.

In the presence of dominant negative eigenvalues $|\lambda_k| < \lambda_c$, the scheme can however be forced to provide the desired eigen–solution by the simple trick of shifting the origin of linearization in $\lambda_0 > 0$, such that we have $(\lambda_c - \lambda_0) < (\lambda_0 - \lambda_k)$. Equation (2.63) is then rewritten as

$$(\mathbf{K}_0 + \lambda_0\mathbf{K}_1)\,\mathbf{v} = -(\lambda - \lambda_0)\mathbf{K}_1\mathbf{v} \tag{2.65}$$

and the iteration scheme transforms into

$$\begin{cases} \mathbf{v}_{j+1} := -\,(\mathbf{K}_0 + \lambda_0\mathbf{K}_1)^{-1}\mathbf{K}_1\mathbf{v}_j \\ \lambda_{j+1} := \|\mathbf{v}_j\|/\|\mathbf{v}_{j+1}\| + \lambda_0 \end{cases} \tag{2.66}$$

Note however the use of scheme (2.66) in spite of (2.63) introduces a noticeable computational extra–cost due to the construction of matrix $(\mathbf{K}_0 + \lambda_0\mathbf{K}_1)^{-1}$ whose decomposition is not available from before.

Solution of the nonlinear eigenvalue problem Solution $(\mathbf{v}_c, \lambda_c)$ provided by the inverse power algorithm is affected by the error produced by the linearization. It could prove unacceptable in cases where it is used for evaluating derived quantities, such as in asymptotic post–buckling analysis, thus justifying the aim of solving the eigenvalue problem in its general nonlinear form (2.61).

We will discuss some standard approaches to this goal and present a simple and convenient alternative.

Wittrick–Williams bisection process The classical approach to the solution of (2.61), that is the well–known method of Wittrick and Williams (see Wittrick, Williams (1973)), comes from observing that, during the Choleski decomposition of matrix $\mathbf{K}_t$:

$$\mathbf{K}_t[\lambda] = \mathbf{L}[\lambda]\mathbf{D}[\lambda]\mathbf{L}^T[\lambda] \tag{2.67}$$

the eigenvalues of the diagonal matrix $\mathbf{D}$ present the same sign as those of $\mathbf{K}_t$; therefore, the loss of positiveness in matrix $\mathbf{K}_t[\lambda]$ is pointed out by the presence of non–positive diagonal terms in matrix $\mathbf{D}[\lambda]$.

The search for the solution can be performed as follows:

1. Operating at increasing values of the parameter λ, matrix $\mathbf{K}[\lambda]$ is assembled and decomposed, until, for some index $i \in [1 \cdots N]$, we get bounds λ_m and λ_x such that:

$$D_{kk}[\lambda_m] > 0 \ \forall k \ , \ \ D_{kk}[\lambda_x] > 0 \ \forall k \neq i \ , \ \ D_{ii}[\lambda_x] \leq 0 \ , \tag{2.68}$$

2. Once a first interval λ_m—λ_x for the required solution λ_1 has been defined, we proceed by bisection, reducing the interval through further decompositions of $\mathbf{K}$, until the interval amplitude reduces to within a prefixed tolerance.

3. Then the solution is obtained (by the inverse power method) with reference to the secant linearization:

$$\left(\mathbf{K}[\lambda_m] + \frac{\lambda - \lambda_m}{\lambda_x - \lambda_m} \left(\mathbf{K}[\lambda_x] - \mathbf{K}[\lambda_m] \right) \right) \mathbf{v} = 0 \tag{2.69}$$

Some comments and alternatives The Wittrick&Williams algorithm is simple, robust and reliable, even in the case of strongly nonlinear $\mathbf{K}[\lambda]$. It can take some advantages from the fact that the decomposition $\mathbf{LDL}^T$ is only used for testing the sign of their diagonal terms and it is not required at the end of the process. Then specialized decomposition algorithms can be used which avoid the memorization of the full matrix. Furthermore, the bisection process can be improved by using appropriate strategies. However the need to repeat for several times (of the order of tenths) the assemblage and decomposition of matrix $\mathbf{K}$ renders the solution process extremely onerous.

Some alternative solution methods are available. The most interesting one seems to be the implicit restarting Lanczos methods which combine Lanczos iterations with a sequence of successive secant linearization of the function $\mathbf{K}[\lambda]$. However the computational effort does not change. Moreover we do not have relevant advantages of using ad–hoc methods available for particular cases (e.g., $\mathbf{K}[\lambda]$ polynomial function of λ).

A fast iterative scheme　A convenient possible choice for solving the nonlinear bifurcation problem (2.61) is provided by the following iterative scheme.

$$\begin{cases} \mathbf{v}_{j+1} & = \mathbf{v}_j - \omega \mathbf{K}_0^{-1} \mathbf{K}[\lambda_j]\mathbf{v}_j \\ \lambda_{j+1} & = 1/\|\mathbf{v}_{j+1}\| \end{cases} \tag{2.70}$$

where $\omega \leq 1$ is a suitable smoothing factor whose meaning will be shown later. The scheme is initialized by using an assigned vector $\mathbf{y}$:

$$\lambda_0 = \frac{\mathbf{y}^T \mathbf{K}_0 \mathbf{y}}{\mathbf{y}^T \mathbf{K}_1 \mathbf{y}}, \qquad \mathbf{v}_0 = \frac{\mathbf{y}}{\lambda_0 \|\mathbf{y}\|} \tag{2.71}$$

and it terminates when the error becomes less than a prefixed tolerance ε:

$$\|\mathbf{v}_{j+1} - \mathbf{v}_j\| \leq \varepsilon \|\mathbf{v}_j\| \tag{2.72}$$

A convenient initialization $\mathbf{y}$ can be easily obtained as the linear solution due to a loading $\mathbf{f}$ corresponding to a unitary action along a component significant for the expected solution $\mathbf{v}$:

$$\mathbf{y} := \mathbf{K}_0^{-1}\mathbf{f} \tag{2.73}$$

Note that the actual computation of scalar product $\mathbf{y}^T \mathbf{K}_0 \mathbf{y}$ and $\mathbf{y}^T \mathbf{K}_1 \mathbf{y}$ can be easily performed through the direct assemblage of element contributions without needing the assemblage of full matrices $\mathbf{K}_0$ and $\mathbf{K}_1$.

Scheme (2.70) can be viewed as a generalization of the inverse power method. In fact, if assuming linearization (2.62) $\mathbf{K}[\lambda]$ and making $\omega = 1$, it provides

$$\begin{cases} \mathbf{v}_{j+1} := \mathbf{K}_0^{-1}\left(\mathbf{K}_0 - (\mathbf{K}_0 + \lambda_j \mathbf{K}_1)\right)\mathbf{v}_j = -\mathbf{K}_0^{-1}\mathbf{K}_1(\lambda_j \mathbf{v}_j) \\ \lambda_{j+1} := 1/\|\mathbf{v}_{j+1}\| = \|\lambda_j \mathbf{v}_j\|/\|\mathbf{v}_{j+1}\| \end{cases} \tag{2.74}$$

then corresponds exactly to scheme (2.64) apart from the use of a normalized input vector $\lambda_j \mathbf{v}_j$ such that $\|\lambda_j \mathbf{v}_j\| = 1$. The rewriting is convenient for the following reasons:

1. Vector $\mathbf{s}_j := \mathbf{K}[\lambda_j]\mathbf{v}_j$ has a direct mechanical meaning (it represents the nodal force vector corresponding to $\mathbf{v}_j$) and can be computed through standard assemblage of element contributions, without needing overall matrix $\mathbf{K}_1$ to be actually constructed.

2. The use of a smoothing factor $\omega < 1$ allows oscillating sequences to be avoided:

$$\mathbf{v}_{j+1} = \mathbf{v}_j - \mathbf{K}_0^{-1}\mathbf{K}[\lambda_j]\mathbf{v}_j = -\mathbf{v}_j \tag{2.75}$$

corresponding to values of λ_j such as

$$\mathbf{K}[\lambda_j]\mathbf{v}_j = 2\mathbf{K}_0 \mathbf{v}_j \tag{2.76}$$

that the scheme tends to generate in presence of dominant negative eigenvalues. Then the factor ω represents a trick equivalent to the shift of the origin in the power method (see (2.66)), but more simple and efficient to use because it does not require further matrix decompositions other than that of $\mathbf{K}_0$.

3. At convergence, the obtained solution, directly characterized by condition $\mathbf{K}[\lambda_j]\mathbf{v}_j = \mathbf{0}$, is not affected by roundoff errors produced in the decomposition process of matrix $\mathbf{K}_0$.

4. The algorithm is actually relatively independent of the choice for the iteration matrix $\omega\mathbf{K}_0^{-1}$ and accepts quite rough estimates of matrix $\mathbf{K}_t[0]^{-1}$ according that

$$\mathbf{v}^T \frac{1}{\omega}\mathbf{K}_0\mathbf{v} > \mathbf{v}^T\mathbf{K}[\lambda]\mathbf{v} \ , \ \forall \mathbf{v} \neq 0 \, , \lambda < \lambda_b \qquad (2.77)$$

which is needed for avoiding undesired oscillations due to the presence of dominant negative eigenvalues (however small ω values tend to slow the iteration process while stabilizing it).

5. For these reasons the scheme can be conveniently used also within "coarse to fine"–"fine to coarse" multilevel solution techniques, as shown in Lopez et al. (1998).

6. For the same reasons, the scheme presents good convergence characteristics also in the general nonlinear case under quite wide hypotheses. A convergence proof is given in Casciaro (1955) and is not considered here. We can only mention that the main restriction comes from condition (2.77)

The scheme provides a simple and robust method for the solution of nonlinear eigenvalue problems. It was first proposed in Casciaro, Aristodemo (1977) and successively it has been widely used and always proved to be robust, fast and reliable, even in the presence of strong nonlinearities Lopez et al. (1998). A computer implementation is contained in Casciaro et al. (1992), (see App.A).

3 A deeper insight into the asymptotic post–buckling analysis

This section aims to provide a deeper insight into the asymptotic post–buckling analysis. We will start by introducing an improved, quadratic version of the asymptotic algorithm, then we discuss some relevant aspects of the structural modeling and finite element discretization and then extend the analysis to cover multimodal buckling and related topics. The reader can refer to Casciaro et al. (1998), Salerno, Casciaro (1997), Garcea (2001) for further details.

3.1   The quadratic asymptotic algorithm

A summary of the linear algorithm   The linear asymptotic algorithm described in the previous section comes from the idea that, in a neighborhood of the bifurcation point, the equilibrium path could be adequately represented by a linear combination of the fundamental and critical mode, $\hat{u}$ and $\dot{v}$ obtained form eqs. (2.7) and (2.11). Therefore, an evaluation of the equilibrium path of the structure can be obtained by assuming a solution in the form:

$$u[\lambda, \xi] = \lambda\hat{u} + \xi\dot{v} \qquad (3.1)$$

and by deriving the λ–ξ equilibrium relationship by Galerkin solution of the equilibrium equation:

$$(\Phi'[u[\lambda, \xi] - \lambda\hat{p})\dot{v} = 0 \qquad (3.2)$$

We obtain, by some manipulations:

$$(\lambda - \lambda_b)\xi\Phi_b'''\hat{u}_b\dot{v}^2 + \frac{1}{2}\xi^2\Phi_b'''\dot{v}^3 + \frac{1}{2}\lambda^2\Phi_b'''\hat{u}_b^2\dot{v} = 0 \tag{3.3}$$

which can be rewritten in the simple form:

$$\lambda + \lambda^2\mu/\xi = \lambda_b + \xi\dot{\lambda}_b \tag{3.4}$$

where the post–buckling slope $\dot{\lambda}_b$ and the implicit imperfection factor μ are defined by eqs. (2.16) and (2.21), respectively.

The algorithm actually represents a simple extension of the traditional Euler method for linearized stability, implying negligible extra–costs in both computer coding and computational demand. However, in spite of this, it is able to give accurate solutions, especially in the evaluation of the limit load, as has been shown from both theoretical investigations (see Brezzi et al. (1986)) and numerical testing (e.g. see Lanzo et al. (1995), Lanzo, Garcea (1996), Garcea (2001)).

Improving the linear algorithm Note that the linear algorithm reduces the nonlinearity of the problem only to the component $\xi\dot{v}$ parallel to the critical mode and nonlinearly satisfies the equilibrium conditions in that direction only (components of $u[\lambda]$ orthogonal to $\dot{v}$ are linear in λ). As a consequence, the algorithm tends to furnish accurate results for structures that present only one direction of strong nonlinearity.

An improvement of the algorithm requires the path description be enriched with an explicitly nonlinear component orthogonal to the critical mode $\dot{v}$ and, conversely, equilibrium also be satisfied in the orthogonal direction. Then we assume:

$$u[\lambda, \xi] = \lambda\hat{u} + \xi\dot{v} + w[\lambda, \xi] , \quad w \perp \dot{v} \tag{3.5}$$

where the *orthogonal correction* $w[\lambda, \xi]$ is obtained by the equilibrium equation

$$(\Phi'[u[\lambda, \xi]] - \lambda\hat{p})\delta w , \quad \forall \delta w \perp \dot{v} \tag{3.6}$$

It is convenient, for reasons we will explain in section 3.3, to assume orthogonality $w \perp \dot{v}$ be defined by the condition

$$\Phi_b'''\hat{u}\dot{v}w = 0 \tag{3.7}$$

We also denote with $\mathcal{V}$ and $\mathcal{W}$ the critical (i.e. parallel to $\dot{v}$) and orthogonal (i.e. orthogonal to $\dot{v}$) components of the displacement space $\mathcal{U}$.

The quadratic algorithm We can suppose the correction w be slightly nonlinear, otherwise the problem can be treated as a multiple bifurcation case, as we will see later. By accepting a quadratic extrapolation for the correction w and recalling (3.5), we obtain:

$$w[\lambda, \xi] = \frac{1}{2}\lambda^2\hat{\hat{w}} + \frac{1}{2}\xi^2\ddot{w} \tag{3.8}$$

where $\hat{\dot{w}}$ and $\ddot{w}$ correspond to the solutions of the linear equations:

$$\Phi_b'' \hat{\dot{w}} \delta w = -\Phi''' \hat{u}^2 \delta w \,, \quad \Phi_b'' \ddot{w} \delta w = -\Phi''' \dot{v}^2 \delta w \,, \quad \forall \delta w \in \mathcal{W} \tag{3.9}$$

Expanding the equilibrium equation

$$\Phi'[u]\delta u - \lambda \hat{p}\delta u = 0 \tag{3.10}$$

up to the 4th-order terms in the energy when $\delta u \in \mathcal{V}$ and up the 3rd–order ones when $\delta u \in \mathcal{W}$, the path equation reduces to the following algebraic λ–ξ relationship

$$\begin{aligned}
\frac{1}{2}\xi^2 \Phi_b''' \dot{v}^3 + \xi(\lambda - \lambda_b)\Phi_b'''(\hat{u} + \lambda_b\hat{\dot{w}})\dot{v}^2 + \frac{1}{2}\lambda^2 \Phi_b''' \hat{u}\hat{u}\dot{v} \\
+ \frac{1}{2}\lambda_b^2 \xi \Phi_b''' \hat{\dot{w}}\dot{v}^2 + \frac{1}{6}\xi^3(\Phi_b''''\dot{v}^4 - 3\Phi_b''\ddot{w}^2) \\
+ \frac{1}{2}\xi(\lambda - \lambda_b)^2(\Phi_b''''\hat{u}^2\dot{v}^2 - \Phi_b''\ddot{w}\hat{\dot{w}}) + \frac{1}{2}\xi^2(\lambda - \lambda_b)\Phi_b''''\hat{u}\dot{v}^3 \\
+ \frac{1}{6}\lambda^2(\lambda - 3\lambda_b)\Phi_b''''\hat{u}^3\dot{v} = 0
\end{aligned} \tag{3.11}$$

We can expect that this equation, together with expansion (3.3),(3.2), should provide a sensible improvement in accuracy with respect to the simpler one (eqs.(3.3),(3.2)) obtained by the linear algorithm. Note, however, that some computational extra cost is required, essentially due to the solution of the linear equations (3.9), The numerical solution of these equations, written in matrix form as

$$\begin{cases} \delta\mathbf{w}^T\{\mathbf{K}_b\ddot{\mathbf{w}} - \ddot{\mathbf{s}}\} = 0 \\ \delta\mathbf{w}^T\{\mathbf{K}_b\hat{\dot{\mathbf{w}}} - \hat{\dot{\mathbf{s}}}\} = 0 \end{cases} \,, \quad \ddot{\mathbf{w}}, \hat{\dot{\mathbf{w}}} \in \mathcal{W}, \ \forall \delta\mathbf{w} \in \mathcal{W} \tag{3.12}$$

will be discussed in section 4.2.

Simplified (essential) version   In some applications, the fundamental path is characterized by small displacements (rotations). By assuming $\hat{\dot{w}} \approx 0$ (and related assumptions) the algebraic path equation simplifies into

$$\tfrac{1}{2}\lambda^2 \Phi_0''' \hat{u}^2 \dot{v} + \tfrac{1}{2}\xi^2 \Phi_b''' \dot{v}^3 + \xi(\lambda - \lambda_b)\Phi_b'''\hat{u}\dot{v}^2 + \tfrac{1}{6}\xi^3(\Phi_b''''\dot{v}^4 - 3\Phi_b''\ddot{w}^2) = 0 \tag{3.13}$$

Defining $\dot{\lambda}_b$ and μ through (2.16) and (2.21) and introducing the *post–buckling curvature*;

$$\ddot{\lambda}_b = -\frac{1}{3}\frac{\Phi_b''''\dot{v}^4 - 3\Phi_b''\ddot{w}^2}{\Phi_b'''\hat{u}\dot{v}^2} \tag{3.14}$$

it can be rewritten in the simple form

$$\lambda + \lambda^2 \mu/\xi = \lambda_b + \xi\dot{\lambda}_b + \frac{1}{2}\xi^2\ddot{\lambda}_b \tag{3.15}$$

which is very simple and expressive and better suited than (3.4) for the discussion.

Note that the differences from the linear formula reduce to the introduction of the *post–buckling* curvature $\ddot{\lambda}_b$. Its evaluation is obtained from the difference between two 4th–order terms:

$$\Phi''''\dot{v}^4 - 3\Phi''\ddot{w}^2 \tag{3.16}$$

which is generally expressed as a small difference among large quantities. To express such differences accurately some care in choosing the finite element discretization fields in order to avoid *locking by discretization* effects. Furthermore, an accurate evaluation of 4–th order energy terms can require a more sophisticated modeling than that provided by the usual technical theories. These, while aimed to express 2nd–order, usually also provide an accurate evaluation of 3rd-order terms but fail in representing 4th-order terms. We will show this by reference to the Euler beam simple example.

At–hand example: the Euler beam Let's consider the Euler beam problem shown in fig.1. Assuming the usual expression for the strain energy

$$\Phi = \frac{1}{2}\int_0^L \{EA\varepsilon^2 + EJ\chi^2\}dx; \tag{3.17}$$

$$\begin{cases} \varepsilon := \varepsilon_b + \mathrm{u},_x + \dfrac{1}{2}\mathrm{w},_x^2 \\[2mm] \chi := \mathrm{w},_{xx}^2 \end{cases} \tag{3.18}$$

coming from the so called technical beam theory, we obtain

$$\Phi_b''\dot{v}\delta u = \int_0^l \{EA\ddot{\mathrm{u}},_x \,\delta\mathrm{u},_x + EJ\dot{\mathrm{w}},_{xx}\,\delta\mathrm{w},_{xx} + N_b\dot{\mathrm{w}},_x\,\delta\mathrm{w},_x\}dx$$

$$\Phi_b'''\dot{v}^2\delta u = \int_0^l EA(2\ddot{\mathrm{u}},_x\,\dot{\mathrm{w}},_x\,\delta\mathrm{w},_x + \dot{\mathrm{w}},_x^2\,\delta\mathrm{u},_x)dx \tag{3.19}$$

$$\Phi_b''''\dot{v}^4 = \int_0^l 3EA\dot{\mathrm{w}},_x^4\,dx$$

Using these expressions, eqs. (2.2) and (2.3) provide the well known solution

$$\hat{\mathrm{w}},_x = 0\;,\quad EA\hat{\mathrm{u}},_x = -F \tag{3.20}$$

$$\lambda_b = -\frac{N_b}{F} = \frac{\pi^2 EJ}{F\ell^2}\quad,\quad \dot{\mathrm{u}},_x = 0\quad,\quad \dot{\mathrm{w}},_x = cos\frac{\pi x}{\ell} \tag{3.21}$$

We have

$$\Phi_b'''\dot{v}^3 = \int_\ell EA\dot{\mathrm{w}},_x^3\,dx = 0 \tag{3.22}$$

so we obtain $\dot{\lambda}_b = 0$. To evaluate $\ddot{\lambda}_b$ we need the orthogonal secondary mode $\ddot{u}$ from (3.9) and (3.19). We obtain

$$\int_0^l EA\ddot{\mathrm{u}},_x\,\delta\mathrm{u},_x\,dx = -\int_0^l EA\dot{\mathrm{w}},_x^2\,\delta\mathrm{u},_x\,dx\;,\quad \forall\delta\mathrm{u},_x \tag{3.23}$$

that provides

$$\ddot{u},_x = -\dot{w},_x^2 \quad , \quad \ddot{w},_x = 0 \tag{3.24}$$

We have

$$\Phi_b''''\dot{v}^4 = 3\int_0^l EA\dot{w},_x^4\,dx \ , \qquad \Phi_b''\ddot{v}^2 = \int_0^l EA\dot{w},_x^4\,dx \tag{3.25}$$

then we obtain

$$\ddot{\lambda}_b := -\frac{1}{3}\frac{\Phi_b''''\dot{v}^4 - 3\Phi_b''\ddot{v}^2}{\Phi_b'''\hat{u}\dot{v}^2} = 0 \tag{3.26}$$

Note that the result $\ddot{\lambda}_b = 0$ is not correct. In fact we know, from the classical "elastic" theory, that the exact result is $\ddot{\lambda}_b = \lambda_b/4$. The error is due to the use of an approximate beam model where kinematical relationships (3.18) are unable to reproduce finite rigid rotations exactly and then provide inexact expressions for the strain energy.

The Euler beam, using Green axial strain Eqs. (3.18) are often justified as an approximation for

$$\begin{cases} \varepsilon := \varepsilon_b + u,_x + \dfrac{1}{2}u,_x^2 + \dfrac{1}{2}w,_x^2 \\[2mm] \chi := w,_{xx}^2 \end{cases} \tag{3.27}$$

where, in the expression of ε, derived from the classical Green strain measure applied to the axis fiber, the "small" term $u,_x^2/2$ is neglected with respect to $u,_x$. We will see however that (3.27) is neither rational nor 4th–order accurate.

Using eqs. (3.20) in spite of (3.18), eqs. (3.19) transform into

$$\Phi_b''\dot{v}\delta u = \int_0^l \{EA(1+\varepsilon_b)\dot{u},_x\,\delta u,_x + EJ\dot{w},_{xx}\,\delta w,_{xx} + N_b\dot{w},_x\,\delta w,_x\}dx$$

$$\Phi_b'''\dot{v}^2\delta u = \int_0^l EA(2\dot{u},_x\,\dot{w},_x\,\delta w,_x + \dot{w},_x^2\,\delta u,_x + 3\dot{u},_x\,\delta u,_x)dx \tag{3.28}$$

$$\Phi_b''''\dot{v}^4 = \int_0^l 3EA(\dot{w},_x^4 + \dot{u},_x^4)dx$$

Fundamental and critical modes are again described by eqs. (3.20)–(3.22), however (3.23) becomes

$$\int_\ell EA\ddot{u},_x\,\delta u,_x\,dx = \int_\ell EA\dot{w},_x^2\,\delta u,_x\,dx \ , \quad \forall \delta u,_x \tag{3.29}$$

and provides

$$\ddot{u},_x = -\frac{1}{1+\varepsilon_b}\dot{w},_x^2 \quad , \quad \ddot{w},_x = 0 \tag{3.30}$$

We then have

$$\Phi_b''''\dot{v}^4 - 3\Phi_b''\ddot{v}^2 = 3EA\int_\ell \dot{w},_x^4\,dx - \frac{3EA}{1+\varepsilon_b}\int_\ell \dot{w},_x^4\,dx = \frac{3N_b}{1+\varepsilon_b}\int_\ell \dot{w},_x^4\,dx$$

$$\Phi_b'''\hat{u}\dot{v}^2 = \int_\ell EA\hat{u},_x\,\dot{w},_x^2\,dx \tag{3.31}$$

that provides

$$\ddot{\lambda}_b := -\frac{1}{3}\frac{\Phi_b''''\dot{v}^4 - 3\Phi_b''\ddot{v}^2}{\Phi'''\hat{u}\dot{v}^2} = -\frac{N_b \int_\ell \dot{w}_{,x}^4\, dx}{(1+\varepsilon_b)F \int_\ell \dot{w}_{,x}^2\, dx} = -\frac{3}{4(1+\varepsilon_b)}\lambda_b \qquad (3.32)$$

so an even worse result is obtained $\ddot{\lambda}_b \approx -3\lambda_b/4$ (see Pignataro et al. (1982) for a deeper insight into this topics)

Some comments The Euler beam case, in spite of its simplicity, allows a real insight into the quadratic algorithm and suggests some useful considerations.

Note from (3.31), (3.32) that the post–critical curvature $\ddot{\lambda}_b$ is obtained from a small difference (of the order of $EA\varepsilon_b$) between large quantities (of the order of EA). An accurate evaluation of this difference requires the reference continuous model be accurate to (at least) the $4th$ order in u. The usual technical models do not satisfy this requirement (in the example we get $\ddot{\lambda}_b = 0$ or the even worse value $\ddot{\lambda}_b \approx -3\lambda_b/4$ in spite of the expected value $\ddot{\lambda}_b = \lambda_b/4$).

It is worth mentioning that our experience, as users or scientists in structural mechanics, in both continuous formulation and finite element modelling is generally limited by the aim to represent a 2nd–order neighbor of the energy, for deriving the tangent stiffness matrix. Actually, we have very few experience of problems related to 4th-order energy neighbor, required by the (quadratic) asymptotic analysis, and in its finite element description. So unusual care has to be taken in structural modelling to avoid 4th–order kinematical incoherences.

Note also that, in slender structures, the critical mode $\dot{v}_b$ essentially corresponds to out-of-plane displacements (i.e., transversal displacements) whereas the correction w essentially develops as plane (axial) displacements. Then an independent discretization of the axial and transversal components of the displacement tends to produce large errors due to the different approximation for these displacements, the difference between the two approximations usually being much bigger than the approximation of the difference. If we do not take to avoid this phenomenon (by an appropriate combined field discretization) it will introduce a spurious nonlinear locking which can invalidate the solution.

For a better understanding of the real relevance of this phenomenon, we can refer to a FEM analysis of the Euler problem where the beam is discretized by one element with linear interpolation for u[x] and cubic for w[x].

Constant u[x] are provided by the discretization, so the fundamental mode is still provided by (3.20). Bifurcation is however obtained as

$$\lambda_b = -\frac{N_b}{F} = \frac{12EJ}{F\ell^2} \quad , \quad \dot{u}_{,x} = 0 \quad , \quad \dot{w}_{,x} = 1 - 2x/\ell \quad , \quad \dot{\lambda}_b = 0 \qquad (3.33)$$

We do not care about this 20% error in the load parameter. We know that it can be reduced and made negligible, by using few elements rather then only one element. The real problem comes from the solution for $\ddot{\lambda}_b$. In fact (3.23) becomes

$$\ddot{u}_{,x} = -\frac{1}{\ell}\int_\ell \dot{w}_{,x}^2\, dx = -\frac{1}{\ell}\int_\ell (1 - 2x/\ell)^2 dx = \frac{1}{3} \qquad (3.34)$$

We have

$$\Phi_b'''' \dot{v}^4 = 3 \int_0^\ell EA(1 - 2x/\ell)^4 dx = \frac{3}{5}EA\ell$$

$$\Phi_b'' \ddot{v}^2 = \int_0^\ell EA(\frac{1}{3})^2 dx = \frac{1}{9}EA\ell \tag{3.35}$$

$$\Phi''' \hat{u}\dot{v}^2 = \int_0^\ell -F(1 - 2x/\ell)^2 dx = -\frac{1}{3}F\ell$$

then we obtain

$$\ddot{\lambda}_b := -\frac{1}{3}\frac{\Phi_b'''' \dot{v}^4 - 3\Phi_b'' \ddot{v}^2}{\Phi_b''' \hat{u}\dot{v}^2} = \frac{4}{15}\frac{EA}{F} = \frac{4}{15}\frac{EA\ell^2}{12EJ}\lambda_b \tag{3.36}$$

which shows an anomalous sensitivity to the axial/flexural rigidity ratio (i.e. a locking effect) and can be really large for slender structures.

We will return on these topics later and we will see how to avoid both nonlinear geometrical incoherences and nonlinear discretization locking sensitivity.

Imperfection sensitivity analysis The presence of small load or geometric imperfections can be taken into account by the same strategy used in the simple linear version of the algorithm. Once the analysis for the reference structural scheme has been performed, the effects of additional imperfections in the structural geometry or in the applied loads can be taken simply by adding some extra–term to the algebraic λ, ξ relationship (3.11) without needing to repeat the analysis and so with a negligible computational extra–cost. Denoting with $\tilde{u}$ the geometrical and with $q = \lambda\tilde{q}$ the load imperfection and referring to the simplified version (3.15) of path equation, we obtain

$$\lambda + (\lambda^2\mu + \lambda\mu_q + \lambda\mu_u)/\xi = \lambda_b + (\xi + \frac{4}{3}\mu_u)\dot{\lambda}_b + \frac{1}{2}\xi^2\ddot{\lambda}_b \tag{3.37}$$

the *additional imperfection factors* μ_q and μ_u being expressed as

$$\mu_u = \frac{\Phi''' \hat{u}\tilde{u}\dot{v}}{\Phi''' \hat{u}\dot{v}^2} \quad , \quad \mu_q = -\frac{\tilde{q}\dot{v}}{\Phi''' \hat{u}\dot{v}^2} \tag{3.38}$$

as in the linear algorithm.

The previous strategy is well suited for imperfection sensitivity analysis where, to obtain statistically reliable values of the collapse safety factor, we need to repeat the analysis for a large number of different possible combinations of geometrical defects and load imperfections. Note however that this treatment of the additional imperfections corresponds to a *3rd*–order expansion of their energy contribution. So it has to be considered as a quite rough procedure.

More accurate results can be obtained by a *4th*-order expansion. However this requires the evaluation of the corresponding orthogonal correction $\tilde{w}$, which is a simple task but not completely inexpensive from the computational point of view. So this improvement appears unsuitable for imperfection sensitivity analysis.

Furthermore note that even better results are, obviously, obtained by keeping imperfections in the external load $\lambda\hat{p}$ and performing a new complete analysis, but this could be done for single imperfections and is impracticable for imperfection sensitivity purposes.

3.2 A beam model suitable for asymptotic analysis

The quadratic algorithm is potentially capable of providing very accurate results. We need however a rational structural model (that is obtained by exact finite kinematics) in the analysis or, at least, models that are accurate up to the 4th–order terms in the strains.

The beam case is discussed here in some detail and a quite simple beam model suitable to that purpose will be described here. For a deeper insight into these topics refer to Pignataro et al. (1982) and Salerno, Casciaro (1997).

The "elastica" theory The classical theory of "elastica", developed by Euler himself in the mid 1700's, assumes that the axial and transversal deformation be null. The displacements are only represented by the rotation $\alpha[x]$ of beam cross section as shown in fig.12

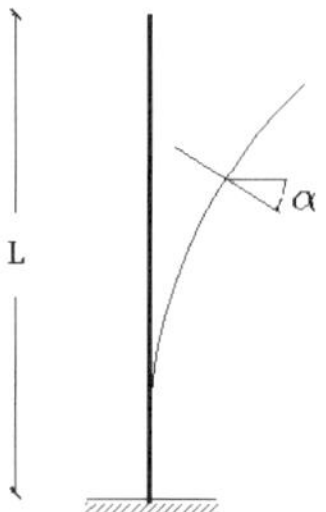

Figure 12. The "elastica" model

The strain energy is assumed as a quadratic function of $\alpha[x]$

$$\Phi[u] := \frac{1}{2} \int_0^L EJ\alpha_{,x}^2 \, dx \tag{3.39}$$

the displacements $u[x]$ and $w[x]$ being derived as functions of $\alpha[x]$ by the nonlinear integral expressions

$$\mathrm{u}[x] := \int_0^x (1 - \cos\alpha)dx \quad , \quad \mathrm{w}[x] := \int_0^x \sin\alpha \, dx \tag{3.40}$$

A very simple while kinetically exact model is obtained by introducing the internal (nonlinear) constraint of axial incompressibility. The whole nonlinearity of the problem is then transferred into the external load work, which becomes nonlinear. So the formulation cannot be considered within the theoretical framework stated in section 2.1.

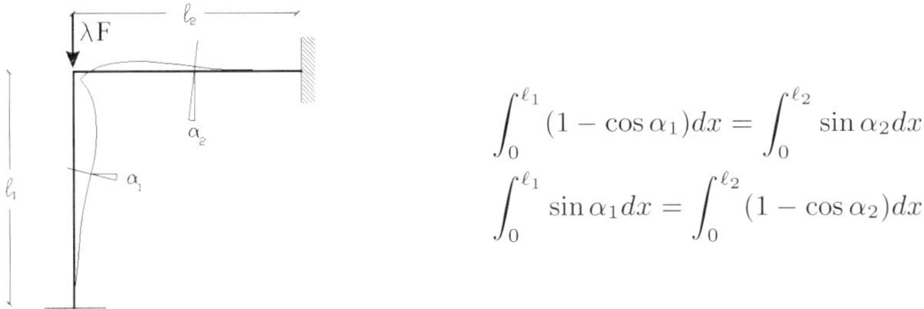

$$\int_0^{\ell_1} (1 - \cos\alpha_1)dx = \int_0^{\ell_2} \sin\alpha_2 dx$$

$$\int_0^{\ell_1} \sin\alpha_1 dx = \int_0^{\ell_2} (1 - \cos\alpha_2)dx$$

Figure 13. Nonlinear inter–element continuity conditions

For the Euler problem in fig.12, where we do not have restrictions on $\alpha[x]$ apart from the linear one $\alpha[0] = 0$, this does not really imply any difficulties related to the dependence of the tangential space $\mathcal{T}$ on the configuration u, but requires some modifications (essentially to refer to the entire potential energy $\Pi[u]$ in spite of the strain energy $\Phi[u]$) of the formulas given in the previous sections. Note however that it can prove to be a serious drawback for structures composed of several elements, the continuity conditions among the elements being expressed by nonlinear integral equations (see fig.??) and practically prevents a FEM analysis process being used. In fact:

1. The construction of the stiffness matrix and of the nodal force vectors can no longer be implemented into the analysis through the usual direct assembly of element contributions.
2. The possible remedies (penalty functions, iterative solutions, etc...) prove to be extremely onerous (see the discussion in the book by Thompson & Hunt).

So, we can conclude that the elastic model is unsuitable for a computational approach.

The Antman beam model The model is based on the following strain measures:

$$\begin{cases} \varepsilon & = \varepsilon_0 + (1 - u_{,x})\cos\varphi - w_{,x}\sin\varphi - 1 \quad \text{(axial strain)} \\ \gamma & = \gamma_0 - (1 + u_{,x})\sin\varphi + w_{,x}\cos\varphi \quad \text{(shear strain)} \\ \chi & = \varphi_{,x} \quad \text{(curvature)} \end{cases} \tag{3.41}$$

and on the elastic laws:

$$N = EA\varepsilon , \quad T = GA^*\gamma , \quad M = EJ\chi \tag{3.42}$$

We obtain the following expression for the strain energy:

$$\Phi[u] := \frac{1}{2}\int_0^\ell \{EA\varepsilon^2 + GA^*\gamma^2 + EJ\chi^2\}dx \tag{3.43}$$

Refer to the Antman paper (Antman (1977)) and to the discussion given in Pignataro et al. (1982), for further detail.

The Antman beam model proves to be rational (i.e. accurate at each order) however their resulting asymptotic formulation is quite complex and difficult to implement within computer codes because of the complexity of the expressions for the strain energy variation. In addition, standard finite element discretizations provide nonlinear discretization locking. So the model does not appear well suited for asymptotic analysis.

A simple rational beam model   We will show here a variant of the Antman beam model which proves to be particularly suitable for asymptotic analysis (the reader can refer to the paper by Salerno, Lanzo (1997), for a deeper insight).

The model can be viewed as derived from the Antman one by a simple change of the strain variables. The following strain components are used:

$$\begin{cases} \tilde{\varepsilon} := \varepsilon \cos\varphi - \gamma \sin\varphi = \tilde{\varepsilon}_0 + \mathrm{u}_{,x} + 1 - \cos\varphi \\ \tilde{\gamma} := \varepsilon \sin\varphi + \gamma \cos\varphi = \tilde{\gamma}_0 + \mathrm{w}_{,x} - \sin\varphi \end{cases} \tag{3.44}$$

and we assume:

$$GA^* = EA \tag{3.45}$$

so, the expression of the strain energy (3.43) becomes

$$\Phi[u] = \frac{1}{2}\int_0^\ell \{EA(\tilde{\varepsilon}^2 + \tilde{\gamma}^2) + EJ\chi^2\}dx \tag{3.46}$$

Assumption (3.45) is allowed for slender structures because the axial and even more the transversal strains are really negligible. In fact they are considered here only for an easier geometric representation. We obtain a rational beam model equivalent to that of Antman but noticeably simpler to use. We also obtain the further useful advantage that, in the case of nodal loading, the exact solution is trivially expressed by the conditions:

$$\tilde{N} := EA\tilde{\varepsilon} = \text{const.} , \quad \tilde{T} := EA\tilde{\gamma} = \text{const.} \tag{3.47}$$

$\tilde{N}$ and $\tilde{T}$ being the axial and transversal strengths associated to $\tilde{\varepsilon}$ and $\tilde{\gamma}$. Then we can take, directly the following expression for the axial and transversal strains:

$$\tilde{\varepsilon} := \tilde{\varepsilon}_0 + \tilde{\mathrm{u}}_{,x} + \frac{1}{\ell}\int_0^\ell (1 - \cos\varphi)dx , \quad \tilde{\mathrm{u}}_{,x} := \frac{u_j - u_i}{\ell}$$
$$\tilde{\gamma} := \tilde{\gamma}_0 + \tilde{\mathrm{w}}_{,x} - \frac{1}{\ell}\int_0^\ell \sin\varphi\, dx , \quad \tilde{\mathrm{w}}_{,x} := \frac{w_j - w_i}{\ell} \tag{3.48}$$

where the indexes i and j refer to the end sections ($x = 0$ and $x = \ell$) of the beam.

Note that eqs.(3.48) do not contain any explicit reference to the displacement field u[x] and w[x] apart from the nodal values $\{\mathrm{u}_i, \mathrm{w}_i\}$ and $\mathrm{u}_j, \mathrm{w}_j\}$. They actually imply an implicit interpolation low for $\{\mathrm{u}[x], \mathrm{w}[x]\}$ which is consequent on that assumed for the rotation $\varphi[x]$. This noticeably simplify the finite element discretization and allows beam elements insensitive to this nonlinear discretization locking to be obtained.

Relevant strain energy variations ¿From eqs.(3.44)–(3.48) we obtain the strain expansions:

$$\tilde{\varepsilon} = \tilde{\varepsilon}_0 + \tilde{u}_x + \frac{1}{\ell}\int_0^\ell \{\frac{1}{2}\varphi^2 - \frac{1}{24}\varphi^4 + \cdots\}dx$$

$$\tilde{\gamma} = \tilde{\gamma}_0 + \tilde{w}_{,x} + \frac{1}{\ell}\int_0^\ell \{-\varphi + \frac{1}{6}\varphi^3 + \cdots\}dx \tag{3.49}$$

Ignoring, for an easier presentation, negligible terms related to the shear ($\tilde{T}_0 = 0$, $\dot{\tilde{T}} \approx 0$), we then get the following expressions for the relevant energy variations:

$$\Phi_0''\dot{u}^2 = \int_0^\ell \{EA\dot{\tilde{u}}_{,x}^2 + EJ\dot{\varphi}_{,x}^2 + N_0\dot{\varphi}^2\}dx$$

$$\Phi_0'''\dot{u}^3 = 3\dot{\tilde{N}}\int_0^\ell \dot{\varphi}^2 dx \tag{3.50}$$

$$\Phi_0''''\dot{u}^4 = \frac{3EA}{\ell}\left(\int_0^\ell \dot{\varphi}^2 dx\right)^2 - \tilde{N}_0\int_0^\ell \dot{\varphi}^4 dx$$

By comparing eqs.(3.49) with the analogous ones (3.19) obtained from the technical beam model, we can see that the expressions for the second and third strain energy variation coincide apart from the presence of $\dot{\varphi}$ in stead of $\dot{w}_{,x}$. Even if these quantities have different meanings (φ in (3.49) is the rotation angle of the cross section and $w_{,x}$ in (3.19) is the derivative of the transversal displacement), they practically coincide at the first order ($\dot{\varphi} \approx \dot{w}_{,x}$). Then the two model, the present (rational) one and that derived from the technical theory, are completely equivalent up to the third–order in the energy and only differ at the 4th–order level.

In the forth–order energy variations the extra–term

$$N_b\int_0^\ell \varphi^4 dx \tag{3.51}$$

is now present and we have

$$\frac{3EA}{\ell}\left(\int_0^\ell \dot{\varphi}^2 dx\right)^2 \quad \text{in spite of} \quad 3EA\int_0^\ell \dot{\varphi}^4 dx \tag{3.52}$$

The first accounts for the 4th–order terms of the strain expansion and renders the model 4th–order accurate, the technical model being only 3rd–order accurate. The second plays an important role in FEM discretization by allowing nonlinear interpolation locking in the evaluation of $\ddot{\lambda}$ to be avoided.

We can easily show this by referring to the Euler beam problem. From (3.50) we obtain

$$\hat{\varphi} = 0 \,, \quad \hat{\tilde{N}} = -F$$

$$\lambda_b F = -\tilde{N}_b = \frac{\pi^2 EJ}{\ell^2} \,, \quad \dot{\tilde{N}} = 0 \,, \quad \dot{\varphi} = \cos\frac{\pi x}{\ell} \,, \quad \lambda_b = 0 \tag{3.53}$$

as in the technical model (see eqs. (3.17)–(3.21)), apart from referring to $\dot\varphi$ in spite of $\dot{w}_{,x}$. The second order displacement is however defined by

$$\ell \ddot{\tilde{u}}_{,x}\,\delta\tilde{N} = -\int_0^\ell \dot\varphi^2\,dx\,\delta\tilde{N} \quad\Longrightarrow\quad \ddot{\tilde{u}}_{,x} = -\frac{1}{\ell}\int_0^\ell \dot\varphi^2\,dx \tag{3.54}$$

and becomes

$$\ddot{\varphi} = 0 \tag{3.55}$$

As a consequence, we have

$$\Phi_b''''\dot{v}^4 = 3EA\ell\left(\frac{1}{\ell}\int_0^\ell \dot\varphi^2\,dx\right)^2 - \tilde{N}_b\int_0^l \dot\varphi^4\,dx \quad,\quad \Phi_b''\ddot{v}^2 = -EA\ell\left(\frac{1}{\ell}\int_0^l \dot\varphi^2\,dx\right)^2 \tag{3.56}$$

Then we obtain a correct evaluation for the post–critical curvature:

$$\ddot{\lambda}_b := -\frac{1}{3}\frac{\Phi_b''''\dot{v}^4 - 3\Phi_b''\ddot{v}^2}{\Phi_b'''\hat{u}\dot{v}^2} = -\frac{\tilde{N}_b\int_0^\ell \dot\varphi^4\,dx}{F\int_0^\ell \dot\varphi^2\,dx} = \frac{1}{4}\lambda_b \tag{3.57}$$

Note that, in the difference $\Phi''''\dot{v}^4 - 3\Phi''\ddot{w}^2$, the two terms containing the axial stiffness EA cancel each other identically and that also holds when using a finite element interpolation for $\dot\varphi$, so providing nonlinear locking insensitivity.

3.3   A plate model suitable for asymptotic analysis

As shown before, an accurate evaluation of the fourth–order energy terms required by the full version of the asymptotic algorithm require a rational or, at least 4th–order accurate, modelling. However, developing suitable rational plate models is really difficult and they are too computationally cumbersome for practical use. In this case it is better to compromise and refer to simplified models if they are accurate enough for the technical demand.

The possibility of referring to approximate (at least, 3rd–order accurate) models arises from observing that the geometrical 4th–order energy terms due to the 4th–order terms in the strain expansion are really very small ($\ddot{\lambda}_b = \lambda_b/4$ obtained for the Euler beam problem provides a 1% load increase for a transversal displacement of 10% of the beam length) and is really appreciable only when the other terms containing the elastic stiffness (i.e. terms containing EA in the beam case) identically cancel each other. However this happens only if the buckling evolution does not imply stress redistribution, that is only for statically determined structures.

In the presence of high indeterminacies, such as in plates or shells, we usually have a strong 2nd–order stress redistribution which produces strong 4th–order energy residuals in the difference $\Phi''''\dot{v}_b^4 - 3\Phi''\ddot{w}^2$ surpassing by some order of magnitude the geometrical ones. So, an exact evaluation of the latter, while desirable, is not strictly required for practical purposes.

The Karman–Marguerre model   A simple plate model can be derived from the classical Karman–Margherre theory for slender plates. Assuming a cartesian coordinate system $\{x_1, x_2\}$ in the plate middle–plane and using commas for representing derivatives, the model is based on the following strain measures:

$$
\begin{cases}
\varepsilon_{11}[u] = u_{1,1} + u_{3,1}^{\,2}/2 \\
\varepsilon_{22}[u] = u_{2,2} + u_{3,2}^{\,2}/2 \\
\varepsilon_{12}[u] = (u_{1,2} + u_{2,1} + u_{3,1}\, u_{3,2})/2
\end{cases}
\quad , \quad
\begin{cases}
\chi_{11}[u] = u_{3,11} \\
\chi_{22}[u] = u_{3,22} \\
\chi_{12}[u] = u_{3,12}
\end{cases}
\tag{3.58}
$$

u_1, u_2 and u_3 being the in–plane and out–of–plane displacements of the middle–plane, and it is defined by the following expression for the strain energy:

$$
\Phi[u] := \frac{1}{2} \int_A \{ N_{ij}\varepsilon_{ij} + M_{ij}\chi_{ij} \}\, dA
\tag{3.59}
$$

where A is the plied domain and the stresses N_{ij} and M_{ij} are related to ε_{ij} and χ_{ij} through the linear constitutive equations:

$$
\begin{cases}
N_{ij} = C_{ijhk}\varepsilon_{hk} \\
M_{ij} = D_{ijhk}\chi_{hk}
\end{cases}
\quad , \quad
i, j, h, k = \{1, 2\}
\tag{3.60}
$$

the elastic tensors C_{ijhk} and D_{ijhk} being totally symmetric and positive defined.

The model assumes that in plane rotations $u_{1,2}$ and $u_{2,1}$ are negligible and, within this assumption, can be considered as a direct extension of the technical beam mode and then it is only 3rd–order accurate.

Note that, while 3rd–order accuracy could be considered acceptable in practical applications, the real drawback consists in neglecting any nonlinear effect of the in–plane rotations. For a single plate the in–plane nonlinear effect plays a minor role and can be considered negligible. This is not true for more complex structures of technical interest, modeled as plate assemblages (thin–walled structures, aeronautical panels etc.) for which the in plane rotation plays the same role as the out–of–plane ones.

Accounting for in–plane rotations   A simple way to account for nonlinearities due to in–plane rotations can be derived from the classical Green Lagrange strain tensor, that in the following we denote as *Complete Green–Lagrange* strain tensor (CL)

$$
\varepsilon_{ij} := u_{i,j} + \frac{1}{2} u_{k,i}\, u_{k,j} \quad , \quad i, j, k = 1, 2, 3
\tag{3.61}
$$

by assuming

$$
\begin{cases}
\varepsilon_{11}[u] = u_{1,1} + \dfrac{1}{2}(u_{1,1}^{\,2} + u_{2,1}^{\,2} + u_{3,1}^{\,2}) \\[2mm]
\varepsilon_{22}[u] = u_{2,2} + \dfrac{1}{2}(u_{1,2}^{\,2} + u_{2,2}^{\,2} + u_{3,1}^{\,2}) \\[2mm]
\varepsilon_{12}[u] = \dfrac{1}{2}(u_{1,2} + u_{2,1}) + \dfrac{1}{2}(u_{1,1}\, u_{1,2} + u_{2,1}\, u_{2,2} + u_{3,1}\, u_{3,2})
\end{cases}
\tag{3.62}
$$

χ_{11}, χ_{22} and χ_{12} already being defined as in (3.58).

The resulting plate model will be rational for in–plane displacements, but the presence of terms such as $u_{1,1}^2$, $u_{2,2}^2$ and $u_{1,1}\,u_{1,2}+u_{2,1}\,u_{2,2}$ provides an underestimate of the post–buckling curvature if considering out–of–plane displacements (for the Euler beam we obtained $\ddot{\lambda}_b \approx -3\lambda_b/4$). Buckling in plates essentially develops as out–of plane displacements, so it is generally convenient to neglect these terms and refer to the *simplified Green–Lagrange strains*(SL)

$$\begin{cases} \varepsilon_{11}[u] = u_{1,1} + \dfrac{1}{2}(u_{2,1}^2 + u_{3,1}^2) \\[2mm] \varepsilon_{22}[u] = u_{2,2} + \dfrac{1}{2}(u_{1,2}^2 + u_{3,1}^2) \\[2mm] \varepsilon_{12}[u] = \dfrac{1}{2}(u_{1,2} + u_{2,1}) + \dfrac{1}{2}(u_{3,1}\,u_{3,2}) \end{cases} \tag{3.63}$$

This makes the description of the plane deformation less accurate but provides a better description of the out-of-plane displacements. So it can be considered to be more suitable in plate analysis and, if considering the cited stress redistribution effect, adequate for technical purposes.

It is worth remembering, however, that the model is not 4th–order accurate and some errors occur in the evaluation of the post–critical curvature $\ddot{\lambda}_b$, which is slightly underestimated. Can we do better than this? The answer is obviously yes. However it is not that easy to obtain a good compromise between accuracy, formulation complexity and finite element implementation practicability, especially when considering the need to avoid nonlinear discretization locking phenomena. We do not consider these topics here, but can only cite that a possible convenient way is obtained by the use of so called *co-rotational approach*.

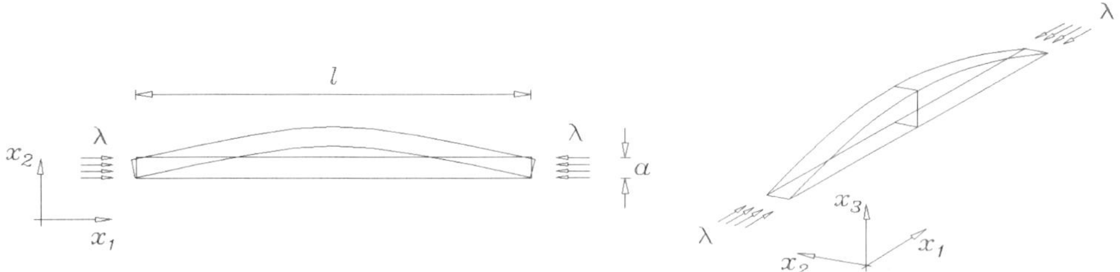

Figure 14. Rectangular plate uniformly compressed, in– and out–of–plane buckling

Differencies between complete and simplified Lagrangian strain measures
The different behaviour when using (3.62) or (3.63) can be shown by the examples in fig.24, relative to the in–plane and out–of–plane buckling of a rectangular plate uniformly compressed in its longitudinal direction. The stress redistribution does not occur in this case so the differences are particularly emphasized. In both cases the results for $\ddot{\lambda}_b/\lambda_b$, reported in tab. 1 and 2, have to be compared with the expected theoretical result $\ddot{\lambda}_b/\lambda_b = 1/4$ obtained for the Euler beam case.

Note that, the CL (complete Lagrangian) strain measure (3.62) recovers the exact result ($\ddot{\lambda}/\lambda_b \approx 0.25$) for the in–plane case, but gives $\ddot{\lambda}/\lambda_b \approx -0.75$ for the out–of–plane case. Conversely the SL (simplified Lagrangian) measure (3.63) provides the values $\ddot{\lambda}/\lambda_b \approx 2.0$ and $\ddot{\lambda}/\lambda_b = 0.0$, respectively. The effects of these errors however are small (about 1% of load for transverse displacements 3% of the beam length, in the worst case). They can be considered as quite negligible for typical cases of thin–walled structures where, due to their high internal statical redundancy, buckling is followed by a strong stress redistribution, as widely discussed in Lanzo, Garcea (1996), and Garcea (2001).

Table 1. In plane plate buckling.   **Table 2.** Out of plane plate buckling.

<table>
<tr><td></td><td colspan="2" align="center">$\ddot{\lambda}_b/\lambda_b$</td></tr>
<tr><td>grid</td><td>SL</td><td>CL</td></tr>
<tr><td>16 × 3</td><td>2.05200</td><td>0.33144</td></tr>
<tr><td>32 × 3</td><td>2.00060</td><td>0.25126</td></tr>
<tr><td>64 × 3</td><td>2.00000</td><td>0.25072</td></tr>
</table>

<table>
<tr><td></td><td colspan="2" align="center">$\ddot{\lambda}_b/\lambda_b$</td></tr>
<tr><td>grid</td><td>SL</td><td>CL</td></tr>
<tr><td>16 × 3</td><td>0.04098</td><td>-0.70902</td></tr>
<tr><td>32 × 3</td><td>0.00062</td><td>-0.74938</td></tr>
<tr><td>64 × 3</td><td>0.00000</td><td>-0.75000</td></tr>
</table>

Some remarks about finite element discretization   Finite element discretization should account for the following considerations:

1. "Exact" shape functions, like that used for beams, are not available for the plate problem.
2. While the buckling is usually characterized by a high degree of displacement continuity, we might need to use fine grids for describing the kinematic complexity of the buckling phenomenon.
3. The finite element model should be insensitive to nonlinear discretization locking in the evaluation of terms $\Phi_b''''\dot{v}_b^4 - 3\Phi_b''\ddot{w}^2$. That requires a careful correlation between in–plane and out–of–plain interpolation. The use of fine grids can help in this.
4. Particular care is also required to avoid high—order kinematical incoherences in the inter–element continuity conditions.
5. Nonlinear internal or external constraints have to be avoided or explicitly taken into account.

The so called **HC** elements (see Aristodemo (1985)) seem a good possible choice for rectangular plates, being characterized by the following features: i) high continuity (C^1) of the overall displacement field; ii) reduced number of description parameters (one node per element); iii) same shape functions for both in–plane and out–of–plane displacements; element construction through a simple and computationally fast algebra.

Numerical experience using KASP code   A general purpose numerical code implementing the post–buckling analysis of slender panels assemblies was developed a few years ago at the University of Calabria within a Brite–Eurom project (see Bilotta et al. (1997), Garcea et al. (1998)). The code called **KASP**, i.e. "**K**oiter's **A**nalysis of

Slender **P**anels", uses a **HC** finite element discretization of each panel and a partitioned description for the overall assembly.

Two successive versions of the code have been implemented and tested successfully: the first one assumes small pre–buckling rotations and uses the compatible formulation described before; the second includes nonlinear pre–buckling behaviour and uses a mixed formulation. Both include multimodal post–buckling analysis.

This is not discussed in detail here. The reader can refer to Lanzo et al. (1995) and Lanzo, Garcea (1996) for a general overview of modelling and implementation choices and to Garcea (2001) for a discussion of nonlinear pre–buckling and mixed formulation. The papers cited also report a series of numerical tests showing the accuracy and effectiveness of the asymptotic approach. Further technical detail about the implementations are reported in Bilotta et al. (1997), Garcea et al. (1998) (the reports can be downloaded from "www.labmec.unical.it").

4 An overview of multimodal buckling

In many cases the buckling phenomenon arises from the interaction between several possible buckling modes. A complex behaviour is produced in that case, usually accompained by strong imperfection sensitivity, which is very difficult to be recovered correctly by standard path–following analysis.

Asymptotic analysis is particularly well suited for multimodal buckling and appears to be the only approach suitable for performing effective imperfection sensitivity analysis in this context. An overview of this and related topics will be given in this section.

4.1 Multimodal analysis

Buckling modes interaction   We have assumed that structural behaviour is dominated by one buckling mode. In some cases, however, due to the high optimization of the structural design, several nearly simultaneous modes $\dot{v}_i$ participate in the buckling phenomenon.

The monomodal analysis discussed before is easily generalized to the multimodal case by assuming that m bifurcation points, defined by the conditions

$$\Phi''[\lambda_i]\dot{v}_i\delta u = 0 \ , \ \forall \delta u \ , \ (\lambda_1 \leq \lambda_2 \leq \cdots \leq \lambda_m) \tag{4.1}$$

are present on the fundamental path $u^f[\lambda]$, and they are sufficiently near to allow the following linearization:

$$\Phi''_b\dot{v}_i\delta u + (\lambda_i - \lambda_b)\Phi'''_b\hat{u}_b\dot{v}_i\delta u = 0 \ , \ \forall \delta u \tag{4.2}$$

λ_b being an opportune reference value of λ (e.g. the first of $\{\lambda_i\}$ or their mean value). Obviously we can assume

$$\Phi'''_b\hat{u}_b\dot{v}_i\dot{v}_j = \left\{ \begin{array}{ll} -1 & \text{if } i \neq j \\ 0 & \text{if } i = j \end{array} \right. \tag{4.3}$$

Then, the path expansion is directly generalized into

$$u[\lambda, \xi_i] := u^f[\lambda] + \sum_{i=1}^{m} \xi_i \dot{v}_i + w[\lambda, \xi_i] \, , \quad w \perp \dot{v}_i \, , \ i = 1 \cdots m \tag{4.4}$$

where, accepting a quadratic expansion for the orthogonal correction $w[\lambda, \xi_i]$, we can assume:

$$w[\lambda, \xi_i] := \frac{1}{2} \sum_{i,j=0}^{m} \xi_i \xi_j \ddot{w}_{ij} \, , \ (\xi_0 := \lambda - \lambda_b \, , \dot{v}_0 := \hat{u}) \tag{4.5}$$

The multimodal algorithm   Using the following orthogonality condition:

$$w \in \mathcal{W} : \Phi_b''' \hat{u} \dot{v}_i w = 0, \ i = 1 \ldots m \tag{4.6}$$

and expanding the Galerkin equilibrium equations

$$\begin{cases} (\Phi'[u[\lambda, \xi] - \lambda \hat{p}) \delta w = 0 \, , \delta w \in \mathcal{W} & \text{(3rd–order expansion)} \\ (\Phi'[u[\lambda, \xi] - \lambda \hat{p}) \dot{v}_i = 0 \, , \ i = 1, \cdots, m & \text{(4th–order expansion)} \end{cases} \tag{4.7}$$

we obtain

$$\Phi_b'' \ddot{w}_{ij} \delta w + \Phi_b''' \dot{v}_i \dot{v}_j \delta w = 0 \, , \ \ \forall \, \delta w \in \mathcal{W} \, , \ \ \ddot{w}_{ij} \in \mathcal{W} \tag{4.8}$$

The simplified form for the algebraic path equation (3.13) generalizes into:

$$\xi_k(\lambda_b - \lambda) + \frac{1}{2} \sum_{i,j=1}^{m} \xi_i \xi_j \mathcal{A}_{ijk} + \frac{1}{6} \sum_{i,j,h=1}^{m} \xi_i \xi_j \xi_h \mathcal{B}_{ijhk} = \mu_i[\lambda] \tag{4.9}$$

where:

$$\begin{cases} \mathcal{A}_{ijk} & := \Phi_b''' \dot{v}_i \dot{v}_j \dot{v}_k \\ \mathcal{B}_{ijhk} & := \Phi_b'''' \dot{v}_i \dot{v}_j \dot{v}_h \dot{v}_k - \Phi_b''(\ddot{w}_{ij} \ddot{w}_{hk} + \ddot{w}_{ih} \ddot{w}_{jk} + \ddot{w}_{ik} \ddot{w}_{jh}) \\ \mu_i[\lambda] & := -\frac{1}{2} \lambda^2 \Phi_0''' \hat{u}_0^2 \dot{v}_i - \lambda(\Phi_b''' \tilde{u} \hat{u}_b \dot{v}_i - \tilde{q} \dot{v}_i) \end{cases} \tag{4.10}$$

$\tilde{u}$ and $\lambda \tilde{q}$ being the additional geometrical and load imperfections.

Some comments are useful:

1. The nonlinear algebraic system (4.9) has a low number of unknowns (the number of considered buckling modes) with respect to the total variables of the problem. It however presents a high nonlinearity because it describes the nonlinear behaviour of the whole structure.

2. The analysis for different imperfections does not require the complete analysis process to be repeated but only the solution of the algebraic equation in λ and $\{\xi_i\}$ for different μ_k has to be actually repeated.

3. The representation trrough buckling modes (and associated orthogonal corrections) permits the recover of both snapping and bifurcation phenomena.

4. Because of the use of the quadratic algorithm, we need a rational model. Furthermore, we have a possible locking in the calculation of the terms $\mathcal{B}_{ijhk}$.

Furthermore note that, by omitting the 4th–order terms, we get a simpler, albeit slightly less accurate, formula corresponding to the multimodal extension of the linear algorithm.

$$\xi_k(\lambda_b - \lambda) + \frac{1}{2} \sum_{i,j=1}^{m} \xi_i \xi_j \mathcal{A}_{ijk} = \mu_i[\lambda] \tag{4.11}$$

Relations with the modal reduction method   The previous equation corresponds to a highly nonlinear system in the m unknowns ξ_i and can be solved using standard Riks methods by reproducing the complete post–buckling behaviour of the structure, including *modal interactions* and *jumping–after–bifurcation* phenomena.

We can notice analogies between the proposed asymptotic method and the modal reduction methods (see Noor (1981), Noor, Peters (1983), Noor (1994)) based on expansions of the displacement field in terms of a combination of significant modes and of the Galerkin solution of the equilibrium equation on these modes. It has however the following advantages:

1. the asymptotic expansion derives from a coherent theoretical framework;

2. the modal reduction method is noticeably more expensive than the solution of the explicit algebraic system in λ and ξ_i whose coefficients are computed once and for all;

3. the use of a "finite–step" equilibrium condition in spite of its (coherent) asymptotic expansion and the omission of the orthogonal correction $\frac{1}{2}\xi_i\xi_j w_{ij}$, generally introduces spurious terms which tend to produce locking in the solution.

An example of multimodal buckling analysis   Some results of multimodal analysis are reported in figs.15, 16, related to an uniformly compressed Channel–section beam and obtained using KASP code. Eleven buckling modes have considered racing in the range $\lambda = 1.4 \div 1.6$ collecting flexural, torsional and local modes with different numbers of half–waves. The analyses have been performed by the KASP code, considering two different imperfections: local–flexural for *case A* and local–torsional for *case B*. The results are compared with solutions reported in Ali and Sridharan (1988) and with numerical values obtained by path following analysis using MSC–NASTRAN. More details on the analises can be found in Lanzo, Garcea (1996).

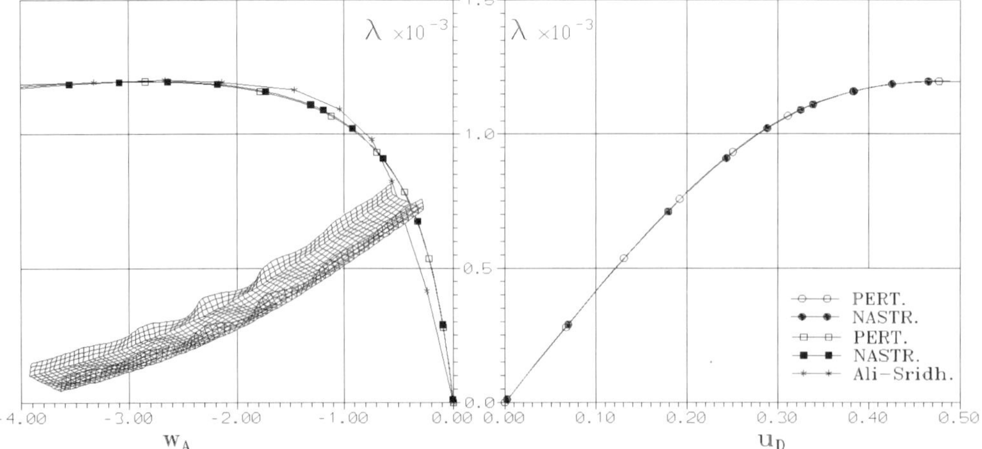

Figure 15. *Example 1.* Load vs displacements for case A.

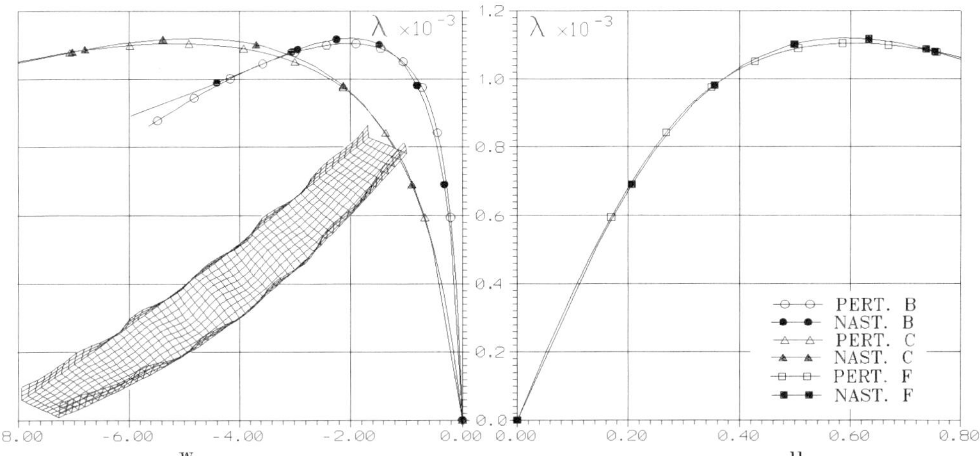

Figure 16. *Example 1.* Load vs displacements for case B

A direct comparison with the NASTRAN results and with those reported in Ali and Sridharan (1988) is carried out in figs.15 and 16. No noticeably different behavior is observed in using the complete Lagrangean (eq.(3.62)) and the simplified plate models (eq.(3.63)), the first giving only a reduction of about 0.8% of the limit load.

4.2 Attractive path theory and imperfection sensitivity analysis

Imperfection sensitivity for multimodal buckling The aim of the imperfection sensitivity analysis is to link the presence of geometrical and load imperfections to the reduction in the limit load. The analysis has a relevant technical interest in the presence of modal interaction, which generally provides high imperfection sensitivity.

The sensitivity largely depends on the initial post–critical coefficients, that is on both the initial slope $\dot{\lambda}_b$ and curvature $\ddot{\lambda}_b$ of the bifurcated path. However, exhaustive results can only be obtained in a statistical context linking the distribution probability of the imperfection to that of the load.

In multimodal buckling, this implies the repeated solution of the algebraic path equation (4.9) for all possible imperfection forms $\tilde{u}$ and $\tilde{q}$. The number of repetitions needed to obtain statistically reliable results increases (quite) exponentially with the number m of the buckling modes and for large m can become very expensive. We could however perform the imperfection sensitivity analysis in a simple and efficient way if we choose a criterion for defining the (few) "significant" imperfection forms, so reducing to some (few) monomodal analyses.

Multiple bifurcation The modal interaction complicates the bifurcation process and increases the imperfection sensitivity.

To simplify the discussion, let's consider the case of perfect structures presenting simultaneous modes ($\lambda_1 = \cdots = \lambda_m$) and refer to the linear algorithm (4.11). The basic equation becomes:

$$\frac{1}{2} \sum_{i,j=1}^{m} \xi_i \xi_j \mathcal{A}_{ijk} - \xi_i(\lambda - \lambda_b) = 0 \tag{4.12}$$

and, making

$$\begin{cases} \lambda &= \lambda_b + \alpha \dot{\lambda}_b \\ \xi_i &= \alpha \dot{\xi}_i \end{cases} , \quad \sum_{i=1}^{m} \dot{\xi}_i \dot{\xi}_i = 1 \tag{4.13}$$

it can be rewritten in variational form as

$$\dot{\lambda}_b \equiv \frac{1}{2} \sum_{i,j,k=1}^{m} \mathcal{A}_{ijk} \dot{\xi}_i \dot{\xi}_j \dot{\xi}_k = \text{stationary} , \quad \dot{\xi}_i \dot{\xi}_i = 1 \tag{4.14}$$

Condition (4.14) provides, in general, a high number of solutions (until $2^n - 1$) each of which characterizes a possible branch defined as a radial path in the plane $\{\xi_i\}$. We will see that paths corresponding to relative minima, are particularly relevant in sensitivity analysis.

Symmetric systems In the case of symmetric problems, we have

$$\mathcal{A}_{ijk} \dot{\xi}_i \dot{\xi}_j \dot{\xi}_k := \Phi_b''' \dot{v}_i \dot{v}_j \dot{v}_k = 0 \; \forall i, j, k \tag{4.15}$$

and the equation in (λ, ξ_i) becomes trivial when considering the linear algorithm. Referring to the quadratic algorithm and proceeding like before, we obtain:

$$\ddot{\lambda}_b \equiv \frac{1}{3} \sum_{i,j,h,k=1}^{m} \mathcal{B}_{ijhk} \dot{\xi}_i \dot{\xi}_j \dot{\xi}_h \dot{\xi}_k = \text{stationary} , \quad \dot{\xi}_i \dot{\xi}_i = 1 \tag{4.16}$$

Also in this case, each solution corresponds to a possible radial branch and those corresponding to relative minima are of special interest.

Determining relative minimum branches The evaluation of all the stationary solutions of the cubic form (in the linear algorithm) or of the quartic form (in the quadratic algorithm) can be expensive. Nevertheless, relative minimum solutions can be (quite easily) obtained by using the iterative scheme suggested in Salerno (1992). Denoting the current estimate by the apex (n) and its iterate by $(n + 1)$, we obtain

$$\begin{cases} \tilde{\xi}_k := \mathcal{A}_{ijk} \dot{\xi}_i^{(n)} \dot{\xi}_j^{(n)} \\ \dot{\xi}_k^{(n+1)} := \tilde{\xi}_k / \|\tilde{\boldsymbol{\xi}}\| \end{cases} \tag{4.17}$$

for the cubic systems and

$$\begin{cases} \tilde{\xi}_k := \mathcal{B}_{ijhk} \dot{\xi}_i^{(n)} \dot{\xi}_j^{(n)} \dot{\xi}_h^{(n)} \\ \dot{\xi}_k^{(n+1)} := \tilde{\xi}_k / \|\tilde{\boldsymbol{\xi}}\| \end{cases} \tag{4.18}$$

for quartic ones. Both schemes can be considered as a generalization of the classical power method scheme for solving standard eigenvalue problems $\mathbf{Kx} = k\,\mathbf{x}$:

$$\begin{cases} \tilde{x}_j := K_{ij} x_i^{(n)} \\ x_j^{(n+1)} := \tilde{x}_j / \|\tilde{\mathbf{x}}\| \end{cases} \tag{4.19}$$

and have the same convergence behaviour of this latter.

Note that both schemes tend to provide solutions characterized by a maximum absolute value for the cubic or quartic forms and so fail to our scope if they provide a positive value for the form. However we can easily avoid this inconvenience: by simply taking the opposite direction, in case of cubic form, and by introducing an appropriate shift in case of quartic form

$$\tilde{\xi}_k := (\mathcal{B}_{ijhk} - c)\dot{\xi}_i^{(n)}\dot{\xi}_j^{(n)}\dot{\xi}_h^{(n)} \tag{4.20}$$

c being a suitable large value.

Attractive paths theory A detailed discussion of the solutions, for both the cubic and the quartic systems, in the presence of external imperfection $\mu_i \neq 0$ has been given in Salerno, Casciaro (1997). The reader is referred to this paper for a deeper insight in this topic.

We only recall here the main results that where obtained.

1. Radial paths $\xi_i = \alpha\dot{\xi}_i$, satisfying stationarity for the cubic or quartic form, are solutions of the imperfect equilibrium equation for $\mu_i = t\dot{\xi}_i$.

2. Radial paths where the stationarity is attained as absolute minimum have a limit point as the first critical point. The corresponding load multiplier provides a lower bound for critical loads (limit or bifurcation) obtained along generically imperfect paths under the same imperfection norm, assuming that $\|\mu\| := (\sum \mu_i^2)^{1/2}$. This is actually an old result, coming from the so called Ho's theorem (see Ho (1972) and Koiter (1974)).

3. Radial paths corresponding to a relative (local) minimum do not present bifurcation points.

4. Radial paths corresponding to a local minimum are attractive with respect to generically imperfect but sufficiently close paths.

5. Radial paths not corresponding to a local minimum present secondary bifurcations, that is, in the current terminology, *modal jumping after bifurcation*.

Sensitivity analysis by monomodal reduction Imperfection sensitivity analysis, while being very easy in cases of single buckling mode, can become much more complex and time consuming when multiple buckling modes are considered. So a strategy which allows the multimodal sensitivity analysis to be reduced to one or few monomodal analyses is convenient in practice.

The general idea is to select a–priori significant buckling shapes, i.e. directions in the $\{\xi_i\}$ plane, and take them within separate monomodal sensitivity analyses. This selection should contain:

1. with reference to the collapse safety: shapes providing the maximum reduction in the limit load with the same imperfection probability.

2. with reference to the predictability of the mechanical phenomenon: shapes that are unaffected by secondary buckling phenomenon and so represent equilibrium configurations the structure could achieve.

Ho's theorem seems to provide a good answer to the first requirement: when considering shapes that correspond to an absolute minimum for the cubic form (or for the quartic form in cases of symmetric or nearly symmetric problems) we actually obtain a lower bound for the collapse load for any given imperfection size. Consider however that "equal size" (actually defined according to a Cartesian norm based on normalization (4.3)) does not generally imply "equal imperfection probability" so this choice, in itself, is of minor interest.

The whole post–critical behaviour of the structure is dominated by the directions of relative minimum for the cubic form (or for the quartic form in cases of symmetric or nearly symmetric systems). Paths that start in different directions, after a complex sequence of modal jumping tend to follow one of these dominant trajectories.

Once all attractive radial paths have been determined, an efficient sensitivity analysis can be performed as follows:

1. Using a MonteCarlo process we generate a sequence of data vectors modelling load and geometrical imperfections with the required random characteristics.

2. Each imperfect path so generated will be attracted by one of the minimum radial paths. We do not have, at the moment, theoretical information for defining this path univocally. However, a lower bound evaluation of the limit load can be obtained considering all minimum directions and referring to the smallest value obtained by performing a monomodal analysis in each of these directions. This implies simple operations and is extremely fast.

3. Repeating the analysis for the different imperfection vectors, we can derive the probability density function for the collapse multiplier and then obtain a meaningful reference value for the safety factor of the structure.

Several numerical tests have been performed for evaluating the practical effectiveness of this simplified procedure and its accuracy in comparison with full asymptotic analysis and with a much more time consuming process based on separate path–following analysis (see Salerno et al. (1996)). Excellent agreement has always been obtained in the results from the three different strategies, so confirming the effectiveness and accuracy of the simplified analysis.

4.3 Local buckling

Local modes The word *local mode* is generally used for buckling shapes confined to a part of the structure: for example, in assemblies of slender panels, buckling confined to a single panel or, in pin-jointed trusses, the Eulerian buckling of a single rod is said to be local. The word is mainly used, as qualitative description terminology, in the context of analytical solutions where the concept of localization could help to obtain approximate solutions.

In a general FEM context, where ad hoc approximations are not possible, a qualitative distinction between local and global modes is ambiguous and without interest.

A precise (quantitative) definition of local mode is then required. Let's assume the following

Proposition 4.1 (Local modes). *The critical modes $\dot{v}_i$ e $\dot{v}_j$ are called* local *if the following conditions of energetic uncoupling hold:*

$$
\begin{aligned}
\Phi_b''' \dot{v}_i \dot{v}_j \delta u &= 0 , \quad \forall \delta u \\
\Phi_b'''' \dot{v}_i \dot{v}_j \delta u_1 \delta u_2 &= 0 , \quad \forall \delta u_1 , \ \delta u_2
\end{aligned}
\tag{4.21}
$$

Proposition 4.2 (Symmetric modes). *The mode $\dot{v}_i$ is called symmetric if it renders null all variations in which it is present an odd number of times:*

$$
\begin{aligned}
\Phi_b''' \dot{v}_i \delta u_1 \delta u_2 &= \Phi_b''' \dot{v}_i^3 = 0 , \quad \forall \delta u_1, \ \delta u_2 \\
\Phi_b'''' \dot{v}_i \delta u_1 \delta u_2 \delta u_3 &= \Phi_b'''' \dot{v}_i^3 \delta u_1 = 0 , \quad \forall \delta u_1, \ \delta u_2, \ \delta u_3
\end{aligned}
\tag{4.22}
$$

Local modes are then characterized by an energetic mutual uncoupling. Global modes, if present, are not necessarily uncoupled. A typical example of symmetric local mode is the Eulerian displacement of an axial loaded rod in a reticular system.

The distinction is useful because it can make the solution process computationally less expensive, as we can see in the next subsection.

A global mode and local symmetric modes   The phenomenon can be modeled by a cubic system because the cubic form $\mathcal{A}_{ijk}$ is not completely null.

$$
\dot{\lambda}_b = \frac{1}{2} \mathcal{A}_{ijk} \dot{\xi}_i \dot{\xi}_j \dot{\xi}_k = \min , \qquad \dot{\xi}_i \dot{\xi}_i = 1
\tag{4.23}
$$

The hypothesis of symmetry and localization for the modes $\dot{v}_i$ for $i = 2 \cdots m$ leads to the following simplification:

$$
\dot{\lambda}_b = \frac{1}{2} \Phi_b''' \dot{v}_1^3 \dot{\xi}_1^3 + \frac{3}{2} \sum_{i=2}^{m} \Phi_b''' \dot{v}_1 \dot{v}_i^2 \dot{\xi}_1 \dot{\xi}_i^2 = \min , \qquad \dot{\xi}_i \dot{\xi}_i = 1
\tag{4.24}
$$

which allows the solution in closed form

$$
\dot{\lambda}_b = \frac{1}{2} \Phi_b''' \dot{v}_1^3 \dot{\xi}_1^3 + \frac{3}{2} \Phi_b''' \dot{v}_1 \dot{v}_{j*}^2 \dot{\xi}_1 \dot{\xi}_{j*}^2 = \min , \qquad \dot{\xi}_1^2 + \dot{\xi}_{j*}^2 = 1
\tag{4.25}
$$

where:

$$
j^* = j \in [2 \cdots m] : \| \Phi_b''' \dot{v}_1 \dot{v}_j^2 \| = \max
\tag{4.26}
$$

Note that the evaluation of the global mode requires an eigenvalue analysis of the overall structure, while the local ones are usually a priori known in closed form. The evaluation of the dominant buckling mode requires the simple minimization of a two variable polynomial function. The problem is particularly well suited for monomodal imperfection analysis.

In figs. 17 the numerical results obtained for the simple Euler beam truss are reported. The analysis has been performed with the radial path theory, a multimodal asymptotic analysis taking into account only the local and the global mode and with a Riks analysis in which the compression axial stresses have been constrained not to overcome the Euler bifurcation load of the single truss.

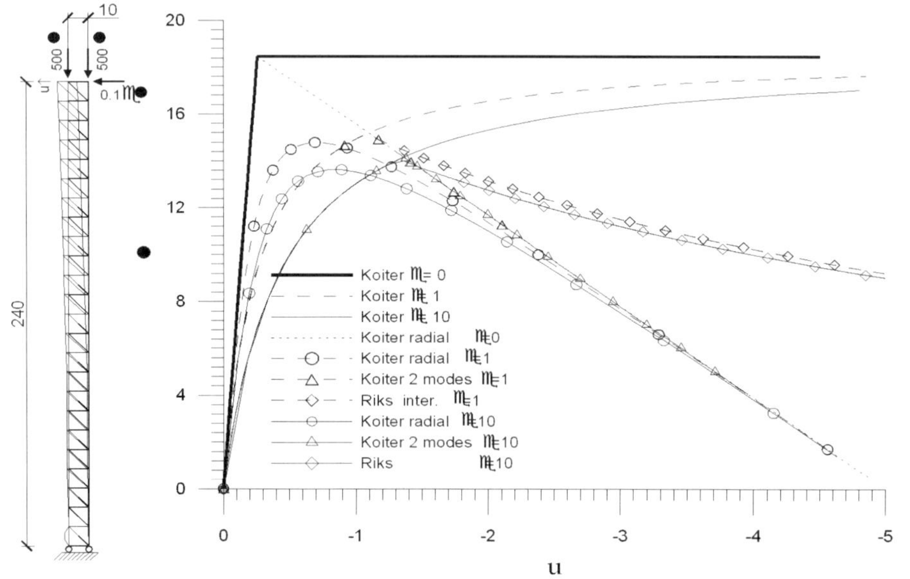

Figure 17. Equilibrium path: global mode and symmetric local modes

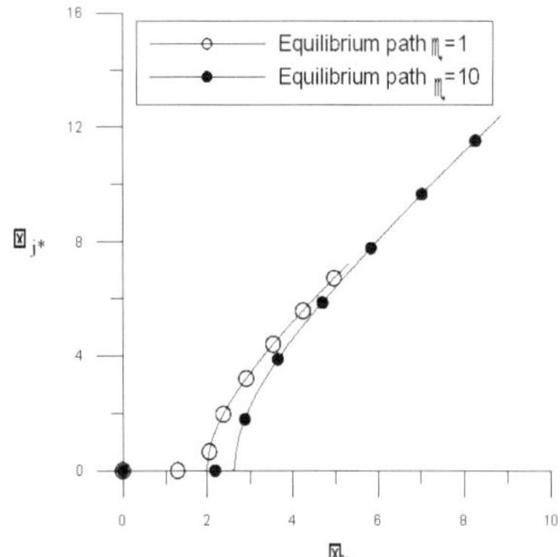

Figure 18. Path in x_1–ξ_j plane

λ_{lim}	Radial	TwoModal	Riks
$\epsilon = 1$	14.787	15.188	15.068
$\epsilon = 10$	13.6224	14.155	13.992
$\epsilon = 100$	8.925	9.821	9.449

Figure 19. Limit loads for Euler beam

Symmetric local modes Since all the critical modes are symmetric, the cubic form $\mathcal{A}_{ijk}$ is null. Then the problem can be modeled as a quartic system.

$$\ddot{\lambda}_b = \frac{1}{3}\mathcal{B}_{ijhk}\dot{\xi}_i\dot{\xi}_j\dot{\xi}_h\dot{\xi}_k = \min , \qquad \dot{\xi}_i\dot{\xi}_i = 1 \tag{4.27}$$

The symmetry and localization hypothesis for all the $\dot{v}_i$ modes leads to the following simplification:

$$\ddot{\lambda}_b = \frac{1}{3}\sum_{i=1}^{m}\Phi_b''''\dot{v}_i^4\dot{\xi}_i^4 - \sum_{i=1}^{m}\sum_{j=1}^{m}\Phi_b''\dot{w}_{ii}\dot{w}_{jj}\dot{\xi}_i^2\dot{\xi}_j^2 = \min , \qquad \dot{\xi}_i\dot{\xi}_i = 1 \tag{4.28}$$

Making:
$$x_i = \dot{\xi}_i^2 , \qquad F_{ij} = \begin{cases} \frac{1}{3}\Phi_b''''\dot{v}_i^4 - \Phi_b''\dot{w}_{ii}^2 & \text{if } i = j \\ -\Phi_b''\dot{w}_{ii}^2 & \text{if } i \neq j \end{cases} \tag{4.29}$$

the minimum condition for the quartic form can be rewritten as

$$\ddot{\lambda}_b = F_{ij}x_i x_j = \min , \quad x_i \geq 0 , \qquad \sum_{i=1}^{m}x_i = 1 \tag{4.30}$$

and corresponds to a quadratic programming minimization problem.

Note that local modes are usually a–priori known in closed form. The evaluation of the corrective displacements w_{ii} requires the solution of the m linear systems

$$\Phi_b''w_{ii}\delta u = -\Phi'''\dot{v}_i^2\delta u , \quad i = 1\cdots m \tag{4.31}$$

where the known right term is also usually known in closed form. The solution of the minimization problems, if we want to obtain all relative minima, is less obvious, matrix **F** being in general not positive definite, and requires specialized solution algorithms such as that presented in Casciaro et al. (2003).

The algorithm was implemented in a finite element computer code for the analysis of slender 3D-Truss. In fig.20 we present a test case. Due to the polar symmetry of the structure also the orthogonal corrections w_{ii} (fig.21) and the 12 "dominant byckling shapes", corresponding to the minimum directions, are symmetric. They are be grouped in two significant sets (fig.22). The sensitivity analysis performed with the MonteCarlo technique for 2000 different imperfections allows statistical informations on the collapse load (fig.23) to be obtained, in an efficient way (2 seconds are necessary to complete the analysis).

Multiple bifurcations: symmetric local modes

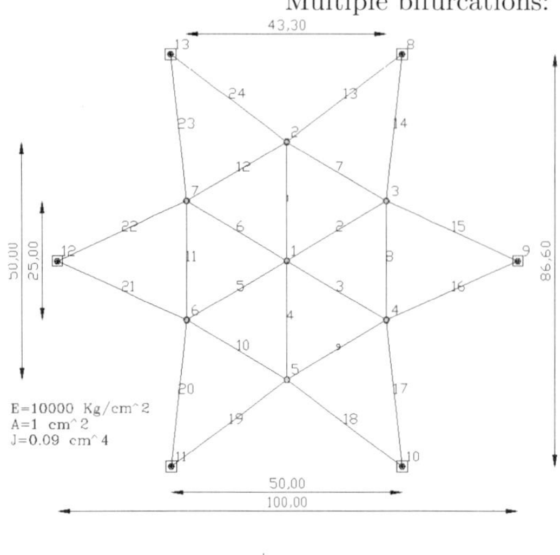

Figure 20. Structure geometry

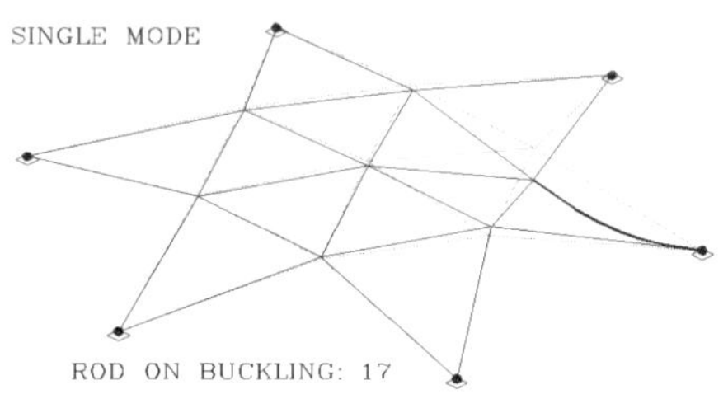

Bar	24	Buckled bars	13-24
Nodes	13	Loaded nodes	1-7
λ_b	3.00	λ_g	11.15

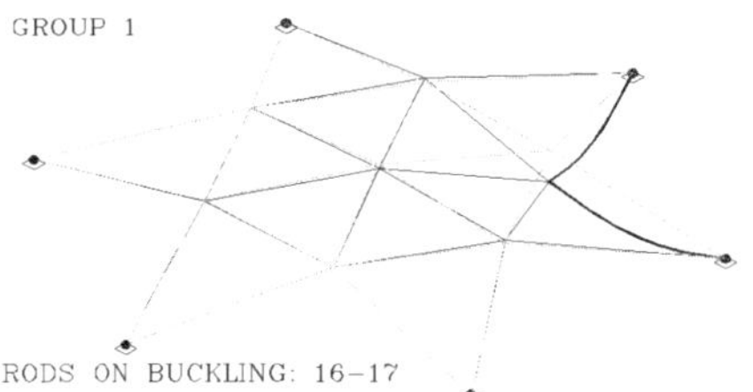

Figure 21. Single mode

Combinations	4095	Minimum founded	12	Time analysis	0.09 sec
λ_b^{**} Group 1	-1.265			λ_b^{**} Group 2	-0.345

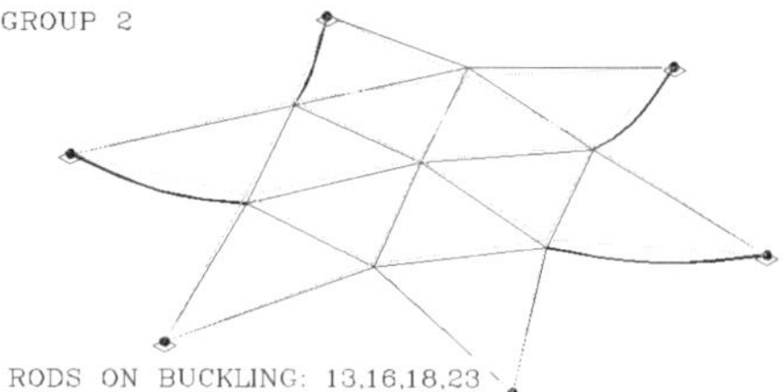

Figure 22. Overall modes

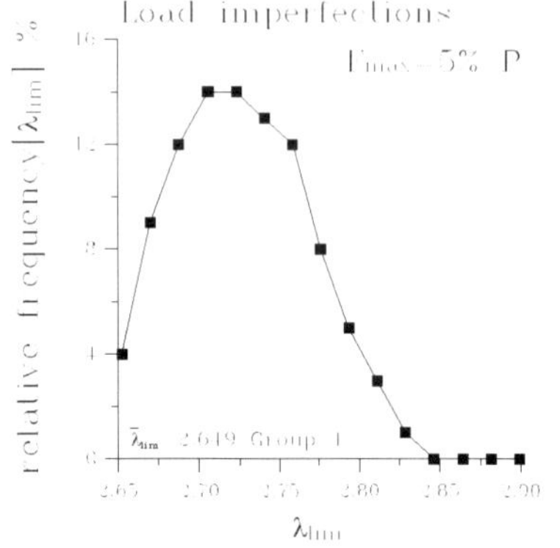

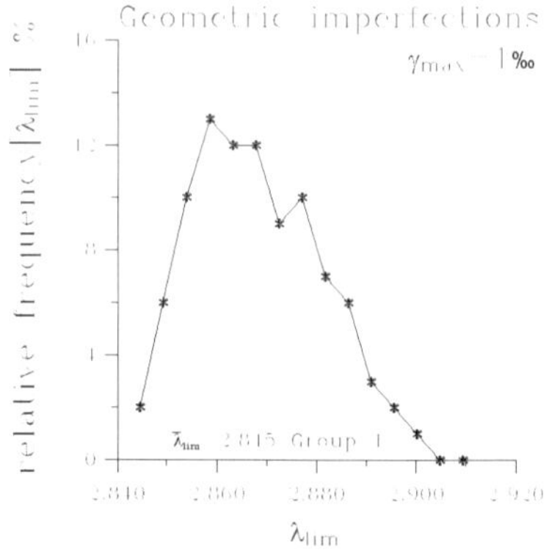

Figure 23. Relative frequency of collapse load

5 Accuracy, reliability effectiveness and related topics

5.1 Reliability of the asymptotic method

Methods that prove to be accurate sometimes but unreliable in general are completely useless. Such as obvious statement is particularly important in the asymptotic analysis because its careless (we can say incorrect) use can lead to very large errors that can destroy the expected accuracy.

An overview of this and related topics will be presented in this section. The reader can refer to Casciaro et al. (1998) and Garcea et al. (1999) for a deeper insight.

Reasons for using asymptotic approach   Safety analysis of slender structures characterized by complex buckling and post–buckling phenomena and by a strong imperfection sensitivity, is largely penalized by a lack of adequate computational tools. In fact, standard algorithms, based on a path–following approach, are extremely expensive in terms of both hardware and computational time. As a consequence, due to the high costs of the single run, it becomes practically impossible to perform the large number of successive runs necessary to the sensitivity analysis.

The asymptotic approach gives a potentially convenient alternative for that purpose. In fact, this method is characterized by several relevant potential advantages: i) a fully automated finite element analysis can be implemented; results can be qualitatively and quantitatively accurate; ii) a complete synthetic description of the structural behaviour is obtained, including the deterioration due to load imperfections, geometrical defects and damage; iii) the analysis is very fast, of the same order as a linearized stability analysis and of some orders less than path–following analysis; iv) re–analysis for different imperfections requires only a small fraction of the first analysis time.

Potential drawbacks of using asymptotic approach   We have already seen that a careless implementation of the asymptotic approach can introduce large errors in the analysis: for instance, using a kinetically incoherent modelling or neglecting nonlinear constraints can noticeably affect the post–buckling behaviour; discretizing without considering nonlinear correlation between the different displacement components can introduce a strong interpolation locking leading to unreliable results. Equally, the usual "at hand" approximations or the omission of "negligible" terms, while being irrelevant in analysis based on the $2nd$–order energy variations, can strongly affect the evaluation of the higher order energy variations preventing a correct recovery of post–buckling behaviour.

The first implementations of the Koiter asymptotic approach within finite element analysis paid insufficient attention to these topics and so unsatisfactory results were obtained. As a consequence, for a long time, the asymptotic method has usually been considered as a powerful tool for obtain analytical "at hand" solutions, in particular cases, but unsuitable to be implemented within a general finite element context that would provide reliable results. The general opinion, within the computational mechanics community, actually was: "the asymptotic method works very well in cases where it works well and that is: in not many cases".

Let's consider, however, that the drawbacks experienced are due to mistakes in implementation more than to intrinsic defects of the asymptotic approach itself, and given adequate care to some, usually considered, minor details, its finite element implementations can be both accurate and reliable.

Locking by extrapolation Some subtle locking phenomena originate from the extrapolation process itself. For a better understanding of the reasons for the problem, consider that the use of asymptotic expansions always provides a correct local description. However it furnishes, at a finite distance, an extrapolation error that is a function of the particular representation used. That is, the finite–distance error depends on the primary variables to be extrapolated (e.g. a linear extrapolation is sufficient to reproduce a circle described in polar coordinates).

The asymptotic method actually uses two different extrapolations:
1. the extrapolation of the fundamental path between (u_0, λ_0) and (u_b, λ_b);
2. the extrapolation of the bifurcated path emerging from (u_b, λ_b).

The first of these, corresponds to a finite–distance extrapolation due to the use of quantities evaluated in both the points (u_0, λ_0) and (u_b, λ_b), an so, quite paradoxically, requires more care.

We will see, in the next section, that the choice of primary variables to be used for describing the structure configurations can significanty affect the final accuracy of results, locking phenomena usually being produced by bad extrapolation choices. We can anticipate here that the use of mixed models, where both tensions and displacements are independently extrapolated, provides, in general, the best choice.

5.2   Extrapolation locking

Now we consider a strong locking phenomenon which arises, in some subtle way, from the extrapolation process itself. For a deeper insight, the reader is referred to Garcea et al. (1999) and to Garcea et al. (1998) where the same topic is discussed in the different context of path–following analysis.

The extrapolation locking phenomenon Even if the asymptotic algorithm is theoretically characterized by a very small error (at least from an asymptotic point of view), its implementation in problems presenting non negligible pre–critical nonlinearities can, sometimes, furnish unexpectedly wrong results.

The phenomenon is produced by the interaction between even small pre–critical rotations and high axial/flexural stiffness ratio of the elements. As a result of this, the accuracy of the asymptotic expansion in recovering the fundamental path noticeably deteriorates.

We can easily see this phenomenon in the simple tests of figs. 24 and 25. In both cases, a small transversal load $\epsilon\lambda$ ($\epsilon << 1$) is added inducing rotations on the fundamental path and large (but quite realistic) values of the stiffness ratio $k := EAl^2/(EJ)$ are assumed.

The results provided by the implementation of the asymptotic method are plotted in figs.26 and 27. We can see that, even for a very small value of ϵ), the bifurcation

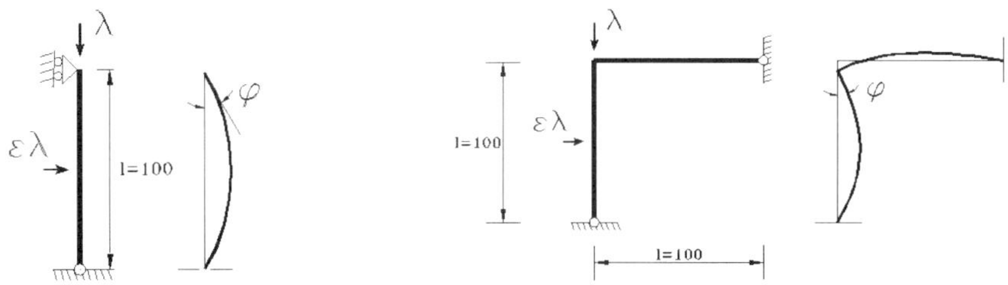

Figure 24. Euler beam.   **Figure 25.** Roorda frame.

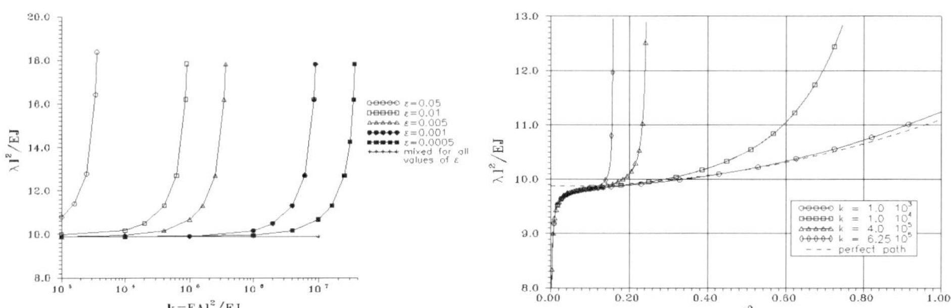

Figure 26. Bifurcation and post–critical locking for the Euler beam

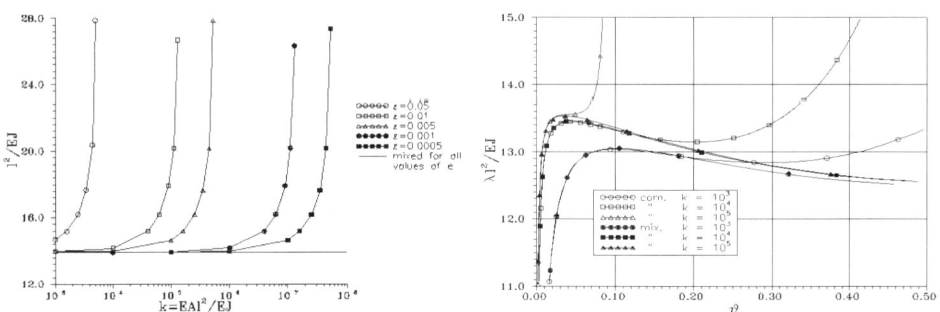

Figure 27. Bifurcation and post–critical locking for the Roorda frame

load is significantly greater than expected and, for increasing values of k, the bifurcation disappears. We can also see that, even in cases where the estimate for λ_b is still apparently unaffected by errors, the evaluation of the post-critical behaviour becomes highly inaccurate. A locking free element discretization has been used, so these results are only related to the extrapolation errors in the evaluation of the fundamental path which affect the estimate for $\Phi''[\lambda]$ (causing an increase in the bifurcation load), and even more its successive derivatives used for recovering the postcritical behaviour.

It is worth mentioning that a possible cure for the problem is obtained, very simply, by treating the horizontal load as an additional imperfection, even if, by this quite rough treatment we should theoretically expect a decrease in accuracy. Another possible trick is that of assuming rotations to be negligible along the fundamental path so the structure can be considered as frozen to the initial geometry. Note, however, that both the tricks cited cannot be implemented in FEM codes intended to analyze general structures, maybe characterized by strong pre–critical nonlinearities.

The role of the primary extrapolation variables   For a better understanding of the reasons for the apparently strange and unexpected behaviour of the asymptotic method shown before, let's consider how extrapolations work.

Remember that asymptotic expansions always provide correct local descriptions, but do not furnish any a–priori quantitative information on the approximation error in the path evaluation at a finite distance from the extrapolation point. The actual error will not only depend on the order of the extrapolation, but also on the particular representation of the equilibrium path, that is, on the choice of the primary variables subject to extrapolation. As an example, consider the circle in fig.28 and try to extrapolate the curve from its upper point. Using a linear extrapolation, we obtain the exact recover when using polar coordinates, and only the initial tangent with very large errors at finite distance when using cartesian coordinates.

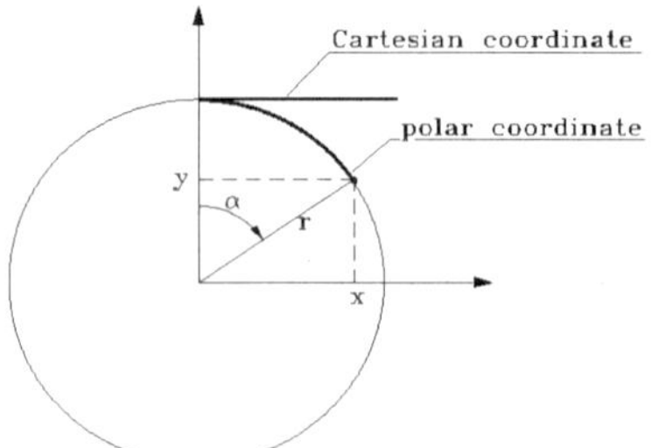

Figure 28. Extrapolation sample

The asymptotic method starts from extrapolating the fundamental path in the rest configuration $(u = 0, \lambda = 0)$. Then, this evaluation for the path is used in the neighborhood of the bifurcation configuration (u_b, λ_b), which is at a finite distance from the initial one, for obtaining the other quantities (the second, third and fourth variations of the strain energy) needed for extrapolating the bifurcated branch and then recovering the post–buckling behaviour. We can expect that the error increases when referring to successive derivatives, so we cannot really be surprised for obtaining large errors in the evaluation of the buckling load (using the second derivatives of the strain energy) and larger in the evaluation of the post–critical slope and curvature (using the third and fourth–order derivatives).

For these reasons, the actual way of defining extrapolations represents a crucial point in the asymptotic approach and an apparently minor but actually very important detail as the appropriate choice for the primary extrapolation variables plays a fundamental role. We will try to show this by a simple implementation example.

A simple test showing extrapolation locking   Consider the simple test–problem in fig.29: a bar of unitary length, characterized by an axial stiffness E and a flexural spring of stiffness K ($K\ell \ll E$ and subjected to horizontal compressive force Q and a transversal force F ($F \ll Q$).

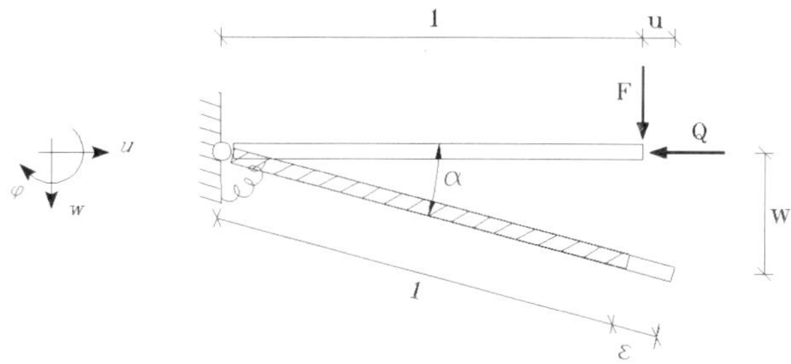

Figure 29. A simple test showing extrapolation locking

The natural choice for describing this problem (within an at–hand procedure) could be the use of the angle of rotation α and the elongation ε. Assuming $u := \{\alpha, \varepsilon\}$, we obtain:

$$\Pi[u] := \frac{1}{2}(K\alpha^2 + E\varepsilon^2) - \lambda Q\left[1 - (1+\varepsilon)\cos\alpha\right] - \lambda F(1+\varepsilon)\sin\alpha \qquad (5.1)$$

By introducing

$$N := \lambda(F\sin\alpha - Q\cos\alpha) \quad , \quad T := \lambda(F\cos\alpha + Q\sin\alpha) \qquad (5.2)$$

we obtain

$$\Pi''[u]\dot{u}\delta u = \left[K + (1+\varepsilon)N\right]\dot{\alpha}\delta\alpha + E\dot{\varepsilon}\delta\varepsilon - T(\dot{\varepsilon}\delta\alpha + \dot{\alpha}\delta\varepsilon) \qquad (5.3)$$

Constitutive relations imply

$$N = E\varepsilon \quad , \quad T = K\alpha/(1+\varepsilon) \qquad (5.4)$$

The critical condition $\Pi''[u]\dot{u}\delta u = 0$ provides

$$N_b = -\frac{K - T_b^2/E}{1 + \varepsilon_b} \quad \Longrightarrow \quad \lambda_b = \frac{K}{Q\cos\alpha - F\sin\alpha}\left(\frac{E - T_b^2/K}{E + N_b}\right) \qquad (5.5)$$

Then, for $K \ll E$ and $F \ll Q$ we have the simplified expression

$$\lambda_b \approx K/Q \tag{5.6}$$

which can be considered as reference value for the bifurcation multiplier.

The use of problem–specialized variables, while convenient within an at–hand analysis, is unsuitable for a computer oriented numerical analysis aiming to construct the relevant vectors and matrices by a standard assemblage procedure. We are then forced to describe the configuration by the free–end displacements u and w or, if using a mixed formulation, by um w and the axial force N.

Assuming small–displacement kinematics

$$\alpha \approx w \quad , \quad \varepsilon \approx u + \frac{1}{2}w^2 \tag{5.7}$$

the potential energy, in the compatible (Π_c) and mixed (Π_m) forms, can be written

$$\Pi_c[u, w] := \frac{1}{2}(Kw^2 + E(u + \frac{1}{2}w^2)^2) - \lambda(Fw - Qu)$$
$$\Pi_m[u, w, N] := \frac{1}{2}(Kw^2 - N^2/E) + N(u + \frac{1}{2}w^2) - \lambda(Fw - Qu) \tag{5.8}$$

From the first, taking u and w as extrapolation variables, we obtain the following linear extrapolation for the fundamental path

$$\begin{cases} u[\lambda] := \lambda\hat{u} = -\lambda Q/E \\ w[\lambda] := \lambda\hat{w} = \lambda F/K \end{cases} \implies N[\lambda] = -\lambda Q + \frac{1}{2}\lambda^2 \frac{F^2}{K}\frac{E}{K} \tag{5.9}$$

where N is treated as dependent variable. From the second, taking both u, w and N as extrapolation variables, we obtain

$$\begin{cases} u[\lambda] := \lambda\hat{u} = -\lambda Q/E \\ w[\lambda] := \lambda\hat{w} = \lambda F/K \\ N[\lambda] := \lambda\hat{N} = -\lambda Q \end{cases} \tag{5.10}$$

Both estimates (5.9) and (5.10) are 1st–order accurate and they are completely equivalent from an asymptotic point of view. In order to evaluate the bifurcation multiplier λ_b, the following, asymptotically equivalent, choices for $\Phi''[u[\lambda]]\dot{u}\delta u$ appear equally possible:

1. *Frozen geometry approximation:* assuming $w \ll 1$ and neglecting all nonlinear terms in w.
$$\Phi''\dot{v}\delta u \approx K\dot{w}\delta w + E\dot{u}\delta u + N\dot{w}\delta w \quad , \quad N \approx Eu \tag{5.11}$$

2. *Direct compatible extrapolation:*
$$\Phi''[\lambda] \approx \Phi''_{c0} + \lambda\Phi'''_{c0}\hat{u}_0 \tag{5.12}$$

3. *Compatible expression:*
$$\Phi''[\lambda]\dot{u}\delta u = \Phi''_c[\lambda\hat{u}, \lambda\hat{w}]\dot{u}\delta u \tag{5.13}$$

4. *Direct mixed extrapolation:*

$$\Phi''[\lambda] \approx \Phi''_{m0} + \lambda \Phi'''_{m0} \hat{u}_0 \qquad (5.14)$$

5. *Mixed expression:*

$$\Phi''[\lambda]\dot{u}\delta u = \Phi''_m[\lambda\hat{u}, \lambda\hat{w}, \lambda\hat{N}]\dot{u}\delta u \qquad (5.15)$$

We will see that quite different results are however obtained by the different choices.

Frozen geometry assumption   This is the usual choice in technical applications and is obtained by assuming rotations small enough to neglect all changes in geometry along the fundamental path. We can equivalently use both compatible and mixed formulations.

From the compatible formulation, we obtain

$$\Phi''_c \dot{v}^2 = K\dot{w}^2 + E\dot{u}^2 + Eu\dot{w}^2 + 2Ew\dot{u}\dot{w} \quad , \quad N = Eu + \frac{1}{2}Ew^2 \qquad (5.16)$$

then the assumption $w \approx 0$ implies

$$\Phi''[\lambda]\dot{v}^2 \approx K\dot{w}^2 + E\dot{u}^2 + N\dot{w}^2 \quad , \quad N \approx -\lambda Q \qquad (5.17)$$

Equally, from the mixed formulation we obtain

$$\Phi''_m \dot{v}^2 = K\dot{w}^2 - \dot{N}^2/E + 2\dot{N}\dot{u} + N\dot{w}^2 + 2w\dot{N}\dot{w} \qquad (5.18)$$

and, with $w \approx 0$ and $\dot{N} = E\dot{u}$, we again obtain (5.17).

We can write, in matrix notations that

$$\Phi''[\lambda]\dot{v}^2 \approx \left\{ \begin{array}{c} \dot{u} \\ \dot{w} \end{array} \right\}^T \left[\begin{array}{c|c} E & \cdot \\ \hline \cdot & K - \lambda Q \end{array} \right] \left\{ \begin{array}{c} \dot{u} \\ \dot{w} \end{array} \right\} \qquad (5.19)$$

and, by zeroing the matrix determinant, we obtain the bifurcation load multiplier λ_b as a solution of

$$E(K - \lambda Q) = 0 \quad \Rightarrow \lambda_b = K/Q \qquad (5.20)$$

It is worth noting that we obtain a reliable (1st–order accurate) approximation for λ_b which is completely uninfluenced by the E/K and F/Q ratios.

Direct compatible extrapolation   Referring to (5.12), we have

$$\Phi''_{c0}\dot{v}^2 = E\dot{u}^2 + K\dot{w}^2 \; , \quad \Phi'''_{c0}\hat{u}\dot{v}^2 = E\hat{u}\dot{w}^2 + 2E\hat{w}\dot{w}\dot{u} \qquad (5.21)$$

so, using eqs.(5.9), we obtain

$$\Phi''[\lambda]\dot{v}^2 \approx (\Phi''_0 + \lambda\Phi'''_0\hat{u}_0)\dot{v}^2 = E\dot{u}^2 + K\dot{w}^2 + 2\lambda F\dot{w}\dot{u}E/K - \lambda Q\dot{w}^2 \qquad (5.22)$$

that is, in matrix notations,

$$\Phi''[\lambda]\dot{v}^2 \approx \left\{ \begin{array}{c} \dot{u} \\ \dot{w} \end{array} \right\}^T \left[\begin{array}{c|c} E & \lambda FE/K \\ \hline \lambda FE/K & K - \lambda Q \end{array} \right] \left\{ \begin{array}{c} \dot{u} \\ \dot{w} \end{array} \right\} \qquad (5.23)$$

From this, by zeroing the matrix determinant, we obtain the bifurcation load multiplier λ_b as a solution of

$$E(K - \lambda Q) - \lambda^2(FE/K)^2 = 0 \tag{5.24}$$

Note that a spurious term $\lambda^2(FE/K)^2$ affects the expected solution $\lambda_b = K/Q$ by negative locking. We also get a spurious buckling at $\lambda = \frac{K}{F}\sqrt{\frac{K}{E}}$ which tends to 0 for $E/K \to \infty$.

Compatible extrapolation By differentiating the strain energy twice, we obtain

$$\Phi''[u]\dot{v}^2 = E\dot{u}^2 + K\dot{w}^2 + Eu\dot{w}^2 + 2Ew\dot{u}\dot{w} + \frac{3}{2}Ew^2\dot{w}^2 + \cdots \tag{5.25}$$

Expansion (5.9) furnishes

$$\Phi''[\lambda]\dot{v}^2 = E\dot{u}^2 + K\dot{w}^2 - \lambda Q\dot{w}^2 + 2\lambda F\dot{w}\dot{u}E/K + \frac{\lambda^2}{2}3E(F/K)^2\dot{w}^2 \tag{5.26}$$

that is

$$\Phi''[\lambda]\dot{v}^2 = \left\{\begin{array}{c} \dot{u} \\ \dot{w} \end{array}\right\}^T \left[\begin{array}{c|c} E & \lambda FE/K \\ \hline \lambda FE/K & K - \lambda Q + \frac{3}{2}E(\lambda F/K)^2 \end{array}\right] \left\{\begin{array}{c} \dot{u} \\ \dot{w} \end{array}\right\} \tag{5.27}$$

By zeroing the matrix determinant, we then obtain

$$E(K - \lambda Q) + \frac{1}{2}\lambda^2(EF/K)^2 = 0 \tag{5.28}$$

where a spurious term $\lambda^2(FE/K)^2/2$ affects the expected solution $\lambda_b = K/Q$ by positive locking. For large values of the aspect ratio E/K there is no bifurcation at all.

Mixed extrapolation The potential energy can be written in mixed form as

$$\Pi[u] := \frac{1}{2}(K\phi^2 + 2N\varepsilon - N^2/E) - \lambda(Fw - Qu), \quad u := \{u, w, N\} \tag{5.29}$$

where $N = E\varepsilon$ is the axial force. Consequently we derive

$$\Phi''[u]\dot{v}^2 = (K + N)\dot{w}^2 + 2\dot{N}\dot{u} + 2w\dot{N}\dot{w} - \dot{N}^2/E \tag{5.30}$$

and, using (5.10) we obtain

$$\Phi''[\lambda]\dot{v}^2 = (K - \lambda Q)\dot{w}^2 + 2\dot{N}\dot{u} + 2\lambda F/K\dot{N}\dot{w} - \dot{N}^2/E \tag{5.31}$$

that is:

$$\left\{\begin{array}{c} \dot{N} \\ \dot{u} \\ \dot{w} \end{array}\right\}^T \left[\begin{array}{c|c} -1/E \;\; 1 & \lambda F/K \\ 1 & \\ \hline \lambda F/K & K - \lambda Q \end{array}\right] \left\{\begin{array}{c} \dot{N} \\ \dot{u} \\ \dot{w} \end{array}\right\} \tag{5.32}$$

By making the determinant of the matrix zero, we obtain

$$K - \lambda Q = 0 \tag{5.33}$$

which furnishes the expected results $\lambda_b \approx K/Q$

Some remarks Compatible and mixed extrapolations essentially differ in the evaluation of the axial strength N. In the mixed formulation, it is directly extrapolated:

$$N = \lambda \hat{N} = -\lambda Q \tag{5.34}$$

and unaffected by the stiffness ratio E/K (frozen extrapolation, albeit for different reasons, gives the same results). In the compatible formulation, it is obtained from the extrapolation of the displacements:

$$N = E\varepsilon = E(\lambda \hat{u} + \frac{1}{2}\lambda^2 \hat{w}^2) = -\lambda Q + \frac{1}{2}\lambda^2 \frac{F^2}{K}\frac{E}{K} \tag{5.35}$$

A term in λ^2 now appears in the expression of N, that produces an overestimate of the strength and so the occurrence of locking.

It is worth mentioning that, when using a more than linear extrapolation, this "spurious", quite large, quadratic term is naturally canceled by some other quadratic contributions in the expansion. So mixed extrapolation (5.34) is actually more accurate.

A criterion for measuring nonlinearity Previous discussion reinforces our conjecture that the good behaviour of the mixed extrapolation is strictly related to its ability in providing a much less nonlinear expression for the fundamental path. To investigate the problem more deaply we need a criterion for evaluating the path nonlinearity and some further comments. A suitable criterion for evaluating the nonlinearity in the description of the fundamental path is that of examining the eigenvalue of the matrix

$$[\mathbf{I} - \mathbf{K}_0^{-1}\mathbf{K}[\lambda]] \tag{5.36}$$

in the work range $0 \le \lambda \le \lambda_b$ (see Garcea et al. (1998), Garcea et al. , 1999). Obviously, matrix $\mathbf{K}[\lambda]$ having been singular at $\lambda = \lambda_b$, one of its eigenvalues will tend to 1 when λ approaches λ_b. However all the others eigenvalues have to be kept small enough (possibly much smaller than 1 in absolute value) within the work range in order to obtain a "good" extrapolation.

Results obtained from the two different extrapolation choices for the simple test in fig.29 are shown in fig, 30 for a fixed value of the transversal load $F = 10^{-3}Q$ and different values of the stiffness ratio $k = E/K$. Note that mixed extrapolation gives the expected results for the eigenvalues of matrix $\mathbf{K}_0^{-1}\mathbf{K}[\lambda]$ for all k: one decreases quite linearly to 0, the others are quite constant at the value 1. Conversely, compatible extrapolation provides incorrect results and shows a strong locking effect.

A geometrical insight Reasons for that different behaviour, can be better understood by looking at the problem from a geometrical point of view.

For the compatible extrapolation, matrices $\mathbf{K}_0$ and $\mathbf{K}[\lambda]$ can be geometrically represented by ellipses corresponding to the level lines $\frac{1}{2}\mathbf{x}^T\mathbf{K}_0^{-1}\mathbf{x} = 1$ and $\frac{1}{2}\mathbf{x}^T\mathbf{K}[\lambda]^{-1}\mathbf{x} = 1$. Both ellipses present high form ratio, due to the high stiffness ratio E/K, and, while being quite similar in form have a slight relative rotation, according to the rigid rotation of the element (see fig.31). The relative difference between the two matrices is emphasized by the different radius which the two ellipses show for the same direction. For high

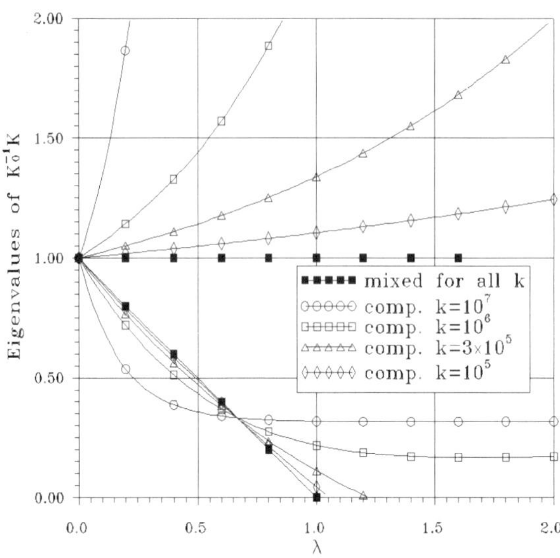

Figure 30. Simple test, eigenvalues of the matrix $\mathbf{K}_0^{-1}\mathbf{K}[\lambda]$.

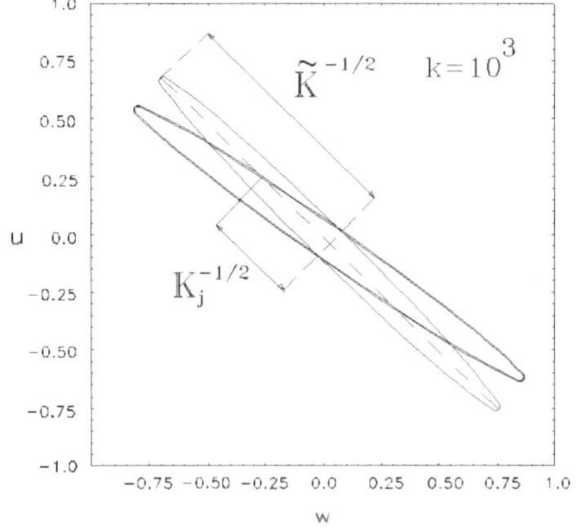

Figure 31. Level curves of $\mathbf{K}_0$ and $\mathbf{K}[\lambda]$

form ratios, even a very small rotation can imply a great difference in the hard direction associated to the axial stiffness.

When the same geometrical interpretation is applied to the mixed extrapolation, the overall figures change considerably, the hard direction being associated to the strength N which is unaffected by the rotation of the element.

A simple test showing nonlinear precritical behaviour Mixed formulation provides noticeable improvements in both accuracy and reliability. Its ability in describing nonlinear precritical behaviours is shown by results reported in fig. 33 for the shallow arc test problem in fig. 32. The analysis has been performed using frozen (*FC-Kasp*) and mixed (*M-Kasp*) versions of KASP code, and frozen (*FC-Beam*) and mixed (*M-Beam*) versions of the beam code described in Garcea et al. (1999). The latter proves to be forth–order accurate. Two buckling modes are considered and results for the bifurcation multipliers and for the third– and fourth–order energy terms are reporteed in table 3.

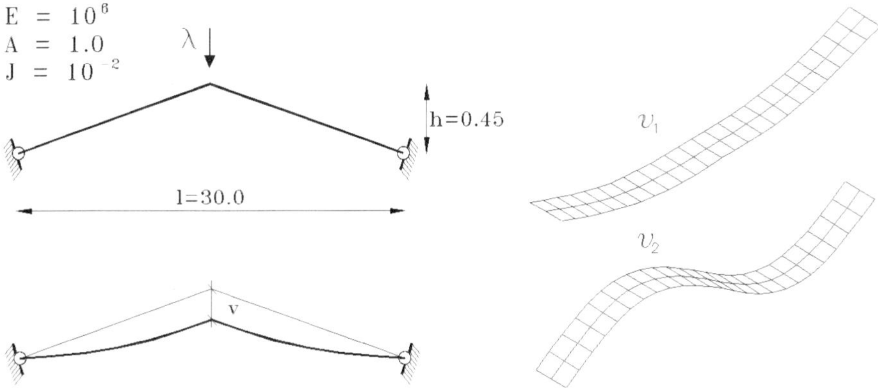

Figure 32. Shallow arc and buckling modes.

Table 3. Shallow arc: comparison of relevant energy perturbation quantities

	M-Beam	*FC-Beam*	*M-Kasp*	*FC-Kasp*
λ_1	22.0202	30.1738	21.9752	30.2122
λ_2	30.1738	46.6956	30.2122	46.7597
$\mathcal{A}_{001}$	0.01934	0.01838	0.01934	0.01841
$\mathcal{A}_{111}$	48.4907	177.138	48.4745	177.655
$\mathcal{B}_{0011}$	-6.26 10^{-4}	0	-6.34 10^{-4}	0
$\mathcal{B}_{0111}$	-3.51 10^{-4}	0	0	0
$\mathcal{B}_{1111}$	8.52431	58.0400	8.51369	60.7125

Note that the frozen geometry model detects two buckling modes: the first (skew–symmetric) at $\lambda_b \approx 30$ and the second (symmetric) at $\lambda_b \approx 46$. In contrast mixed formulation obtains $\lambda_b \approx 22$ for the first (symmetric) critical load. The frozen approach overestimates the limit load value, as reported in Fig. 33. The same figure shows the good agreement between the results obtained by the both KASP and Beam mixed analysis and those obtained with the beam path–following scheme (*Riks*) described in Garcea et al. (1998).

Some full–scale numerical results   Some numerical results are reported here, referring to the compressed channel beam in fig. 34 and assuming initial imperfections in order to emphasize the effects of a nonlinear precritical behaviour. The results have been obtained using both frozen and mixed version of the Kasp code and are compared

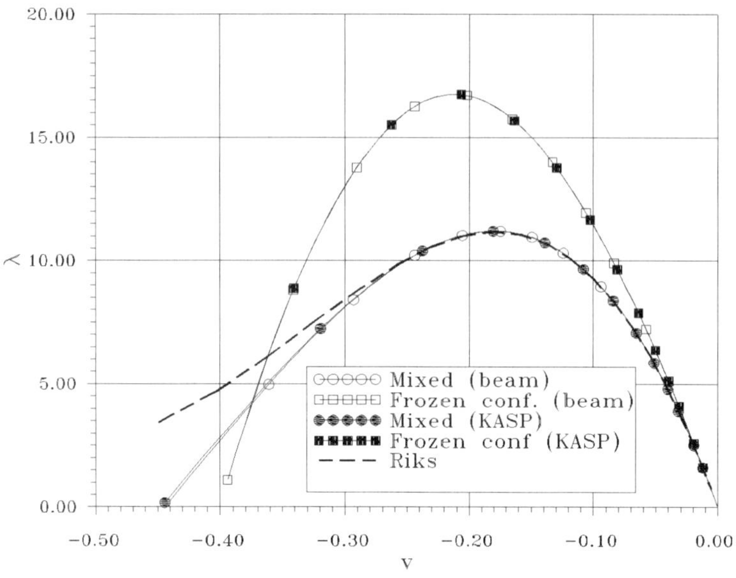

Figure 33. Shallow arc - equilibrium paths.

with results of path–following analysis provided by MSC–NASTRAN. Two different load conditions have been considered, assuming the different initial imperfections (torsional and flexural) shown in fig.34.

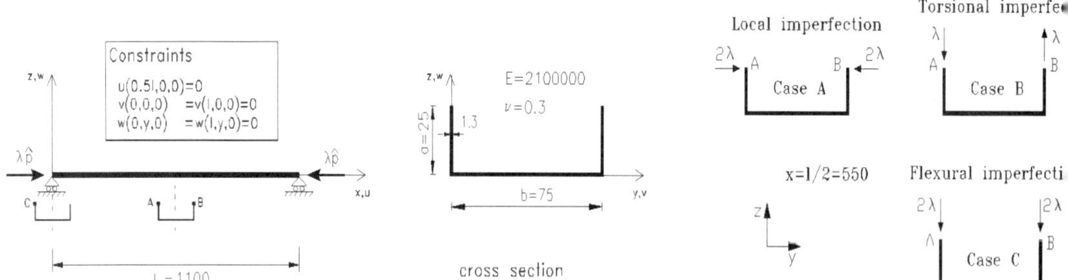

Figure 34. Channel section and force imperfection.

Channel beam with torsional imperfection The first 4 buckling modes obtained by the mixed formulation are drafted in Fig. 35. The buckling loads and a description of the mode shape obtained with the three different codes are listed in Tab. 4. The values and shapes of the buckling modes obtained with NASTRAN, **M** and **FC** are similar and the differences are essentially due to the different finite element implementations. It is worth noting that NASTRAN code uses a linearized buckling analysis neglecting

| mode | $\lambda_{\text{crit.}}$ | | | mode type |
	M	FC	NASTRAN	
$\dot{v}_1$	1264.9	1289.9	1296.0	flexural
$\dot{v}_2$	1833.9	1711.8	1715.6	torsional
$\dot{v}_3$	2808.5	2912.1	2842.4	7 half–waves
$\dot{v}_4$	2823.9	2930.0	2863.5	8 half–waves

Table 4. Buckling values: torsional imperfection.

| λ_{lim} | | |
M	FC	NASTRAN
1128.67	1204.17	1116.96

Table 5. Limit loads: torsional imperfection.

the prebuckling deformations. In this case the results of the buckling analysis must be compared with those of the **FC** ones. The differences between **M** and **FC** are due to the effect of the nonlinear precritical behaviour.

The accuracy of the mixed strategy in the evaluation of the limit load and of the initial postcritical behaviour is highlighted in Figs.36 and 37. In particular the evaluation of the limit load of the proposed formulation agrees perfectly with the NASTRAN values (see Table 5) proving the reliability of the mixed approach. The equilibrium paths, obtained by both **M** and **FC** strategies are based on an expansion of the first two or the first four buckling modes. As highlighted in Figs.36 and 37 the strong effect of modal interaction between the third local mode and the first two flexural–torsional global modes leads to unstable behaviour. The NASTRAN equilibrium path has a different evolution in the postbuckling range in u_C and w while the $\lambda - v_B$ equilibrium path displays a good agreement also in the post–buckling range. These differences could be due to the presence of high order terms neglected in the asymptotic analysis.

Channel beam with flexural imperfection In Fig. 38 the first 4 buckling modes detected by mixed formulation are depicted. In Tab. 6 the buckling loads and a description of the the mode shape obtained with the three different codes are reported.

Note the agreement between **FC** and NASTRAN results with respect to the buckling loads, the differences on the local modes being essentially due to the different finite element technologies. A greater difference with respect to the **M** approach, that detects as the first mode the torsional one (the second for **FC** and MSC–NASTRAN (see Tab.6)) can also be observed.

There is a good agreement between **M** and NASTRAN in the evaluation of the limit load as shown in Tab.7. The equilibrium paths (see Figs. 39 and 40) show the best approximation given by the mixed formulation with respect to the frozen one. In this case, however, the differences are of minor relevance. The effect of neglecting prebuckling deformation is highlighted in Fig. 40 where the solutions near the bifurcation point are amplified.

The postcritical behaviour is characterized by the interaction between the first (flexural) and the third (local) critical modes. The results show a good agreement with those obtained with NASTRAN especially in the v_B displacement component. A few differences can be observed in the u_C and w_A components with the usual deeper descending path depicted by the perturbation analysis.

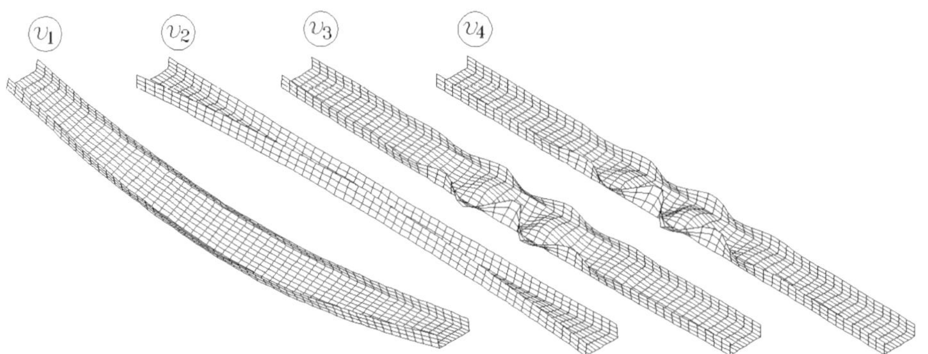

Figure 35. Channel section: Buckling modes for the torsional imperfection.

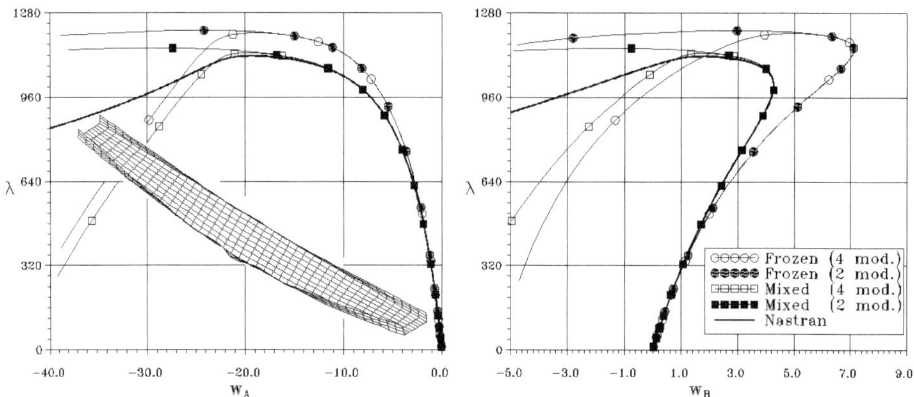

Figure 36. Case A: Equilibrium path $\lambda - w_A$, $\lambda - w_B$

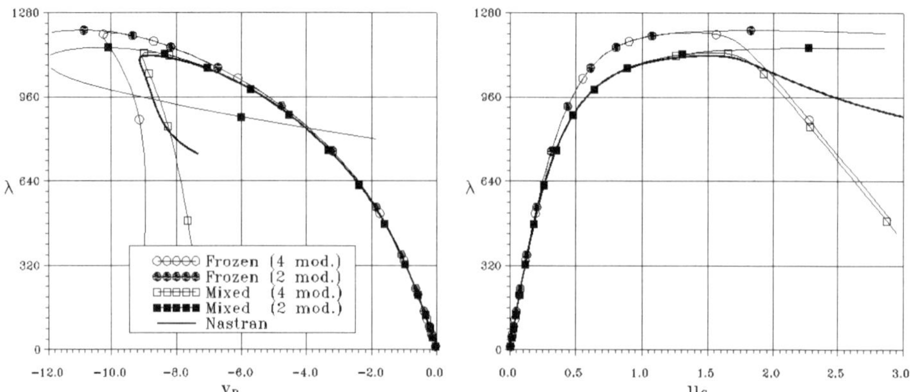

Figure 37. Case A: Equilibrium path $\lambda - v_B$, $\lambda - u_C$.

mode	$\lambda_{\text{crit.}}$ Mixed	Frozen	**NASTRAN**	mode type
$\dot{v}_1$	1291.6	1292.2	1298.2	flexural
$\dot{v}_2$	1396.0	1192.5	1195.2	torsional
$\dot{v}_3$	1994.4	1992.8	1919.4	3 half–waves
$\dot{v}_4$	2046.4	2044.7	2094.5	4 half–waves

Table 6. Buckling values: flexural imperfection.

$\lambda_{\max}$		
Mixed	Frozen	NASTRAN
900.3	908.4	896.3

Table 7. Limit loads: flexural imperfection.

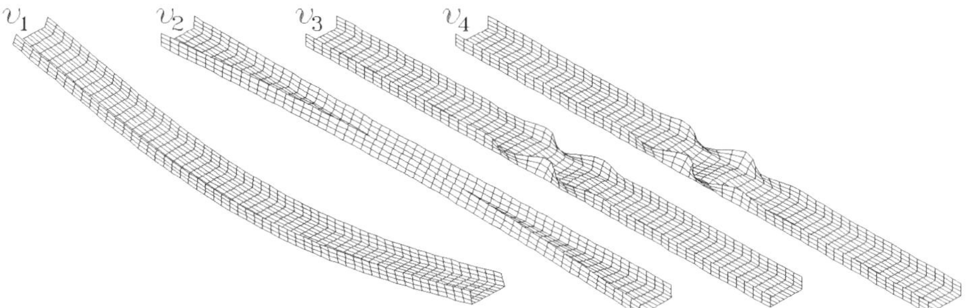

Figure 38. Channel section: Buckling modes for the flexural imperfection.

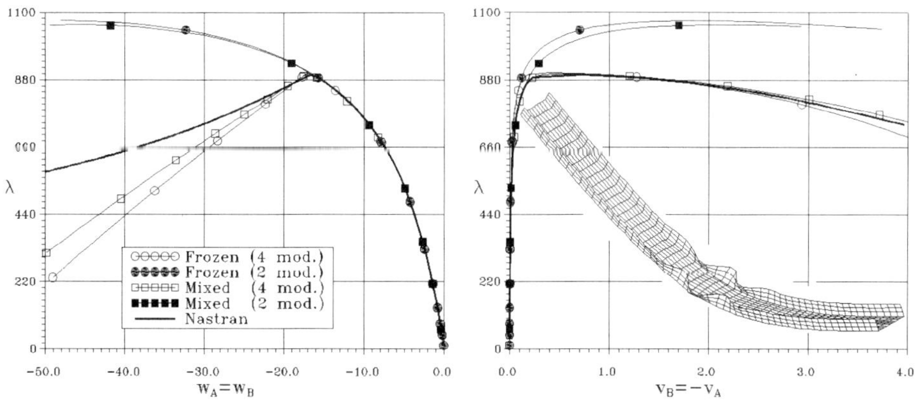

Figure 39. Flexural imperfection: equilibrium path $\lambda - w_A$, $\lambda - v_B$

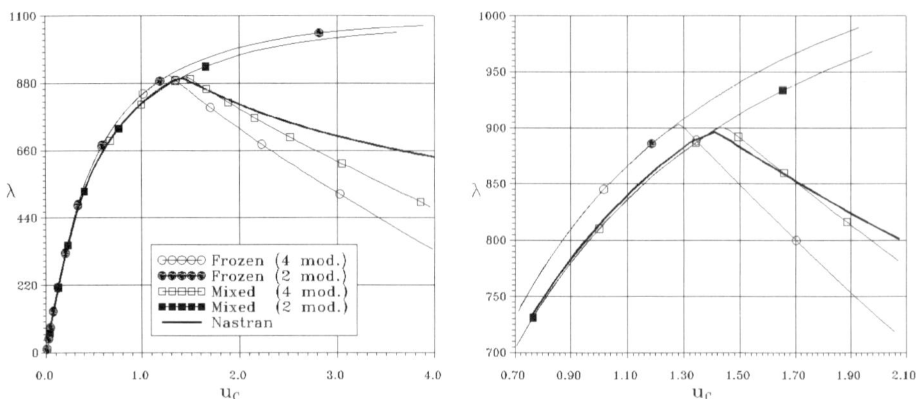

Figure 40. Flexural imperfection: Equilibrium path $\lambda - u_C$, zoom on the prebuckling path.

5.3 Effective implementation in a finite element code

We will try to answer here the following question: *is it possible to implement asymptotic analysis in a general purpose code able to provide reliable results suitable in practical structural design ?*

We will find that it is possible, trough some care has to be taken in the implementation.

Initial remarks The implementation of the asymptotic approach into a general purpose finite element code implies defining computational strategies for the following operations:

1. Assembly and triangular decomposition of the tangent stiffness matrix $\mathbf{K}_0$ defined by the energy equivalence

$$\delta\mathbf{u}^T\mathbf{K}_0\delta\mathbf{u} \equiv \Phi_0''\delta u^2 \ , \quad \forall \delta u \in \mathcal{T} \tag{5.37}$$

2. Solution of the linear problem
$$\mathbf{K}_0\hat{\mathbf{u}} = \hat{\mathbf{p}} \tag{5.38}$$

   defining the fundamental path $\mathbf{u}^f[\lambda]$.
3. Solution of the eigenvalue problem

$$\mathbf{K}[\mathbf{u}[\lambda]]\dot{\mathbf{v}} = 0 \tag{5.39}$$

   providing the critical multipliers λ_{bi} and the modes $\dot{v}_i$.
4. Evaluation of the orthogonal corrections w_{ij} by solving the linear systems

$$\Phi_b''w_{ij}\delta w = -\Phi'''\dot{v}_i\dot{v}_j\delta w \tag{5.40}$$

5. Numerical evaluation of the third– and fourth–order energy variations required by the analysis.
6. Determination of the equilibrium curve defined by the asymptotic equations in λ and ξ_i.

Points 1 and 2 are full standard operations, always available within general purpose finite element codes. The other points require some additional comments.

Bifurcation search The bifurcation multiplier λ_b and the critical mode $\dot{\mathbf{v}}$, are defined by the eigenvalue problem
$$\mathbf{K}[\mathbf{u}[\lambda]]\dot{\mathbf{v}} = 0 \tag{5.41}$$

as the first positive eigenvalue and the corresponding eigenvector of the problem, respectively.

Finite element codes usually include the solution of this problem as a standard option. However they provide the solution through a linearization of the functional dependence of $\mathbf{K}[\lambda]$ from λ, usually obtained by what we called "frozen geometry assumption". Bifurcation condition is then reduced to a linear eigenvalue problem:

$$(\mathbf{K}_0 + \lambda\mathbf{K}_1)\dot{\mathbf{v}} = 0 \tag{5.42}$$

which can be solved by standard methods such as the power method, the subspace iterations method or the Lanczos method.

Note, however, that this approximation can be unsuitable if aiming to obtain reliable results for the post–buckling behaviour. In fact, the error due to the linearization, while it could be irrelevant with respect to the evaluation of the bifurcation point, can imply some kinetical incoherences in the extrapolation and can appreciably affect the evaluation of the higher order energy terms $\Phi'''\dot{v}^3$ e $\Phi''''\dot{v}^4$ used to define the post–buckling behaviour of the structure. Furthermore, when shape functions dependent on λ are used in the finite element discretization (as, in beams systems, the "exact" shape functions described in section 2) the linearization is impossible.

It is worth remembering that this is not a real problem. We have already seen in section 2 a simple and efficient solution method for the nonlinear eigenvalue problem. The scheme (2.70) can be conveniently implemented in the form

$$\begin{cases} \dot{\mathbf{v}}_{j+1} = \dot{\mathbf{v}}_j - \tilde{\mathbf{K}}_0^{-1}\dot{\mathbf{s}}_j \\ \lambda_{j+1} = 1/\|\dot{\mathbf{v}}_j\| \end{cases} \tag{5.43}$$

where vector $\dot{\mathbf{s}}_j$ is obtained by

$$\dot{\mathbf{s}}_j := \mathbf{K}[\mathbf{u}^f[\lambda_j]]\dot{\mathbf{v}}_j \tag{5.44}$$

The iteration scheme requires an approximation $\tilde{\mathbf{K}}_0^{-1}$ of the inverse of the initial tangent matrix $\mathbf{K}_0$. It is already available at the end of step 1 so this operation does not require any computational extra–cost. The scheme also requires the availability of the incremental vector of internal forces $\dot{\mathbf{s}}_j$ defined by (5.43). Numerical procedures for its construction are usually present, as standard features, in codes that include the option for path–following analysis. Vector $\dot{\mathbf{s}}$ is generally assembled, element by element, directly from the energy equivalence

$$\dot{\mathbf{s}}^T \delta\mathbf{u} \equiv \Phi''[u^f[\lambda]]\dot{v}\delta u \ , \quad \forall \delta u \in \mathcal{T} \tag{5.45}$$

without actually needing the assembly of the full matrix $\mathbf{K}[\mathbf{u}^f[\lambda]]$. Then it is obtained at a small computational cost. Note that for $\tilde{\mathbf{K}}_0^{-1}$ we can assume even a very rough estimate of $\mathbf{K}_0^{-1}$ without noticeably affecting the convergence of the iteration process (e.g. in Lopez et al. (1998) the matrix implicitly defined by a rough coarse multigrid solution is used).

Multimode eigenvalue search A subspace iteration strategy can be used, owing to the closeness of the required eigenvalues which implies:

$$\Phi''[\lambda] \approx \Phi_b''\dot{v}_i\delta u + (\lambda_i - \lambda_b)\Phi_b'''\hat{u}_b\dot{v}_i\delta u = 0 \tag{5.46}$$

λ_b being an appropriate reference value (e.g. an estimate of the average value of the cluster).

The solution is obtained by repeating until convergence the 2–step iterative process:
1. Perform some (usually $2 \div 4$) iterative steps according to the scheme

$$\begin{cases} \dot{\mathbf{v}}_{i,j+1} & = & \dot{\mathbf{v}}_{i,j} - \mathbf{K}_0^{-1}\mathbf{K}[\lambda_{i,j}]\dot{\mathbf{v}}_{i,j} \\ \lambda_{i,j+1} & = & 1/\|\dot{\mathbf{v}}_{i,j+1}\| \end{cases} \quad i = 1..m \tag{5.47}$$

extended to all m modes $\dot{v}_i$ to be determined.

This phase filters not–critical directions and provides a better approximation of the critical subspace $\mathcal{V}$.

2. Assemble the matrices $\mathbf{A}$ and $\mathbf{B}$ defined by the coefficients

$$A_{ik} = \Phi_b'' \dot{v}_i \dot{v}_k \quad , \quad B_{ik} = \Phi_b''' \hat{u}_b \dot{v}_i \dot{v}_k \quad , \quad i, k = 1..m \tag{5.48}$$

Then solve, by a standard Jacobi routine, the linear eigenvalue problem:

$$\mathbf{A}\mathbf{x} + (\lambda - \lambda_b)\mathbf{B}\mathbf{x} = 0 \tag{5.49}$$

providing the new estimate for the required buckling modes

$$\tilde{\mathbf{v}}_i = x_{ik}\dot{\mathbf{v}}_k \tag{5.50}$$

Evaluation of scalar high–order terms   The asymptotic algorithm needs some 3rd– and 4th–order (scalar) energy terms, like $\Phi'''\dot{v}^3$ or $\Phi''''\dot{v}^4$, be computed in order to define the equilibrium path of the structure. Such quantities are not directly provided by standard functions in finite element codes, so we need to implement the proper routine to compute these quantities by recovering the necessary information from the host code.

Note that the generic 3rd–order variation

$$\Phi_b''' u_1 u_2 u_3 \tag{5.51}$$

can be computed as the integral, extended to all elements of the structure, of the product of three known functions defined, through the assumed shape functions of the element, by the three incremental displacement vectors $\mathbf{u}_1$, $\mathbf{u}_2$ and $\mathbf{u}_3$. So the computational cost reduces to the evaluation of a scalar quantity that is given in explicit form and can be computed, in the worst case, by Gauss–point integration.

The same holds for the evaluation of the generic 4th–order variation

$$\Phi_b'''' u_1 u_2 u_3 u_4 \tag{5.52}$$

Evaluation of vector–valued 3rd–order terms   In order to compute the orthogonal correction $w[\lambda, \xi_i]$ we require the evaluation of vectors $\ddot{\mathbf{s}}_{ij}$ defined by the energy equivalence

$$\ddot{\mathbf{s}}_{ij}^T \delta \mathbf{u} \equiv \Phi_b''' \dot{v}_i \dot{v}_j \delta u \tag{5.53}$$

which appears as a known term in the orthogonal equations.

Like the scalar quantities $\Phi_b''' \dot{v}_i \dot{v}_j \dot{v}_k$, the vector $\ddot{\mathbf{s}}$ is not directly provided by standard functions in finite element code. It, however, can be assembled, element by element, from the equivalence

$$\delta \mathbf{u}^T \ddot{\mathbf{s}}_{ij} = \Phi''' \dot{v}_i \dot{v}_j \delta u \tag{5.54}$$

in the same way as $\dot{\mathbf{s}}$. A new routine has to be implemented to compute the element contributions by recovering information from the host code, while both the routine architecture and the computational costs are similar to that required for obtaining $\dot{\mathbf{s}}$.

Evaluation of the orthogonal corrections $\mathbf{w}_{ij}$ The orthogonal correction $\mathbf{w}_{ij}$ is obtained by the solution of a constrained linear system.

The presence of the orthogonality constraints renders this equation unsuitable for a direct implementation. By standard Lagrange multiplier technique, it is conveniently rewritten in an unconstrained form:

$$\begin{cases} \Phi_b'' w_{ij} \delta u + \sum_{k=1}^{m} c_{ijk} \Phi_b''' \hat{u} \dot{v}_k \delta u &= -\Phi''' \dot{v}_i \dot{v}_j \delta u \\ \Phi_b''' \hat{u} \dot{v}_k w_{ij} &= 0 \end{cases} \tag{5.55}$$

where $w_{ij} \in \mathcal{T}$, $\delta u \in \mathcal{T}$ and $k = 1, \cdots, m$, and c_{ijk} are the unknown Lagrange multipliers.

Note that, with

$$\Phi_b'' w_{ij} \dot{v}_k = 0 \ , \quad \Phi_b''' \hat{u}_b \dot{v}_i \dot{v}_j = \begin{cases} -1 & \text{if } i = j \\ 0 & \text{if } i \neq j \end{cases} \tag{5.56}$$

by taking $\delta u = \dot{v}_k$, we obtain

$$\sum_{h=1}^{m} c_{ijk} \Phi_b''' \hat{u} \dot{v}_h \dot{v}_k = -c_{ijk} = -\Phi_b''' \dot{v}_i \dot{v}_j \dot{v}_k \tag{5.57}$$

Then the equation can be rewritten in matrix form as:

$$\mathbf{K}_b \mathbf{w}_{ij} = \mathbf{f}_{ij} \ , \quad \mathbf{f}_{ij} := -\ddot{\mathbf{s}}_{ij} - \sum_{i=1}^{m} \mathcal{A}_{ijk} \ddot{\mathbf{s}}_{0k} \tag{5.58}$$

where $\ddot{\mathbf{s}}_{0k}$ is obtained by assuming $\dot{v}_0 := \hat{u}$ and, as previously defined,

$$\mathcal{A}_{ijk} := \Phi_b''' \dot{v}_i \dot{v}_j \dot{v}_k \tag{5.59}$$

Iterative solution for $\mathbf{w}_{ij}$ Note that, $\mathbf{K}_b$ being near to singular in $\dot{v}_k$ directions, the numerical solution $\bar{\mathbf{w}}$ of the previous equation may have significant errors in $\dot{v}_k$ components, i.e.:

$$\mathbf{w}_{ij} = \bar{\mathbf{w}} + \sum_{k=1}^{m} d_k \dot{\mathbf{v}}_k \tag{5.60}$$

Correct values of $\mathbf{w}_{ij}$ can be obtained by using the orthogonality condition for computing d_k. We obtain

$$d_k = \Phi_b''' \hat{u} \bar{w} \dot{v}_k \tag{5.61}$$

It is worth noting that $\mathbf{K}_0 \approx \mathbf{K}_b$ in the orthogonal subspace $\mathcal{W}$. Then $\mathbf{w}_{ij}$ can be obtained by iterating with the following scheme

$$\begin{cases} \bar{\mathbf{w}} = \mathbf{w}_{ij}^{(n)} - \mathbf{K}_0^{-1}(\mathbf{K}_b \mathbf{w}_{ij}^{(n)} - \mathbf{f}_{ij}) \\ \mathbf{w}_{ij}^{(n+1)} = \bar{\mathbf{w}} - \sum_{k=1}^{m} d_k \dot{\mathbf{v}}_k \ , \quad d_k := \Phi_b''' \hat{u} \bar{w} \dot{v}_k \end{cases} \tag{5.62}$$

The use of this iterative scheme is particularly convenient because matrix $\mathbf{K}_0$ is already available and, vector $\mathbf{K}_b \mathbf{w}_{ij}^{(n)}$ being directly obtained by element–by–element assemblage, we do not need matrix $\mathbf{K}_b$ actually be constructed.

Final evaluation of the equilibrium curve The equilibrium curve is defined by the asymptotic equations, expressed in simplified form by:

$$\lambda + \lambda^2 \mu/\xi = \lambda_b + \xi\dot{\lambda}_b + \frac{1}{2}\xi^2\ddot{\lambda}_b \tag{5.63}$$

or, when considering multiple buckling modes,

$$\xi_k(\lambda_b - \lambda) + \frac{1}{2}\sum_{i,j=1}^{m}\xi_i\xi_j\mathcal{A}_{ijk} + \frac{1}{6}\sum_{i,j,h=1}^{m}\xi_i\xi_j\xi_h\mathcal{B}_{ijhk} = \mu_i[\lambda] \tag{5.64}$$

We obviously need a direct implementation of the equation. Note however that, in the monomodal case, due to the simplicity of the equations, the solution is obtained by points with a neglegeable cost from both the implementational and the computational points of view. In the multimodal case, even if the dimension m of the problem can be considered quite small, very high nonlinearities are involved, then the use of a specific, carefully tuned, path-following solution strategy is required.

Note that, in any case, it is convenient to implement the full versions of the asymptotic equations. Referring to simplified formulas is only convenient for discussing aims and does not provide any real computational advantage within a numerical solution process, while even producing some small decrease in the accuracy of the results.

A final comment The computational time required by the perturbation method is similar to that of an ordinary linearized stability analysis. Many of the required operations are supported as standard options by commercial finite element codes that can then be used as a high optimized host ambient for performing the perturbation analysis.

Some new routines are however needed to implement the bifurcation search algorithm and the iterative solution for the orthogonal corrections, and to compute the scalar and vector quantities expressing the high–order energy terms. In the latter case, an interface is required for recovering information (like values of the tensions and the displacement gradients in Gauss–points) usually available at low level only.

The implementation of the asymptotic approach within a commercial general–purpose FEM code is then quite a simple but not a completely straightforward task. However it should be convenient because careful implementations can provide excellent results.

6 Questions and answers about the asymptotic analysis

We already stated that the asymptotic analysis is still not that popular within the computational mechanics community.

The suspicions about asymptotic methods initiated by the, however excellent, review by Gallagher (see Gallagher (1974)) at the first Ticom Conference in 1974, as a consequence of the really poor results obtained by the earlier implementation of the Koiter approach as a numerical procedure. The main criticisms have been related to the apparent numerical complexity of the approach, to the need for computing 3rd– and 4th–order energy variations, and to its inaccuracy in recovering the post–buckling behaviour.

This general opinion didn't change for two decades, even if some papers where published in this period showing that the method was actually able to provide good results. The main reasons for this may be related to the large widespread of the path–following method as a general purpose approach to the analysis of nonlinear structural problems. However suspicions about the asymptotic method also arise from its logical complexity and apparent unreliability (we can't really be surprised about this because of the high sensitivity of the method to the "minor" implementation details).

We tried to show in the previous sections that the asymptotic method, when implemented with sufficient care, can actually lead to powerful reliable finite element solution procedures. However, while the method is now accepted as a possible technique for the analysis of nonlinear slender structures, the old doubts about its reliability have still not completely disappeared and criticisms, not always well motivated, about the method sometimes reappear in the literature. The most frequent of these are summarized and commented on here.

6.1 Frequent criticisms about asymptotic analysis

Question 1   The asymptotic approach requires an accurate evaluation of buckling loads λ_i and modes ϕ_i, solutions of the nonlinear eigenvalue problem (36)

$$\mathbf{K}[\lambda]\phi = 0 \tag{6.1}$$

Its linearization

$$(\mathbf{K}_0 + \lambda \mathbf{K}_g)\phi = 0 \tag{6.2}$$

as is commonly assumed in the asymptotic approach, can produce excessive inaccuracies.

Question 2   In multimodal analysis, the successive eigenvalues are assumed to be orthogonal to each other. However, they are not exactly orthogonal when using a $\mathbf{K}_0$–based norm or a Euclidian norm i.e. we have

$$\dot{\mathbf{v}}_i^T \mathbf{K}_0 \dot{\mathbf{v}}_j \neq 0 \quad \text{and} \quad \dot{\mathbf{v}}_i^T \dot{\mathbf{v}}_j \neq 0 , \quad \text{for } i \neq j \tag{6.3}$$

For this reason, the buckling mode concept i.e. the existence of a set of orthogonal critical directions seems in itself questionable.

Question 3   In cases of modal interaction, the number of buckling modes to be taken into account is a completely open question. We do not have any coherent criterion for that choice and the selection of the appropriate modes requires a clear insight into the problem at hand.

Question 4   The asymptotic approach require an accurate analytical evaluation for the higher–order energy terms used for computing the coefficients A_{ijk} and A_{ijhk} in the expansion (4.9). This implies a careful choice for both the continuum modeling and its finite element discretization in order to obtain an expression for the energy accurate to the required order. The usual technical theories and finite elements can imply truncation errors not compatible with that purpose.

Question 5   AS the approach is based on a Taylor expansion, its applicability is limited to a neighborhood of the origin. This prevents a deep excursion in the post–buckling regime. For extending the neighborhood more terms have to be considered in the expansion, but that can be impracticable in cases of multimodal analysis.

Question 5   Noticeable differences can sometimes be recognized in recovering the post–buckling behaviour when comparing asymptotic with path–following results. The difference should be related to: i) the fourth-order truncation of the strain energy; ii) the insufficient number of buckling modes considered; iii) the errors introduced by the use of a simplified structural modelling.

Question 6   The direct use of the displacements provided by the asymptotic solution within the equilibrium equations can lead to large equilibrium errors, rather difficult to estimate.

Question 7   As a consequence of this, the asymptotic approach, while very appealing, from a conceptual viewpoint, because of its local nature, can hardly be considered a reliable numerical tool, especially for deep excursions into the post–buckling regime.

6.2   Answers

Answer to question 1   We can easily agree with the need to obtain an accurate solution for the nonlinear eigenvalue problem (2.61) which defines the buckling modes and loads. This point has a certain relevance because a crude linearization of (2.61) can introduce kinematical inconsistencies in the solution that, even if negligible for the evaluation of the critical load, can significantly affect the evaluation of the high–order energy terms involved in the sequel of the analysis. However to solve the nonlinear problem (2.61) exactly is quite easy, e.g. by using the iterative scheme (2.70) which is fast, robust and effective, and represents a convenient alternative to the classical inverse power method also in the linear case.

It is worth mentioning that a linearization for the fundamental path $u^f[\lambda]$ does not imply a linearization for matrix $\mathbf{K}[\lambda]$ (e.g., $\mathbf{K}[\lambda]$ depending on both stresses and displacements it is, at least, quadratic on λ within a compatible formulation). We can also mention that when using linearization in the form

$$(\mathbf{K}_0 + \lambda \mathbf{K}_g)\phi = 0 \tag{6.4}$$

this is usually obtained by assuming a frozen geometry hypothesis along the fundamental path ($u^f[\lambda] \approx 0$) rather than taking $\mathbf{K}_g := d\mathbf{K}[\lambda]/d\lambda|_{\lambda=0}$. The two assumptions provide different values for $\mathbf{K}_g$, the first somewhat being more reliable while maybe inaccurate.

Answer to question 2   We can easily agree with (6.3). However, when considering successive critical modes $\{\phi_i, \phi_2, .., \phi_m\}$ we can refer (see (4.2)) to the linearization

$$\mathbf{K}[\lambda_b]\dot{\mathbf{v}} + (\lambda - \lambda_b)\mathbf{K}'[\lambda_b]\dot{\mathbf{v}} = 0 \ , \ \mathbf{K}' := d\mathbf{K}/d\lambda \tag{6.5}$$

λ_b being a convenient reference value internal to the range $\{\lambda_i..\lambda_m\}$, we have (see (4.3)):

$$\dot{\mathbf{v}}_i^T \mathbf{K}'[\lambda_b]\dot{\mathbf{v}}_j = 0 \ , \ \forall i \neq j \tag{6.6}$$

Usually the interacting cluster of buckling modes presents a sufficiently small excursion in λ ($\lambda_m - \lambda_1$ not more than 10% – 20% of λ_b) to render the linearization acceptable, so the $\{\phi_i\}$ can be assumed as orthogonal with respect to matrix $\mathbf{K}'[\lambda_b]$. In any case, the deviations from the orthogonality only affect the relation (4.9), where some further relevant contribution has to be added, and not the asymptotic approach itself.

It is worth noting that (4.9) represents a highly simplified expression which is obtained when a series of small but not always negligible terms are cut off. It could be useful for discussing the general features of the method, but, in general e.g. for structures characterized by non negligible precritical displacements, we have to refer to the full expression (e.g. see (3.11)).

It is worth mentioning that matrix $\mathbf{K}'[\lambda_b]$ is implicitly defined by the equivalence:

$$\dot{\mathbf{v}}^T \mathbf{K}'[\lambda_b]\dot{\mathbf{v}} = \Phi'''\hat{u}\dot{v}^2 \tag{6.7}$$

then it can be obtained directly by an assemblage of the element contributions to the relevant third–order energy terms without no need to use finite matrix differentiations which are both cumbersome and potentially inaccurate. Furthermore, using the solution process (5.47)–(5.50), we don't really need to construct matrix $\mathbf{K}'[\lambda_b]$ at all.

Finally, we can also note that, in some mixed formulations, the strain energy proves to be cubic in stresses and displacements and therefore, when referring to a linear extrapolation of the fundamental path (which, however, does not imply a linear pre–buckling behaviour) the eigenvalue problem is exactly linear in λ.

Answer to question 3 I can agree with that criticism. While we can adopt as a base criterion that of fixing in advance the interval (10% – 20% of the first critical load) where locating the relevant buckling modes, some different attempts may be necessary to be sure not to neglect anything of importance. So, some insight into the problem may be necessary. However this is rather frequent in computational mechanics (structural modeling, finite element meshes, etc.) so it can be hardly considered an intrinsic defect of the asymptotic approach.

Answer to question 4 I completely agree with this statement. Within the asymptotic approach the (initial) post–buckling regime is derived using information contained in the higher–order energy terms and then requires that these terms be accurately expressed. When, as usual, the analytical expression for the strain energy is defined as quadratic in the strain ε, its expression is 4th–order accurate only if the $\varepsilon[u]$ is also 4th–order accurate, because of the presence of zero–order term ε_0.

It is worth emphasizing the importance of using a rational or, at least, 4th–order accurate modeling. However, while deriving rational beam theories and beam elements is relatively easy, it may be difficult or too cumbersome for plates and shells. Then we could be forced to accept some compromises if the error introduced can be presumed to be quantitatively small as happens, at least, in some applications.

We also have to consider that much larger errors can be introduced by inappropriate finite element discretization. For instance, what we called interpolation locking can introduce errors of more than 100% errors.

It is worth mentioning that, while consolidated experience is already available within the finite element people on how to obtain an accurate description of a 2nd–order neighbor of the strain energy (we have a lot of good finite elements able to provide an accurate evaluation for the equilibrium equations and the tangent stiffness matrix), we know much less about how to accurately describe higher–order neighborhoods, as required by the asymptotic approach.

Answer to question 5   Questions implied in this point are of central relevance in the asymptotic approach because it is essentially based on the use of a Taylor expansion of the potential energy truncated to some order (usually, for a series of reasons, to the 4th–order). Further comment is necessary.

First of all, the equilibrium path is implicitly defined by an expansion of the potential energy, so the accuracy is strictly related to the smoothness of the surface which defines the potential energy in terms of the configuration variables (nodal displacements, in a purely compatible formulation, stresses and displacements in a mixed one), and not on the smoothness of the equilibrium path, which is quite a different thing. In fact, the path tends to follow the low–energy directions then and is little influenced by nonlinearities in the high–energy directions, so we can have smooth paths in the presence of non–smooth energy. Conversely, if the energy surface is very flat, even smooth energy variations can produce strongly nonlinear paths.

Furthermore, the expansion is taken from a reference bifurcation point $\{u_b, \lambda_b\}$ which is itself obtained by an extrapolation taken from the initial point $\{u_0, \lambda_0\}$. The interaction between these two extrapolations, while being responsible for the high level of accuracy which can be potentially provided by the asymptotic approach, can cause a strong sensitivity to the smoothness of the energy surface.

We have, however, to consider that smoothness is not an intrinsic characteristic of the problem but depends only on the analytical representation of the energy surface, that is, it is essentially related to the choice of the configuration "primary" variables which are assumed for describing the energy and are directly subjected to the extrapolation process. While, from a theoretical pont of view, we can always refer to a smooth representation for the energy (we are only subjected to some topological constraints), a crude choice for the primary variables, can lead to very large (actually, much larger than even be expected) unpredictable errors in the path evaluation. However, once recognized, the problem can easily be sanitized by more appropriate choices for the energy representation which lead to a smooth surface and then allow reliable and accurate results to be obtained. These topics have been widely discussed in section 5. We only mention here that, in order to extend the range of accuracy of the asymptotic method, an appropriate parameterization of the energy function is easier and much more effective than the use of high order expansions which has little (or even worse) effect in cases of non–smooth surfaces and rapidly tends to be very cumbersome.

Answer to question 6 Differences between asymptotic results and those obtained by path–following analysis can occur in the recovery of multimodal buckling phenomena. However this is largely due to the occurrence in the asymptotic solution of secondary bifurcations not detected by the path–following analysis. So, we should be careful in relating the responsibilities of the differences only to the asymptotic approach.

Note that the use of asymptotic expansions certainly introduces some errors if very large displacements are considered. However, noticeable accuracy is obtained in evaluating the initial post–buckling behaviour and that is usually sufficient for technical purposes.

Answer to question 7 The asymptotic approach is really very appealing. It can be extremely efficient computationally and provides an insight into the mechanical behaviour of the structure which can allow an effective structural safety analysis.

The fact that the asymptotic approach can lead to larger errors in the equilibrium equations than in the load–displacement curve is really true. However the statement can be reversed. In fact, the most important feature of the approach is its ability to provide evaluations for the load–displacement relationship which are much more accurate than the local representation of the equilibrium, and even better accuracy in the evaluation of the limit load, which is the main technical result.

Contents

Bibliography

Antman S.S., 'Bifurcation problems for nonlinearly elastic structures', Rabinowitz ed., Application of Bifurcation Theory, Academic Press, New York, 1977.

Aristodemo M., 'A High-Continuity Finite Element Model for Two–Dimensional Elastic Problems', *Computer and Structures*, Vol.21, 987–993, 1985.

Bilotta A., Garcea G., Trunfio G.A., Casciaro R., 'Postcritical analysis of thin–walled structures by KASP (Koiter Analysis of Slender Panels) code', Report APRICOS–DE–1.3–03–1/UNICAL, Sept. 1997.

Brezzi F.,Cornalba M., Di Carlo A., 'How to get around a simple quadratic fold', *Numer.Math.*, Vol.48, pp.417-427, 1986.

Budiansky B., 'Theory of buckling and post-buckling of elastic structures', *Advances in Applied Mechanics*, Vol.14, Academic Press, New York, 1974.

Casciaro R., Di Carlo A., Pignataro M., 'A finite element technique for bifurcation analysis', *Proc. 14th IUTAM Congress*, Delft, The Netherlands, august-sept. 1976.

Casciaro R., Aristodemo M., 'Perturbation analysis of geometrically nonlinear structures', *Proc. Int. Conference on Finite Elements Nonlinear Solid and Structural Mechanics*, Geilo, Norway, august-sept. 1977.

Casciaro R., Lanzo A.D., Salerno G.,'Computational problems in elastic structural stability', in Nonlinear Problems in engineering, C.Carmignani, G.Maino eds., World Scientific publ., Singapore, 1991.

Casciaro R., Salerno G., Lanzo A.D.,' Finite Element Asymptotic Analysis of Slender Elastic Structures: a Simple Approach', *Int. J. Numer. Meth. Eng.*, Vol.35, pages 1397–1426, 1992.

Casciaro R.,'Un algoritmo per il problema non lineare agli autovalori' 13th *AIMETA Congress of Theoretical and Applied Mechanics*, Brescia, Italy, November 13–15, 2000.

Casciaro R., Garcea, Attanasio G., Giordano F., 'Pertubative approach to elastic post-buckling analysis', *International Journal of Computer & Structure*, Vol.66, 585-595, 1998.

Casciaro R., Mancusi G., Formica G., 'Analisi post-critica in presenza di instabilità locali accoppiate',16th *AIMETA Congress of Theoretical and Applied Mechanics*, Ferrara, Italy, Sept. 9–13, 2003.

Gallagher R. H., 'Perturbation procedures in nonlinear finite element analysis', *Int. Conference on Computation Methods in Nonlinear Mechanics*, Texas (USA), 1974.

Garcea G., Trunfio G.A., Casciaro R., 'Mixed formulation and locking in path following nonlinear analysis', *Comput. Meth. Appl. Mech.Engrg.*, 165 1-4, pp. 247-272, 1998.

Garcea G., Bilotta A., Trunfio G.A., Casciaro R., 'Mixed implementation in Koiter nonlinear analysis of thin–walled structures: KASP implementation', Report APRICOS–DE–1.3–10/UNICAL, Sept. 1998.

Garcea G., Salerno G., Casciaro R., 'Sanitizing of locking in Koiter perturbation analysis through mixed formulation', *Computer Methods in Applied Mechanics and Engineering*, Vol.180, 137-167, 1999.

Garcea G., 'Mixed formulation in Koiter nonlinear analysis of thin–walled beam', *Computer Methods in Applied Mechanics and Engineering*, Vol. 190/26-27,3369-3399, marzo 2001.

Ho D., 'The influence of imperfections on systems with coincident buckling loads', Int.J. Nonlinear Mech. Vol.**7**, 311-321, 1972.

Ho D., 'Buckling load of nonlinear systems with multiple eigenvalues', Int. J. Solids Struct., Vol.**10**, 1315–1333, 1974.

Koiter W.T., *Current Trends in the Theory of Buckling*, In *Buckling of structures*, B.Budiansky ed., Proc.of IUTAM symposium, Cambridge, Cambridge, June 17-21, 1974, Springer–Verlag, Berlin, 1976.

Koiter W.T., *Elastic stability*, Koiter's course 1978–79, Technologic University of Delft, manuscript transcription by A.M.A. v.d. Heijden.

Lanzo A.D., Garcea G., Casciaro R., 'Asymptotic post-buckling analysis of rectangular plates by HC finite elements', *International Journal of Numerical Methods in Engineering*, Vol.**38**, 2325-2345, 1995.

Lanzo A.D., Garcea G., 'Koiter's analysis of thin-walled structures by s finite element approach', *International Journal of Numerical Methods in Engineering*, Vol.**39**, 3007–303, 1996.

Livesley R.K., **M**atrix Methods of Structural Analysis, 2nd ed., Pergamon Press.

Lopez S., Fortino S., Casciaro R., "An adaptive multigrid solver for plate vibration and buckling problems", *Computer & Structure*, Vol. **69**, pp. 625-637, 1998.

Noor A.K., 'Recent advances in reduction methods for nonlinear problems', *Computers & Structures*, Vol.**13**, pp.31-44, 1981.

Noor A.K., Peters J.M., 'Recent advances in reduction methods for instability analysis of structures', *Computers & Structures*, Vol.**16**, No. 1-4, pp.67-80, 1983.

Noor A.K., 'Recent advances and application of reduction methods', *Appl.Mech. Rev.*, Vol.**5**, 125–146, May 1994.

Pignataro M., Di Carlo A., Casciaro R., 'On nonlinear beam model from the point of view of computational postbuckling analysis', *Int. J. Solids & Structures*, Vol.**18**,pp. 327-347, 1982.

Salerno G.,'How to recognize the order ofinfinitesimal mechanism: a numerical approach', *International Journal of Numerical Methods in Engineering*, Vol.**35**, 1351–1395, 1992.

Salerno G., Colletta G., Casciaro R., 'Attractive post-critical paths and stochastic imperfection sensitivity analysis of nonlinear elastic structures within an FEM context', *2nd ECCOMAS Conference on Numerical Methods in Engineering*, Paris, France, Sept. 9–13, 1996.

Salerno G., Casciaro R., 'Mode jumping and attrative paths in multimode elastic buckling', International Journal for Numerical Methods in Engineering, Vol.**40**, 833-861, 1997.

Salerno G., Lanzo A.D., 'A nonlinear beam finite element for the post-buckling analysis of plane frames by Koiter's pertubation approach', *Comp.Meth.Appl.Engng.*, Vol.**146**, pp. 325-349, 1997.

Timoshenko S.P., Gere J.M., **T**heory of Elastic Stability, Mc Graw Hill.

W.H.Wittrick, F.W.Williams, 'An algorithm for computing critical buckling loads of elastic structures', *J.Struct.Mech.*, 1(4), pp.497–518, 1973.

M. A. Ali and S. Sridharan, 'A versatile model for interactive buckling of columns and beam –columns', *Int. J. Solids Struct.*, **24** (5), 481–496 (1988).

G. M. van Erp, 'Advanced buckling analyses of beam with arbitrary cross section', *Ph.D. Thesis*, Eindhoven University of Technology (1989).

Mechanical Models for the Subclasses of Catastrophes

Zsolt Gaspar[1,2]

[1] Department of Structural Mechanics, Budapest University of Technology and Economics, Hungary
[2] Research Group for Computational Structural Mechanics, Hungarian Academy of Sciences, Budapest, Hungary

Abstract. First some concepts of the structural stability and the elementary catastrophe theory are shown. A short chapter explains which types of the catastrophes are typical at elastic structures. Hence the load parameter has a special role among the parameters, a subclassification is needed in the stability analysis. The main part of the paper shows this subclassification and illustrates almost every type by simple elastic models.

1   Equilibrium, Stability, Imperfection-sensitivity

Only finite-degree-of-freedom structures will be considered, which carry conservative loads. To analyse them we use total potential energy functions. (For the sake of simplicity we will use always the same units for every lengths, forces etc., but we will not print them.)

Example 1. Let us consider the structure shown in Figure 1 comprising a rigid link of length L, supported by two linear springs. c_i denotes the stiffness in both tension and compression, and l_i is the stressfree length of the ith spring (i=1, 2). The structure is loaded by a vertical dead load of magnitude Λ. Let us determine the total potential energy function of the structure.

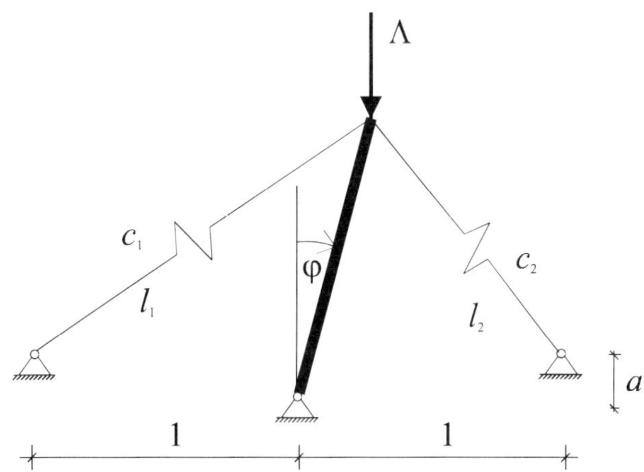

Figure 1. A one-degree-of-freedom structure.

Position of the link can be uniquely described by the angle φ. In this position the lengths of the springs are

$$b_1 = \sqrt{\left(1 + L\sin\varphi\right)^2 + \left(L\cos\varphi - a\right)^2}\,, \qquad b_2 = \sqrt{\left(1 - L\sin\varphi\right)^2 + \left(L\cos\varphi - a\right)^2}\,, \tag{1.1}$$

so the strain energy of the system is

$$U = \frac{1}{2}\left(c_1\left(b_1 - l_1\right)^2 + c_2\left(b_2 - l_2\right)^2\right). \tag{1.2}$$

If we choose the zero level at the height of the bottom of the link, potential of the load is

$$P = \Lambda L\cos\varphi\,. \tag{1.3}$$

The total potential energy of the system is the sum of (1.2) and (1.3):

$$V(\varphi) = U + P\,. \tag{1.4}$$

Definition. The necessary and sufficient condition for statical *equilibrium* is the vanishing of all first derivatives of V, according to the generalized coordinates.
Subscripts of V denote derivations according to the variable, e. g.

$$V_\varphi \equiv \frac{\partial V}{\partial \varphi}\,.$$

Example 2. Let us determine equilibrium states of our earlier structure for the following parameter values:

$$L = 1,\ a = 0.9,\ c_1 = c_2 = 2,\ l_1 = l_2 = \sqrt{1.01},\ \Lambda = 3\,. \tag{1.5}$$

There are three solutions of equation $V_\varphi = 0$ in the interval $\left(-\pi/2, \pi/2\right)$:

$$\varphi_1 = -0.6265878293,\ \varphi_2 = 0,\ \varphi_3 = 0.6265878293\,. \tag{1.6}$$

Definition. Let us consider Λ as a load parameter. Showing the equilibrium solutions as a function of the load parameter we get the *equilibrium paths*.

Example 3. Let us illustrate the equilibrium paths for four cases of parameter values:

a) $L = 1,\ a = 0,\ c_1 = c_2 = 2,\ l_1 = l_2 = \sqrt{2}\,,$ \hfill (1.7a)

b) $L = 1,\ a = 0.9,\ c_1 = c_2 = 2,\ l_1 = l_2 = \sqrt{1.01}\,,$ \hfill (1.7b)

c) $L = 1$, $a = 0$, $c_1 = 2$, $c_2 = 3$, $l_1 = l_2 = \sqrt{2}$,   (1.7c)

d) $L = 1$, $a = 0.9$, $c_1 = c_2 = 2$, $l_1 = \sqrt{1.01}$, $l_2 = 1$.   (1.7d)

Figure 2 shows the very different equilibrium paths. There are bifurcation points in the first three paths, there are local minimum points in the last three paths. At the last case there is also a local maximum point.

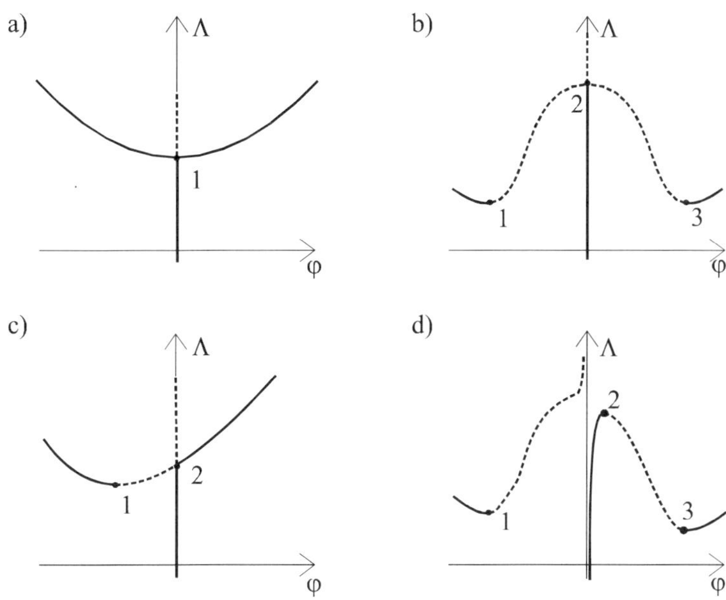

Figure 2. Equilibrium paths.

Definition. An *equilibrium state is stable* if the total potential energy function has an isolated local minimum at the point belongs to this state, otherwise it is *unstable*.
A sufficient but not necessary condition is to have an isolated local minimum, if (the gradient of V is zero and) the Hessian matrix of V is positive definite.

Example 4. Let us determine whether the states (1.6) of the structure in Example 2 are stable or unstable.
Hence

$$V_{\varphi\varphi}(\varphi_1) = V_{\varphi\varphi}(\varphi_3) = -2.932091311 \quad \text{and} \quad V_{\varphi\varphi}(\varphi_2) = 0.960396040,$$ (1.8)

the second state is stable while the other two are unstable states.
Stable states are denoted by solid lines, and unstable states by broken lines in Figure 2. On the border of the stable and unstable states the Hessian matrix must be singular.

Definition. Every equilibrium states at which the Hessian matrix of the total potential function is singular are said to be *critical state*.

Example 5. Let us determine the critical states for the four structures of Example 3.
Λ is considered as a load parameter, so $V(\varphi,\Lambda)$ has two unknowns (a variable and a parameter). The equations

$$V_{\varphi} = 0, \quad V_{\varphi\varphi} = 0 \tag{1.9}$$

determine the following solutions for the four cases

a) $\quad \varphi_1^{cr} = 0, \qquad\qquad\qquad \Lambda_1^{cr} = 2,$

b) $\quad \varphi_1^{cr} = -1.163728287, \quad \Lambda_1^{cr} = 0.4056170347,$

$\qquad \varphi_2^{cr} = 0, \qquad\qquad\qquad \Lambda_2^{cr} = 3.960396040,$

$\qquad \varphi_3^{cr} = 1.163728287, \quad\;\; \Lambda_3^{cr} = 0.4056170347,$

c) $\quad \varphi_1^{cr} = -0.5956588802, \Lambda_1^{cr} = 2.386636825,$

$\qquad \varphi_2^{cr} = 0, \qquad\qquad\qquad \Lambda_2^{cr} = 2.5,$

d) $\quad \varphi_1^{cr} = -1.164460881, \quad \Lambda_1^{cr} = 0.4123088138,$

$\qquad \varphi_2^{cr} = 0.1544489481, \quad \Lambda_2^{cr} = 3.864457731,$

$\qquad \varphi_3^{cr} = 1.162994680, \quad\; \Lambda_3^{cr} = 0.4147724729.$

The critical states are denoted by dots in Figure 2.

The question arises: How many different types of equilibrium paths do exist? This question cannot be answered if we analyse the equilibrium paths globally, hence the type depends on the region where the paths are analysed. But we can try to answer the following: How many different types of neighbourhood of the critical points do exist?

Figure 2 shows the four most important types, which were first analysed by Koiter (1945), later by many others (e.g. Thompson and Hunt, 1973, 1984, Gioncu and Ivan 1984, Gaspar, 1999):

- limit point (Figure 2b 1st and 3rd, 2c 1st, 2d 1st, 2nd, 3rd points),
- asymmetric point of bifurcation (Figure 2c 2nd point),
- unstable-symmetric point of bifurcation (Figure 2b 2nd point),
- stable-symmetric point of bifurcation (Figure 2a 1st point).

Imperfections change the potential energy functions, so the equilibrium paths and critical load parameters will change, too. (E. g. the structure d) in example 3 can be considered as an imperfect version of structure b), and really the corresponding critical points are near to each other.) These structures can be handled together as the energy function will depend on the state variables, the load parameter and the imperfection parameter(s): $V(\varphi,\Lambda,\varepsilon)$.

Definition. Imperfection-sensitivity curves/surfaces shows the critical load parameter against the imperfection parameter(s).

To analyse the different types of imperfection sensitivity surfaces we will use the results of the catastrophe theory.

2   Catastrophe Theory

The most important concepts and results will be shown on the basis of Poston and Stewart (1978).

A *function* or *map* (*f*) consisting of *m* scalar functions with *n* variables will be denoted by
$$f : R^n \to R^m .$$

It is a *smooth function* if all derivatives of the function exist in every point of it. Usually the functions will be analysed *locally*, i. e. in the neighbourhood of the examined point. *Diffeomorphism* is a smooth and locally reversible map
$$R^n \to R^n .$$

A function of the form
$$x \to -x_1^2 - \ldots - x_l^2 + x_{l+1}^2 + \ldots + x_n^2 \tag{2.1}$$

is called a *Morse l-saddle*. Let
$$f : R^n \to R$$

be a smooth function. A point $u \in R^n$ is a *critical point* of f if its gradient is zero at this point, i. e.

$$Df\big|_u = \begin{bmatrix} \dfrac{\partial f}{\partial x_1}\Big|_u \\ \vdots \\ \dfrac{\partial f}{\partial x_n}\Big|_u \end{bmatrix} = 0 . \tag{2.2}$$

We say that f has a *non-degenerate critical point* at u if it is a critical point and the Hessian matrix is nonsingular at this point, i. e. the determinant of

$$Hf\big|_u = D^2 f\big|_u = \begin{bmatrix} \dfrac{\partial^2 f}{\partial x_1^2}\Big|_u & \ldots & \dfrac{\partial^2 f}{\partial x_1 \partial x_n}\Big|_u \\ \vdots & & \vdots \\ \dfrac{\partial^2 f}{\partial x_n \partial x_1}\Big|_u & \ldots & \dfrac{\partial^2 f}{\partial x_n^2}\Big|_u \end{bmatrix} \tag{2.3}$$

is not zero.

Morse Lemma: Let u be a non-degenerete critical point of a smooth function $f : R^n \to R$. Then there is a local co-ordinate system $(y_1 \ldots y_n)$ in the neighbourhood U of u, with $y_i(u) = 0$ for all i, such that $f\text{-}f(u)$ is a Morse l-saddle in U.

The function $f : R^n \to R$ is *structurally stable* if, for all sufficiently small smooth functions $p : R^n \to R$, f and $f + p$ have the same numbers and the same types of critical points. A single function is typically structurally stable, but we want to deal with *r-parameter families of function* $R^n \to R$,

$$f_{u_1,\ldots,u_r}(x_1,\ldots,x_n)$$

or, in an equivalent way:

$$f : R^n \times R^r \to R .$$

Our aim is to classify these, up to suitable co-ordinate changes, in such a way that the critical point structure is not affected qualitatively by the co-ordinate changes involved.

Two smooth *r*-parameter families of function $f, g : R^n \times R^r \to R$ are said to be *equivalent* if there exist a diffeomorphism $e : R^r \to R^r$, a smooth map $y : R^n \times R^r \to R^n$ (such that for each $s \in R^r$ the map $y_s : R^n \to R^n$, $y_s(x) = y(x,s)$ is a diffeomorphism) and a smooth map $\gamma : R^r \to R$, all defined in a neighbourhood of 0, such that

$$g(x,s) = f(y_s(x), e(s)) + \gamma(s) \tag{2.4}$$

for all $(x,s) \in R^n \times R^r$ in that neighbourhood.

Splitting Lemma: Let $f : R^n \times R^r \to R$ be smooth. Denote a point in $R^n \times R^r$ by (x,c). Suppose that the gradient is zero and the rank of the Hessian matrix **H** is $n - m$ at $(x,c)=0$. Then f is equivalent to a family of the form

$$\tilde{f}(y_1(x,c),\ldots,y_m(x,c),c) \pm y_{m+1}^2 \pm \ldots \pm y_n^2 . \tag{2.5}$$

We call $y_1,\ldots,y_m$ the *active variables*, $y_{m+1},\ldots,y_n$ the *passive variables*.

If $f : R^n \times R^r \to R$ is equivalent to any family $f + p : R^n \times R^r \to R$, where p is a sufficiently small family $R^n \times R^r \to R$, then f is *structurally stable*. An *r-unfolding* of a function $f : R^n \to R$ is a function $F : R^{n+r} \to R$, such that $F(x_1,\ldots,x_n,0,\ldots,0) = f(x_1,\ldots,x_n)$. F is a universal unfolding of f if all other unfoldings of f can be induced from F, and r is as small as possible.

The following theorem classifies the typical singularities of families of functions with less than 6 parameters:

Thom's theorem: Typically an *r*-parameter family of smooth functions $R^n \to R$, for any n and for $r \leq 5$, is structurally stable, and is equivalent around any point to one of the following forms:

1. u_1
2. $u_1^2 + \ldots + u_i^2 - u_{i+1}^2 - \ldots - u_n^2 \quad (0 \leq i \leq n)$
3. $u_1^3 + t_1 u_1 + (M)$
4. $\pm(u_1^4 + t_2 u_1^2 + t_1 u_1) + (M)$
5. $u_1^5 + t_3 u_1^3 + t_2 u_1^2 + t_1 u_1 + (M)$
6. $\pm(u_1^6 + t_4 u_1^4 + t_3 u_1^3 + t_2 u_1^2 + t_1 u_1) + (M)$
7. $u_1^7 + t_5 u_1^5 + t_4 u_1^4 + t_3 u_1^3 + t_2 u_1^2 + t_1 u_1 + (M)$
8. $u_1^2 u_2 - u_2^3 + t_3 u_1^2 + t_2 u_2 + t_1 u_1 + (N)$
9. $u_1^2 u_2 + u_2^3 + t_3 u_1^2 + t_2 u_2 + t_1 u_1 + (N)$
10. $\pm(u_1^2 u_2 + u_2^4 + t_4 u_2^2 + t_3 u_1^2 + t_2 u_2 + t_1 u_1) + (N)$
11. $u_1^2 u_2 - u_2^5 + t_5 u_2^3 + t_4 u_2^2 + t_3 u_1^2 + t_2 u_2 + t_1 u_1 + (N)$
12. $u_1^2 u_2 + u_2^5 + t_5 u_2^3 + t_4 u_2^2 + t_3 u_1^2 + t_2 u_2 + t_1 u_1 + (N)$
13. $\pm(u_1^3 + u_2^4 + t_5 u_1 u_2^2 + t_4 u_2^2 + t_3 u_1 u_2 + t_2 u_2 + t_1 u_1) + (N)$

where $(u_1,\ldots,u_n) \in R^n$, $(t_1,\ldots,t_r) \in R^r$ and

$$(M) = u_2^2 + \ldots + u_i^2 - u_{i+1}^2 - \ldots - u_n^2 \quad (1 \le i \le n),$$

$$(N) = u_3^2 + \ldots + u_i^2 - u_{i+1}^2 - \ldots - u_n^2 \quad (2 \le i \le n).$$

The first function has no critical point, the second function is a Morse $(n\text{-}i)$-saddle hence it has a non-degenerate critical point. These two types are not catastrophe forms: they do not change with t. The other functions are catastrophes. They can be put in two groups. Functions 3 to 7 have only one active variable, they are called *cuspoid catastrophes*. Functions 8 to 13 have two active variables, they are called *umbilic catastrophes*. Each catastrophe has a pet name and a symbol in the systematic classification of Arnol'd (1972). These are:

Cuspoid catastrophes:

3. the fold (A_2)
4. the cusp (A_3)
5. the swallowtail (A_4)
6. the butterfly (A_5)
7. the wigwam (A_6)

Umbilic catastrophes:

8. the elliptic umbilic $\left(D_4^-\right)$
9. the hyperbolic umbilic $\left(D_4^+\right)$
10. the parabolic umbilic (D_5)
11. the second elliptic umbilic $\left(D_6^-\right)$
12. the second hyperbolic umbilic $\left(D_6^+\right)$
13. the symbolic umbilic (E_6).

Four functions contain two signs. In the case of a positive sign they are called *standard forms*, those with a negative sign are referred to as *dual forms*.

For each catastrophe a canonical form is shown. *Canonical forms* consist a function with a universal unfolding, and they have some speciality. Later we will use other canonical forms.

A smooth function $f : R^n \to R$ is *k-determinate* at 0 if $f+g$ for any g of order $k+1$ is locally equivalent by a smooth change of co-ordinates to the Taylor series of f up to degree k. *Determinacy* of f is the lowest k for which f is k-determinate. The determinacy of the catastrophes is as follows.

Determinacy	Types
3	A_2, D_4^-, D_4^+
4	A_3, D_5, E_6
5	A_4, D_6^-, D_6^+
6	A_5
7	A_6

If the number of the parameters is less than 5 then the number of the possible catastrophes decreases:

Parameters	Catastrophes
1	1
2	2
3	5
4	7
5	11
6	∞

The last row shows why the number of the parameters was limited in Thom's theorem. Originally Thom (1972) dealt with 4 parameters, and Zeeman (1977) extended (and proved) the theorem to 5 parameters.

3 Catastrophes Arise at Elastic Structures

The behaviour of an engineering structure depends only on a single control parameter, the time (the variation of the loads and imperfections also occur in time). According to Thom's theorem only *fold catastrophe* can arise typically: i. e. a real structure (with its inevitable manufacturing imperfections) always loses its stability at a fold catastrophe, in which the equilibrium path merely reaches a maximum load at a limit point.

The structure and the load are often assumed to be symmetric. Hence in this case a Taylor expansion of the potential function in some (suitable) variables must contain vanishing coefficients of odd powers. In the case of perfect structures, a fold catastrophe cannot arise, but perfect symmetry on the drawing board can give rise to the *cusp catastrophe*. To model the structure successfully, allowing for the qualitative effect of all random perturbations of the system and its environment, we must include symmetry-destroying imperfection in the analysis.

Obviously a designer would like his system to carry loads as efficiently as possible, and this often calls for the simultaneity of failure loads so that the structure loses its stability with respect to two buckling modes simultaneously. In these cases there are two active variables, and umbilic catastrophes arise, typically the *elliptic and hyperbolic umbilic* (hence these need the fewest control parameters).

What is more, if both buckling modes are symmetric, then the perfect system must exhibit one of the *double-cusp catastrophes*, which are not included in Thom's theorem (because they need more than five control parameters).

One can design structures which have other types of catastrophes, but it is not an engineering aim, as was to have symmetrical structures carrying symmetrical loads or the optimization to spare money. Nevertheless, there is some reason why it worth dealing with other types, too:

- In the neighbourhood of catastrophes appear every simpler type of catastrophes, and we have got double-cusp catastrophes, too.
- At optimization one determine special values for different design variables, but the final plan usually somewhat other values, so the critical values will not be equal, only they will be near to each other. This can be handled as an imperfection of the ideal structure. Similarly, if our structure is near to a higher type of catastrophe, we might to consider this special structure as ideal one, and our structure as an imperfect one.
- A near higher catastrophe might give a much better global picture about the behaviour of our structure then the true simpler one.

This is why we will deal briefly with other types of catastrophes, too.

4   The Most Important Catastrophes

4.1   The Fold Catastrophe

The following canonical form will be analysed:

$$f(x,a) = \frac{1}{3}x^3 + ax. \tag{4.1}$$

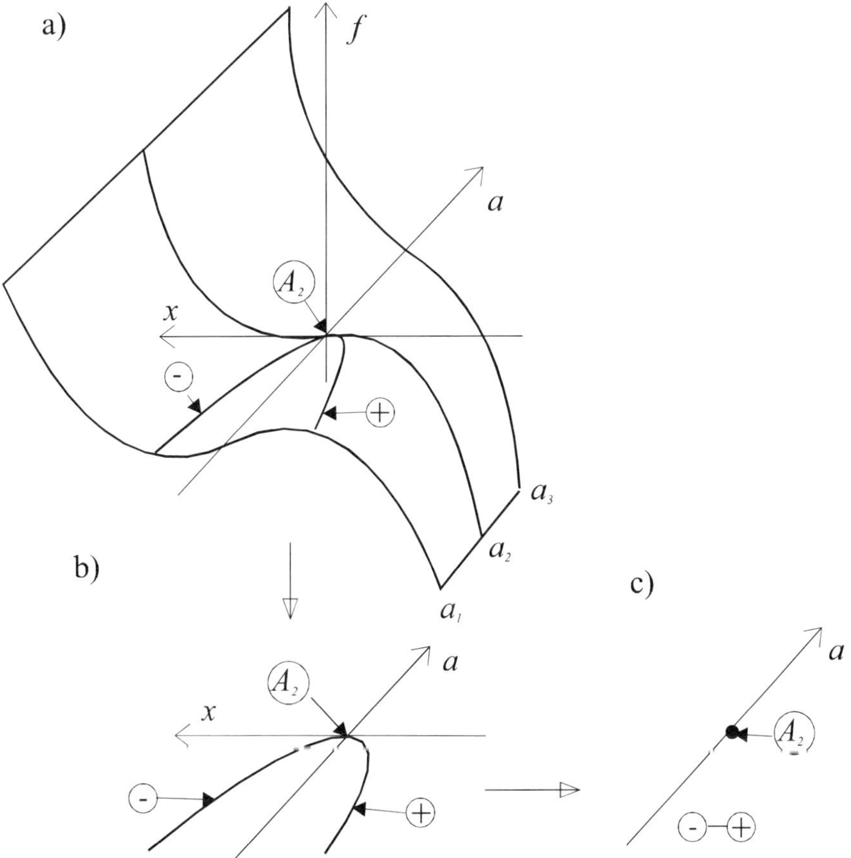

Figure 3. Family of functions (a), equilibrium surface (b) and bifurcation set (c) of the fold catastrophe.

Equation (4.1) can be drawn in a three-dimensional space (Figure 3a), where we have shown also three members of the family.

At a stationary point (critical point in catastrophe theory) the gradient is zero:

$$x^2 + a = 0. \tag{4.2}$$

(In the figures minimum, maximum and saddle points are denoted by a negative, a positive sign and an empty circle, respectively, while for catastrophes we use Arnol'd's notations.) Hence the value of f is usually not important, we map the stationary points into the space (x,a), having the so called equilibrium surface (Figure 3b), it is given by Eq. (4.2).

There is a catastrophe when the Hessian matrix (now: one scalar) is singular:
$$2x = 0.\tag{4.3}$$
Eliminating x from Eqs (4.2) and (4.3) one has
$$a = 0.\tag{4.4}$$

Eqs (4.3) and (4.4) determine the singularity set (now: one point) in Figure 3b, or mapping it into the parameter space (a) we get the bifurcation set (Figure 3c). The *bifurcation set* divides the parameter space into parts. Two functions have the same number and the same types of stationary points if their parameters belong to the same part, but they differ if they belong to different parts.

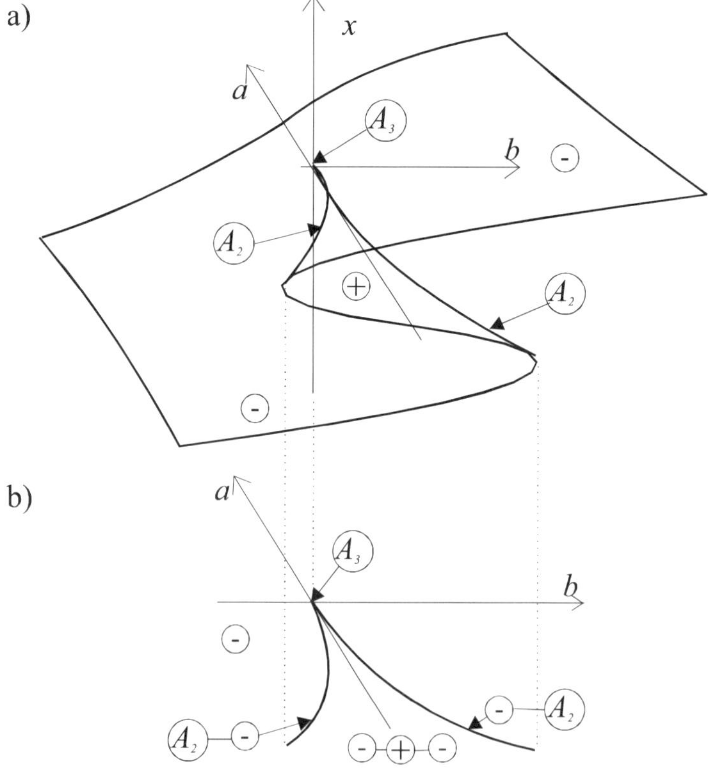

Figure 4. Equilibrium surface (a) and bifurcation set (b) of the standard cusp catastrophe.

4.2 The Cusp Catastrophe

The following canonical form will be analysed in the case of the standard cusp catastrophe (which is marked by A_3^+):

$$f(x,a,b) = \frac{1}{4}x^4 + \frac{1}{2}ax^2 + bx \, ,\tag{4.5}$$

while in the case of the dual cusp (A_3^-) the signs are changed.

To show the family of functions (4.5) we ought to use four dimensions, so the first set we can show is the equilibrium surface (Figure 4a), determined by the gradient:

$$x^3 + ax + b = 0 \, .\tag{4.6}$$

At catastrophe points the Hessian is singular, i. e.

$$3x^2 + a = 0 \, .\tag{4.7}$$

Using Eqs (4.6) and (4.7) the bifurcation set (Figure 4b) can be given in a parametric form:

$$a = -3p^2 \, , \qquad b = 2p^3 \, .\tag{4.8}$$

In the case of the dual cusp we have the same equations, only the maximum and minimum are inverted. Figure 5 shows the bifurcation set.

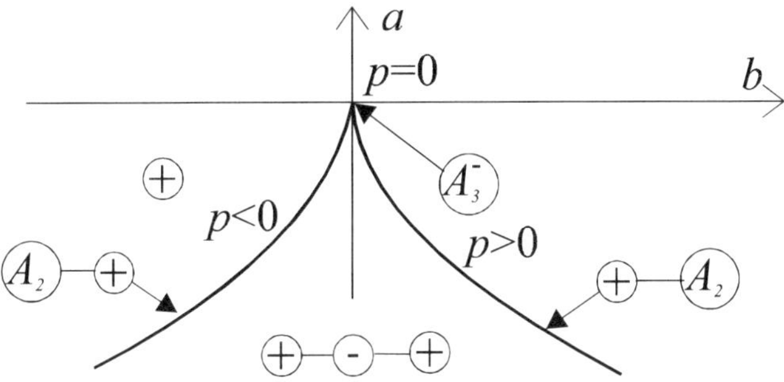

Figure 5. Bifurcation set of the dual cusp catastrophe.

4.3 The Swallowtail Catastrophe

A canonical form of the swallowtail catastrophe is

$$f(x,a,b,c) = \frac{1}{5}x^5 + \frac{1}{3}ax^3 + \frac{1}{2}bx^2 + cx \tag{4.9}$$

and the bifurcation set (Figure 6) in a parametric form is the following:

$$a = 3q - 6r^2 \, ,$$
$$b = -6rq + 8r^3 \, ,\tag{4.10}$$
$$c = 3qr^2 - 3r^4 \, .$$

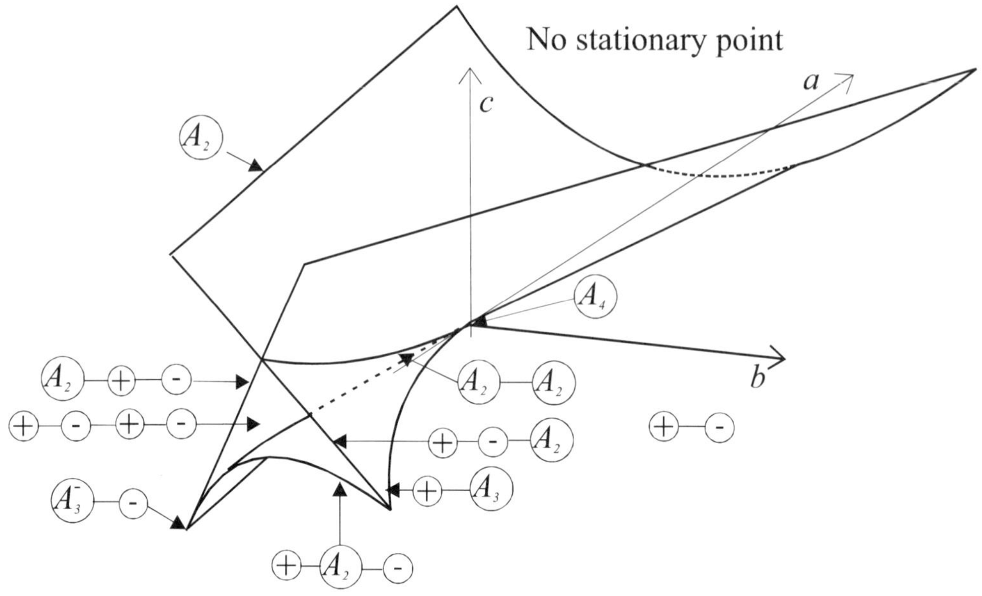

Figure 6. Bifurcation set of the swallowtail catastrophe.

4.4 The Butterfly Catastrophe

A canonical form of the standard butterfly catastrophe (A_5^+) is:

$$f(x,a,b,c,d) = \frac{1}{6}x^6 + \frac{1}{4}ax^4 + \frac{1}{3}bx^3 + \frac{1}{2}cx^2 + dx \qquad (4.11)$$

and the bifurcation set in a parametric form is the following:

$$a = 4r - 10s^2,$$
$$b = 3q - 12rs + 20s^3,$$
$$c = -6qs + 12rs^2 - 15s^4, \qquad (4.12)$$
$$d = 3qs^2 - 4rs^3 + 4s^5.$$

This ought to be shown in a four-dimensional space, so only a characteristic section of it is shown in Figure 7. Other sections are illustrated in Poston and Stewart (1978).

In the case of dual catastrophes the minimum and maximum points invert.

4.5 The Umbilic Bracelet

This section is based on Zeeman (1977). The umbilic catastrophes have two active variables. Homogeneous cubic forms with two-variables

$$f = ax^3 + bx^2y + cxy^2 + dy^3 \qquad (4.13)$$

can be classified by their root structure:

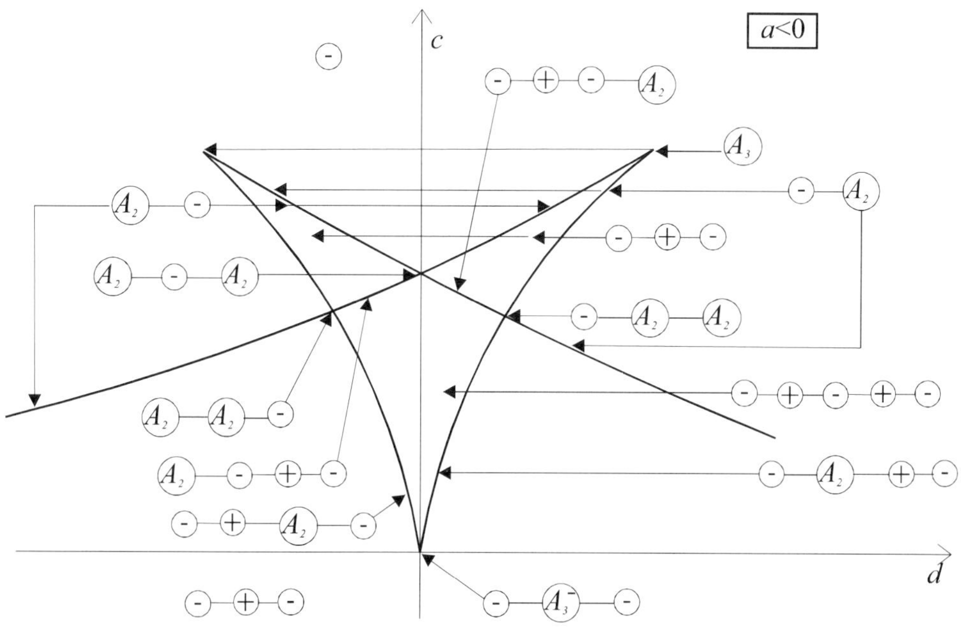

Figure 7. A section $(a < 0, b = 0)$ of the bifurcation set of the standard butterfly catastrophe.

Class	Root structure	Dimension	Example
elliptic	3 distinct lines	4	$x^3 - 3xy^2$
hyperbolic	a single real line	4	$x^3 + y^3$
parabolic	3 lines, 2 coincide	3	$x^2 y$
symbolic	3 coincident lines	2	x^3

To show that in which class the form is for given a, b, c, d coefficients, first we make a linear transformation:

$$\begin{bmatrix} \alpha_1 \\ \alpha_2 \\ \beta_1 \\ \beta_2 \end{bmatrix} = \frac{1}{4} \begin{bmatrix} 1 & 0 & -1 & 0 \\ 0 & -1 & 0 & 1 \\ 3 & 0 & 1 & 0 \\ 0 & -1 & 0 & -3 \end{bmatrix} \begin{bmatrix} a \\ b \\ c \\ d \end{bmatrix}. \tag{4.14}$$

The $\alpha_1 = \alpha_2 = 0$ cases are hyperbolic points. Otherwise we multiply f by a scalar to fulfil the

$$\alpha_1^2 + \alpha_2^2 = 1 \tag{4.15}$$

condition, so the classes can be illustrated in three-dimension by the umbilic bracelet (Figure 8a), where a three-cusped hypocycloid (Figure 8b) is rotated round a circle together with a one-third twist.

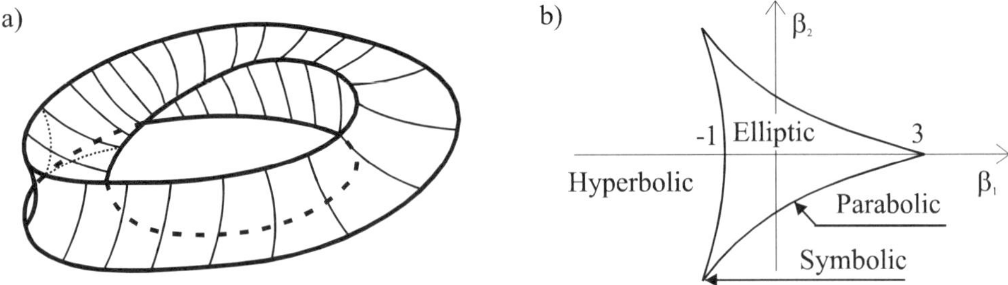

Figure 8. The umbilic bracelet (a) and a section $\left(\alpha_1 = 1, \alpha_2 = 0\right)$ of it (b).

Elliptic points are inside, hyperbolic points outside of the surface; parabolic points lie on the surface, while symbolic points are on the cusped edge.

4.6 The Elliptic Umbilic

The following canonical form will be analysed:

$$f(x, y, a, b, c) = y^3 - 3x^2 y + a(x^2 + y^2) + bx + cy . \tag{4.16}$$

The equilibrium surface is determined by the equations

$$\frac{\partial f}{\partial x} = -6xy + 2ax + b = 0 , \tag{4.17}$$

$$\frac{\partial f}{\partial y} = 3y^2 - 3x^2 + 2ay + c = 0 . \tag{4.18}$$

The elements of the Hessian are:

$$H_{11} = -6y + 2a , \quad H_{12} = H_{21} = -6x , \quad H_{22} = 6y + 2a ,$$

so the matrix is singular if

$$(-6y + 2a)(6y + 2a) - 36x^2 = 0 . \tag{4.19}$$

Eliminating x and y from Eqs (4.17)-(4.19) one obtains the equation of the bifurcation set, but the following parametric form is more useful:

$$a = a , \quad b = \frac{a^2}{3}\left(\sin 2\Theta - 2\cos\Theta\right), \quad c = \frac{a^2}{3}\left(\cos 2\Theta - 2\sin\Theta\right) . \tag{4.20}$$

Apart from the parabolic increasing factor $a^2/3$ the cross sections for a constant a have the same Θ-dependence. The cross section for $a = \sqrt{3}$ is a three-cusped hypocycloid, as shown in Figure 9a. The whole bifurcation set and the types of the stationary points are drawn in Figure 9b. The parameter space is divided into three parts by the bifurcation set, but only one of them contains a minimum. Usually only the border of this part (given by the restriction $a \leq 0$) is important, and it is shown in Figure 9c.

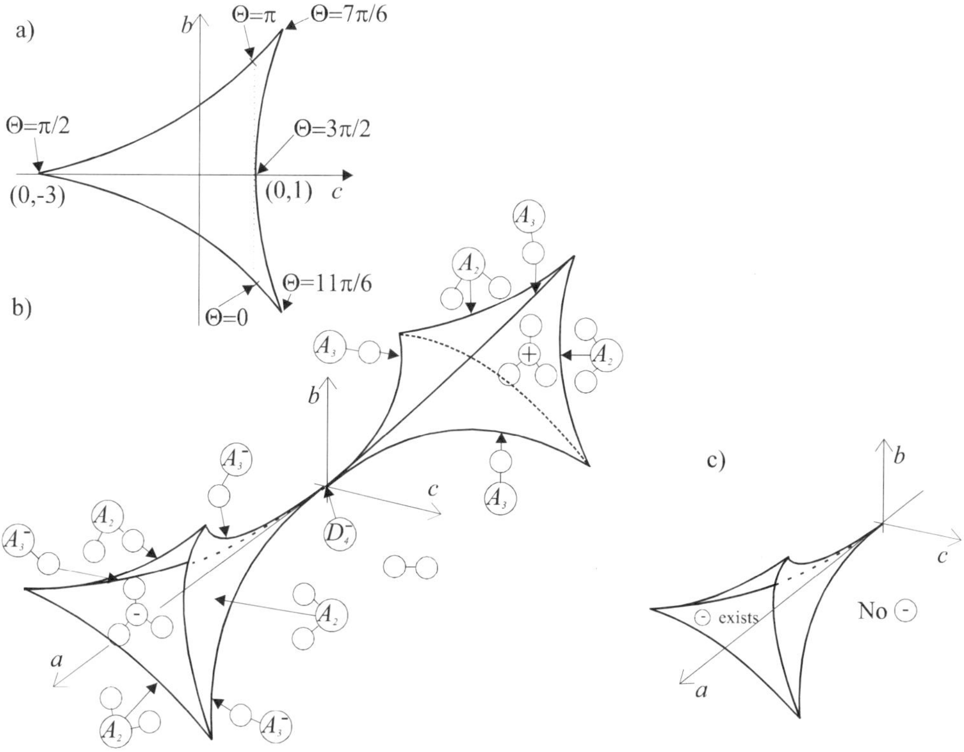

Figure 9. Bifurcation set of the elliptic umbilic catastrophe: a section (a), the whole set (b) and the most important part of it (c).

4.7 The Hyperbolic Umbilic

The following canonical form will be analysed:

$$f(x,y,a,b,c) = x^3 + y^3 + axy + bx + cy.$$ (4.21)

The bifurcation set can be written in the following parametric form:

$$a = 6p,$$
$$b = -3p^2\left(2q^{-1} + q^2\right),$$ (4.22)
$$c = -3p^2\left(2q + q^{-2}\right),$$

but we have to add to it the negative b-axis and the negative c-axis. The cross sections for constant a-s are similar; one of them is shown in Figure 10a. The whole bifurcation set and the types of the stationary points are drawn in Figure 10b. The parameter space is divided into four parts by the bifurcation set, and two of them contain a minimum. So only one half of the bifurcation set has different number of minimum points on its both sides. This half satisfies the condition $pq > 0$ and is shown in Figure 10c.

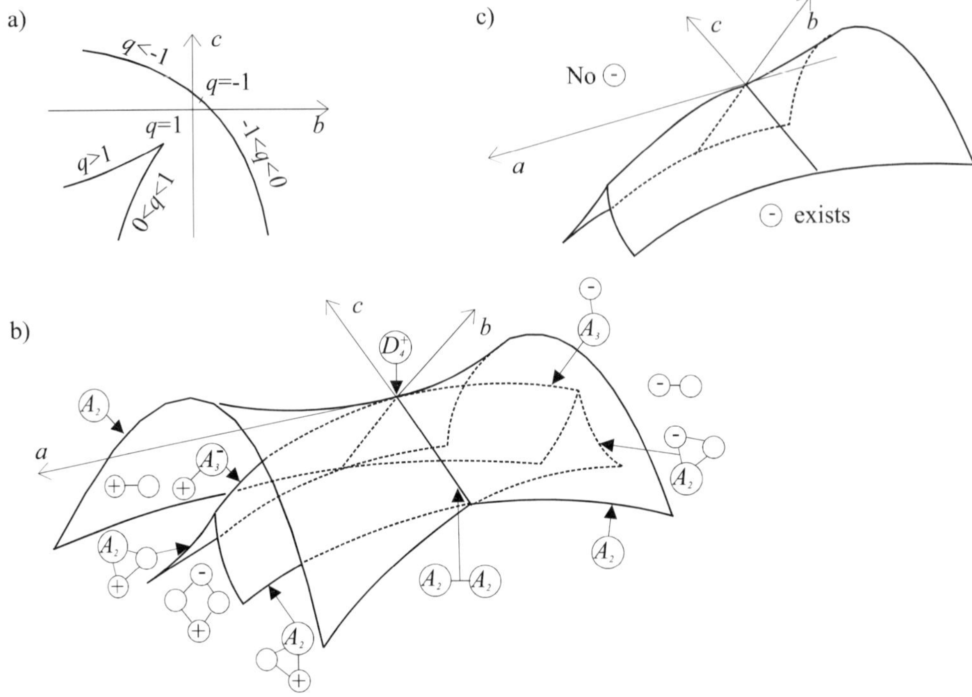

Figure 10. Bifurcation set of the hyperbolic umbilic catastrophe: a section (a), the whole set (b) and the most important part of it (c).

4.8 Double Cusp Catastrophes

This section is based on Poston and Stewart (1976). The double cusp catastrophes have two active variables. Homogeneous quartic forms with two-variables

$$f = ax^4 + bx^3 y + cx^2 y^2 + dxy^3 + ey^4 \qquad (4.23)$$

can be classified by their root structure:

Class	Root structure	Dimension	Example
1.	r_1, r_2, r_3, r_4	5	$x^4 - 6x^2 y^2 + y^4$
2.	r_1, r_2, c_3, c_4	5	$x^4 - y^4$
3.	c_1, c_2, c_3, c_4 +	5	$x^4 + y^4$
4.	c_1, c_2, c_3, c_4 −	5	$-x^4 - y^4$
5.	$r_1 = r_2, r_3, r_4$ +	4	$x^2(x^2 - y^2)$
6.	$r_1 = r_2, r_3, r_4$ −	4	$-x^2(x^2 - y^2)$
7.	$r_1 = r_2, c_3, c_4$ +	4	$x^2(x^2 + y^2)$

8.	$r_1 = r_2, c_3, c_4 \quad -$	4	$-x^2\left(x^2 + y^2\right)$
9.	$r_1 = r_2 = r_3, r_4$	3	$x^3 y$
10.	$r_1 = r_2, r_3 = r_4 \quad +$	3	$x^2 y^2$
11.	$r_1 = r_2, r_3 = r_4 \quad -$	3	$-x^2 y^2$
12.	$c_1 = c_2, c_3 = c_4 \quad +$	3	$\left(x^2 + y^2\right)^2$
13.	$c_1 = c_2, c_3 = c_4 \quad -$	3	$-\left(x^2 + y^2\right)^2$
14.	$r_1 = r_2 = r_3 = r_4 \quad +$	2	x^4
15.	$r_1 = r_2 = r_3 = r_4 \quad -$	2	$-x^4$

(Here r and c denote real and complex roots, respectively.)

The first four classes have essential difference according to the classes in Thom's theorem, hence two functions from the same class are usually not equivalent, they have infinitely many subclasses. If the family of functions has six parameters then also these classes may arise.

Similarly to the way, which led to the umbilic bracelet, connections of the 15 classes can be illustrated, but the number of the dimensions increased by one, so only the new cross section can be shown in three-dimension space (Figure 11). This contains of a tetrahedron with curved edges, two bowls and two whiskers.

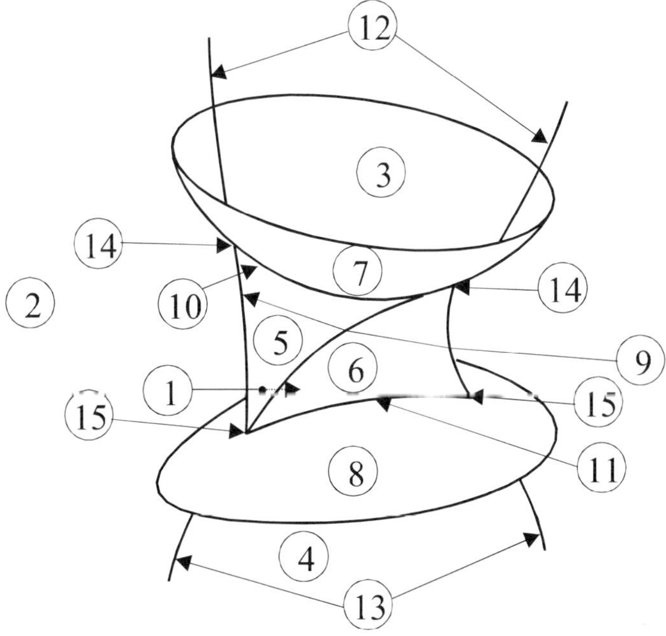

Figure 11. Classes of the double-cusp catastrophes.

5 Illustrative Models

In this section by the help of some illustrative models we show the imperfection-sensitivity surfaces and the equilibrium paths of the different types of loss of stability.

In section 1 we have shown how to write the potential energy function, how to determine its critical points. It was shown in section 3 that usually we may suspect which catastrophe occurs at this critical point. The catastrophe theory (section 2) has given the determinacy of this catastrophe, so one can determine the appropriate Taylor expansion of the energy function, and according to the Splitting Lemma we can separate the active and passive parts of it. Here we will use incremental coordinates (difference from the critical values) both for state variables and load parameter, e. g.:

$$u = \varphi - \varphi^{cr}, \qquad \lambda = \Lambda - \Lambda^{cr}. \tag{5.1}$$

By the help of the active part it is easy to check whether really the suspected catastrophe occurs or not.

In section 4 we have shown the bifurcation sets in the parameter space for the most important catastrophe types. In the case of a given elastic structure, fixing the imperfections and changing the load parameter we will have a route in the parameter space (Thompson and Hunt, 1984). This route can reach or cross the bifurcation set some different ways, this fact result in a subclassification of the catastrophe types. Similar ways lead to similar imperfection-sensitivity surfaces, so it is worth analysing simple models for the different subclasses.

5.1 Fold Catastrophe

According to Figure 3c the parameter space of the fold catastrophe has one dimension and the bifurcation set contains one point. The negative half of axis a belongs to functions having one minimum and one maximum, while the other part gives functions without stationary point. (We always speak about the active part of the function and suppose that the passive part has a minimum point.) The structure may lose its stability if originally it was in stable state, so if $\lambda < 0$, then a must be negative. There are two possibilities (Figure 12):

- the λ route cross the bifurcation set (limit point) or
- it turns back to the negative part of axis a (asymmetric point of bifurcation).

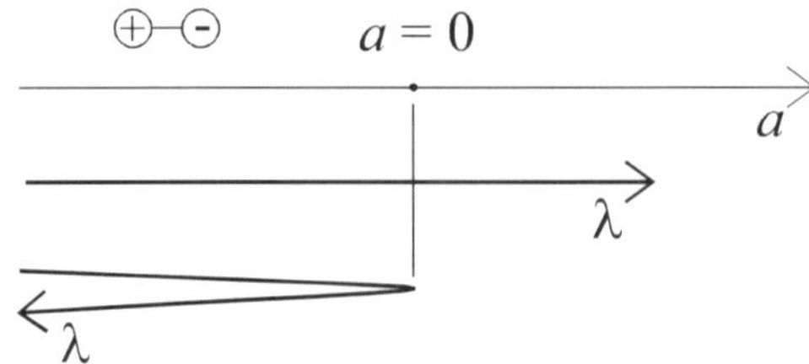

Figure 12. Routes in control space through fold catastrophe.

Example 6. Let as consider again the structure of Example 3d, but we introduce an imperfection for the rigidity of spring 2. So the parameters are:

$$L = 1,\ a = 0.9,\ c_1 = 2,\ c_2 = 2 + \varepsilon,\ l_1 = \sqrt{1.01},\ l_2 = 1. \tag{5.2}$$

We want to determine the equilibrium paths and the imperfection sensitivity curve in the small vicinity of the second critical point of the perfect structure (this point was determined in Example 5).

According to (5.1) we slip the origin into the critical point. Determinacy of the fold catastrophe is 3, so we truncate the Taylor series of the potential energy function at third degree terms in u; even we keep only the lowest degree terms in u, which contains λ and ε. So

$$V(u, \lambda, \varepsilon) = -0.2181503913u^3 - 0.1538356302\lambda u + 0.1490701384\varepsilon u. \tag{5.3}$$

Using its first derivative the equation of the equilibrium paths is

$$\lambda = -4.254223635u^2 + 0.9690221843\varepsilon, \tag{5.4}$$

which is illustrated in Figure 13a for a positive, the zero and a negative ε values.

Hence there maximum points are always at $u = 0$ the imperfection-sensitivity curve is given by

$$\lambda^{cr} = 0.9690221843\varepsilon, \tag{5.5}$$

and is shown in Figure 13b (C substitute a positive constant).

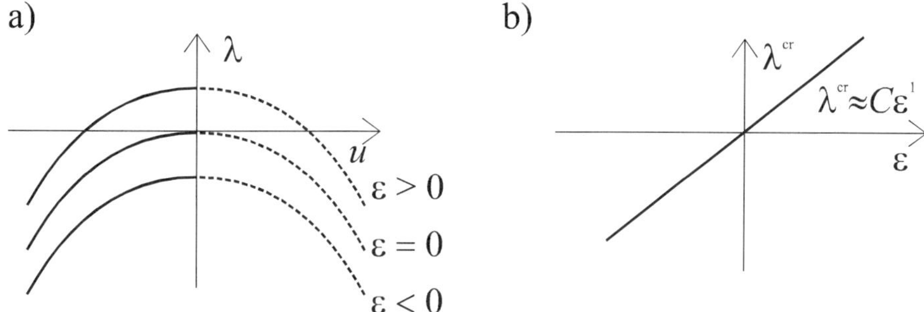

Figure 13. Equilibrium paths for different imperfections (a) and imperfection-sensitivity curve in the case of limit point.

Example 7. Let as consider again the structure of Example 3c, but we introduce an imperfection for the length of spring 2. So the parameters are:

$$L = 1,\ a = 0,\ c_1 = 2,\ c_2 = 3,\ l_1 = \sqrt{2},\ l_2 = \sqrt{2} + \varepsilon. \tag{5.6}$$

We want to determine the equilibrium paths and the imperfection sensitivity curve in the small vicinity of the second critical point of the perfect structure (this point was determined in Example 5).

Similarly to Example 6 we slip the origin into the critical point and determine the truncated Taylor series:

$$V(u, \lambda, \varepsilon) = 0.125u^3 - 0.5\lambda u^2 + 2.121320343\varepsilon u. \tag{5.7}$$

The equilibrium paths can be given as

$$\lambda = \frac{0.375u^2 + 2.1211320343\varepsilon}{u} \tag{5.8}$$

and are illustrated in Figure 14a for a positive, the zero and a negative ε values.

Eliminating u from the following equations

$$V_u = 0, \quad V_{uu} = 0 \tag{5.9}$$

we get the imperfection-sensitivity curve (Figure 14b):

$$\lambda^{cr} = 1.783811970\varepsilon^{1/2} . \tag{5.10}$$

Broken lines show those parts of the imperfection-sensitivity curves, which do not belong to the loss of stability, but some new equilibrium states arise.

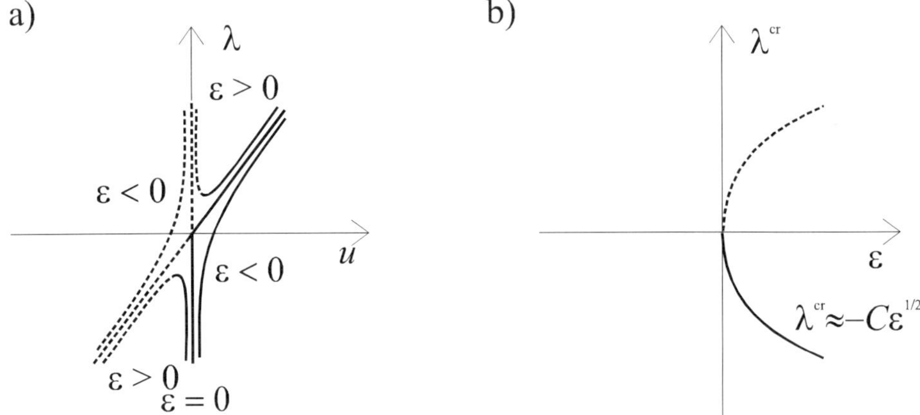

Figure 14. Equilibrium paths for different imperfections (a) and imperfection-sensitivity curve in the case of asymmetric point of bifurcation.

Generalization

The active part of the potential energy function exhibits a fold catastrophe at its critical point if the first few terms of its Taylor series have the form

$$V' = c_1 u^3 + \varepsilon_1 u^2 + \varepsilon_2 u - \lambda\left(c_2 u^2 + c_3 u\right), \tag{5.11}$$

where u is the only active generalized co-ordinate, $c_1 > 0$ (if $c_1 < 0$, then we change the direction of axis u), ε_1 and ε_2 are functions of the imperfections α, and $\varepsilon_1(0) = \varepsilon_2(0) = 0$. For a perfect structure typically $c_3 \neq 0$. If $c_3 = 0$, then this coefficient will be denoted by ε_3 and typically $c_2 \neq 0$. The case $c_2 = c_3 = 0$ will not be considered.

First applying the linear transformation

$$u = \frac{v}{\left(3c_1\right)^{1/3}} - \frac{\varepsilon_1 - \lambda c_2}{3c_1} \tag{5.12}$$

we eliminate the second-order terms from Eq. (5.11):

$$V(v,\lambda,\varepsilon_1,\varepsilon_2) = \frac{1}{3}v^3 + \left(\frac{\varepsilon_2 - \lambda c_3}{(3c_1)^{1/3}} - \frac{(\varepsilon_1 - \lambda c_2)^2}{(3c_1)^{4/3}} \right)v. \tag{5.13}$$

Function (5.13) can be induced from the canonical form (4.1), because the transformations

$$x = v, \qquad\qquad a = \frac{\varepsilon_2 - \lambda c_2}{(3c_1)^{1/3}} - \frac{(\varepsilon_2 - \lambda c_2)^2}{(3c_1)^{4/3}} \tag{5.14}$$

give
$$f(x,a) = V(v,\lambda,\varepsilon_1,\varepsilon_2). \tag{5.15}$$

Substituting the transformation (5.14) into the equation (4.4) of the bifurcation set, one obtains the equation of the imperfection-sensitivity surface:

$$\varepsilon_2 = \frac{(\varepsilon_1 - \lambda c_2)^2}{3c_1} + \lambda c_3. \tag{5.16}$$

A typical form of it is shown in Figure 15.

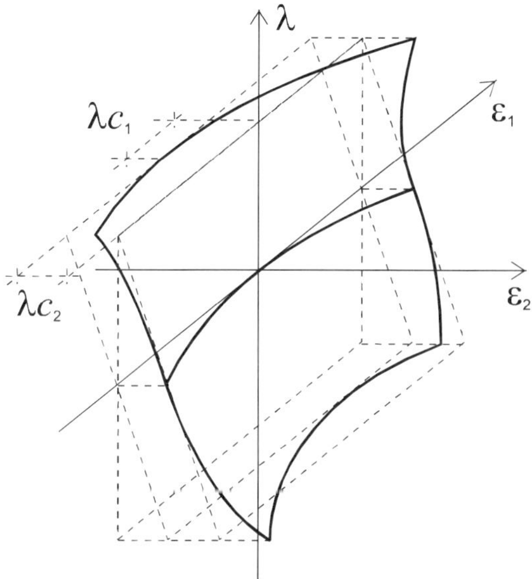

Figure 15. The imperfection-sensitivity surface.

Imperfection-sensitivity surfaces are usually governed by multi-valued functions, so it is not easy to determine the smallest critical load for a described imperfection. This is why the concept of the *critical imperfection territory* (Gaspar, 1982) was introduced. This territory covers all imperfections resulting in a value of the critical load smaller than a prescribed value. To illustrate how the border of the critical imperfection territory depends on the prescribed value of the load parameter, we show always two borders. One for the case $\lambda = 0$ and the other for a $\lambda_0 < 0$. So the space of the imperfections is divided into three parts (and their number will be shown in the figures):

1. Critical imperfection territory belonging to a given load parameter $\lambda_0 < 0$.
2. Imperfections resulting in critical loads that are smaller then the critical load of the perfect system, but greater than λ_0.
3. Imperfections that result in no loss of stability if the load is smaller than Λ^{cr}.

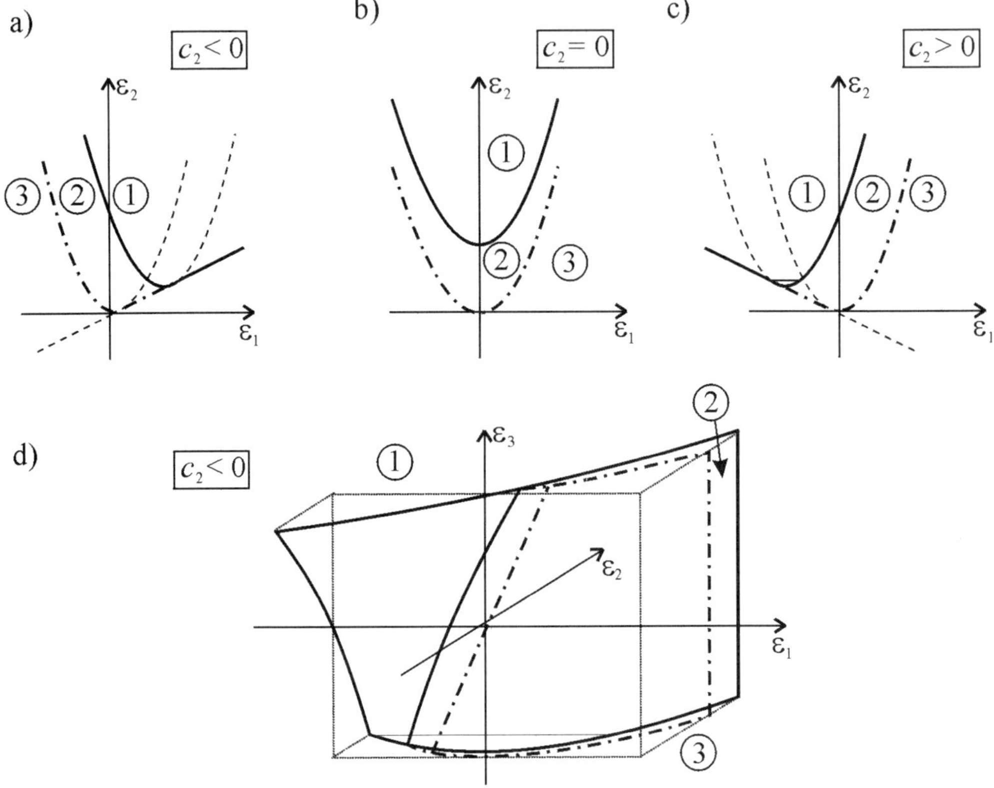

Figure 16. Critical imperfection territories.

The border of the critical imperfection territories contains partly a level line (λ =const.) of the imperfection-sensitivity surface and partly the envelope of the level lines. The envelope of the level lines given by Eq. (5.16) is a straight line and has the form:

$$\varepsilon_2 = \frac{c_3}{c_2}\varepsilon_1 - \frac{3c_1 c_3^2}{4c_2^2}. \tag{5.17}$$

(If $c_2 = 0$, there is no envelope.) The envelope touches the level lines at the points

$$\varepsilon_1^e = c_2\lambda + \frac{3c_1c_3}{2c_2}, \qquad \varepsilon_2^e = c_3\lambda + \frac{3c_1c_3^2}{2c_2^2} . \tag{5.18}$$

Figures 16a-c show the typical shape of the critical imperfection territories for the case $c_3 < 0$ and Figure 16d shows the special case when $c_3 = 0$ for the perfect structure (so we have a third imperfection parameter).

According to Eqs (4.2) and (5.14) the equilibrium paths are given by

$$v = \left(\frac{\lambda c_3 - \varepsilon_2}{(3c_1)^{1/3}} - \frac{(\lambda c_2 - \varepsilon_1)^2}{(3c_1)^{4/3}} \right)^{1/2} \tag{5.19}$$

which can have three different types (Figure 17), depending on the values of its constants. The distance between the two curves in Figure 17a increases when the value of $|c_2|$ decreases, and the upper curve disappears when $c_2 = 0$. Using transformation (5.12) the path can be written in dependence of the original active state variable (see Figure 14a).

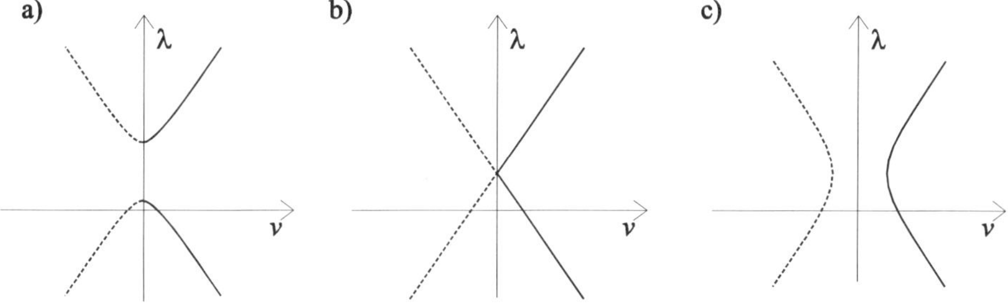

Figure 17. Equilibrium paths of fold catastrophes.

5.2 Dual Cusp Catastrophe

According to Figure 5 the parameter space of the dual cusp catastrophe has two dimensions and the bifurcation set is a curve. Below this curve there are points belong to functions having one minimum and two maxima, while the other part gives functions with one maximum. So if $\lambda < 0$, we must be below of the curve. There are two possibilities for the perfect structure to lose its stability (Figure 18):

- the λ route cross the bifurcation set at the cusp point (unstable-symmetric point of bifurcation) or
- it turns back (unstable-X point of bifurcation).

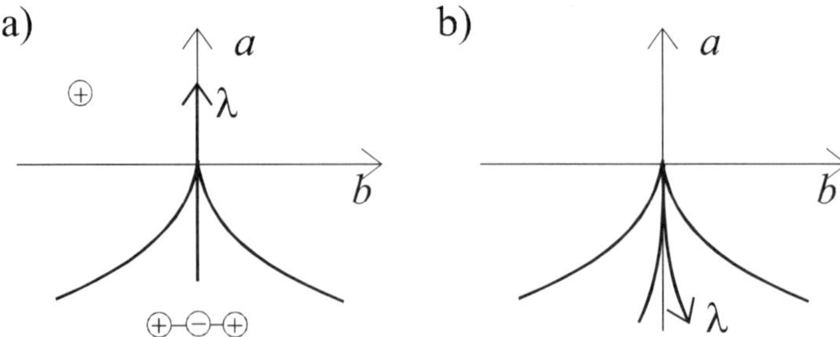

Figure 18. Routes in control space through dual cusp catastrophe.

Example 8. Let as consider again the structure of Example 3b, but we introduce an imperfection for the length of spring 2. So the parameters are:

$$L = 1,\ a = 0.9,\ c_1 = 2,\ c_2 = 2,\ l_1 = \sqrt{1.01},\ l_2 = \sqrt{1.01} + \varepsilon.$$

$$(5.20)$$

We want to determine the equilibrium paths and the imperfection sensitivity curve in the small vicinity of the second critical point of the perfect structure (this point was determined in Example 5).

Similarly to Example 6 we slip the origin into the critical point and determine the truncated Taylor series:

$$V(u, \lambda, \varepsilon) = -0.3143833693u^4 - 0.5\lambda u^2 + 1.990074380\varepsilon u.$$

$$(5.21)$$

The equilibrium paths can be given as

$$\lambda = \frac{-1.257533477u^3 + 1.990074380\varepsilon}{u}$$

$$(5.22)$$

and are illustrated in Figure 19a for a positive, the zero and a negative ε values.

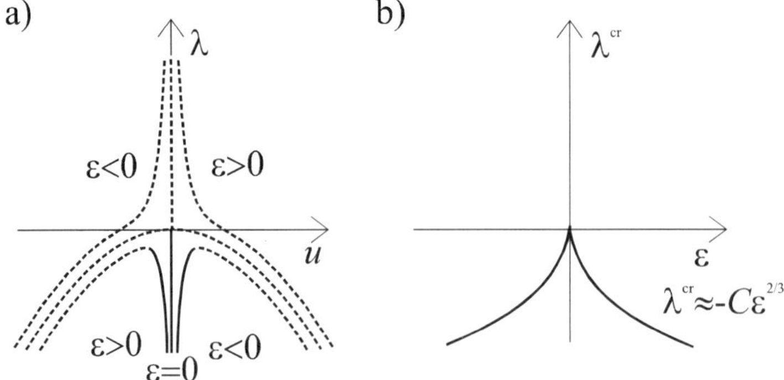

Figure 19. Equilibrium paths for different imperfections (a) and imperfection-sensitivity curve (b) in the case of unstable-symmetric point of bifurcation.

Eliminating u from equations $V_u = 0,\ V_{uu} = 0$ we get the imperfection-sensitivity curve (Figure 19b):

$$\lambda^{cr} = -2.033140156\varepsilon^{2/3}.$$ (5.23)

Example 9. Let as consider a simplified version of the structure analysed by Gaspar and Domokos (1991). The structure is a hinged cantilever (Figure 20a) comprising a link with normal rigidity k, pinned to the rigid foundation and supported by a linear rotational spring of stiffness c. The vertical load acts on the top of the link. A state of the structure can be given by two state variables (φ, h). The unloaded perfect structure is in equilibrium in the state $\varphi = 0,\ h = 1$. We introduce an imperfection, when the rotational spring is unstressed if $\varphi = \varepsilon$.

Let us determine the equilibrium paths and the imperfection sensitivity curve in the small vicinity of their critical point.

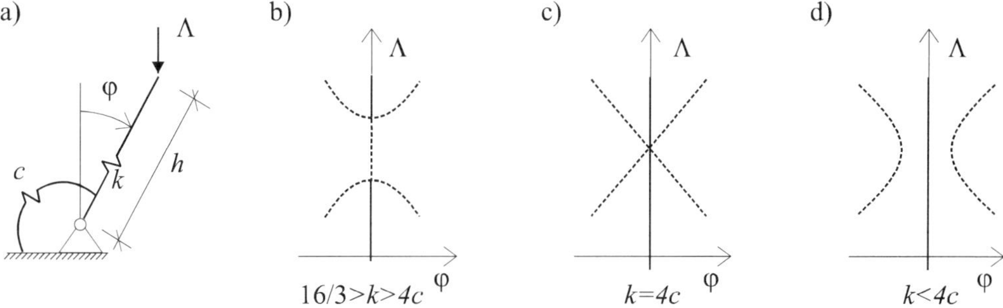

Figure 20. A pinned cantilever (a) and the equilibrium paths of some perfect structures (b-d).

The total potential energy function is

$$V(\varphi, h, \Lambda, \varepsilon) = \frac{1}{2}c(\varphi - \varepsilon)^2 + \frac{1}{2}k(1 - h)^2 + \Lambda h \cos\varphi.$$ (5.24)

The perfect structure is in equilibrium if

$$V_\varphi = c\varphi - \Lambda h \sin\varphi = 0,$$

$$V_h = k(h - 1) + \Lambda \cos\varphi = 0,$$ (5.25)

which are fulfilled when $\varphi = 0$ and $h = 1 - \Lambda/k$. The Hessian matrix on the primary equilibrium path is diagonal:

$$\mathbf{H}_0 = \langle c - \Lambda(1 - \Lambda/k)\quad k \rangle,$$ (5.26)

so it is singular if

$$\Lambda^{cr}_{1,2} = \frac{k \pm \sqrt{k^2 - 4ck}}{2}.$$ (5.27)

There are two critical load parameter when $k > 4c$, one (double) when $k = 4c$, and the structure do not lose its stability if $k < 4c$. The equilibrium paths are shown for these cases in Figure 20b-d. (Sign of the forth degree term changes when k>16/3.)

We should like to illustrate the case shown in Figure 18b, so we choose $k = 4c$, and the critical values are the following:

$$\Lambda^{cr} = 2c, \quad \varphi^{cr} = 0, \quad h^{cr} = 0.5 . \tag{5.28}$$

We slip the origin into the critical point by the linear transformations

$$\Lambda = \Lambda^{cr} + \lambda, \quad \varphi = \varphi, \quad h = h^{cr} + v , \tag{5.29}$$

and determine the truncated Taylor series (we choose $c = 1$):

$$V(\varphi, v, \lambda, \varepsilon) = \frac{1}{24}\varphi^4 - \left(1 + \frac{\lambda}{2}\right)\varphi^2 v - \frac{1}{4}\lambda\varphi^2 + 2v^2 + \lambda v - \varepsilon\varphi . \tag{5.30}$$

Both φ and v appears in the second term, but the diffeomorphism

$$\varphi = \varphi, \quad v = \frac{u\sqrt{2}}{2} + \left(\frac{1}{4} + \frac{\lambda}{8}\right)\varphi^2 - \frac{\lambda}{4} \tag{5.31}$$

splits V into an active and a passive parts:

$$V(\varphi, u, \lambda, \varepsilon) = -\frac{1}{12}\varphi^4 + \frac{1}{8}\lambda^2\varphi^2 - \varepsilon\varphi + u^2 . \tag{5.32}$$

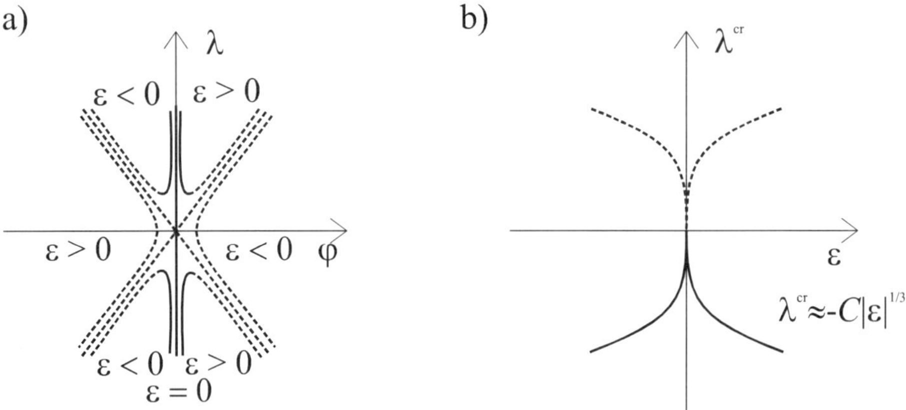

Figure 21. Equilibrium paths for different imperfections (a) and imperfection-sensitivity curve (b) in the case of unstable-X point of bifurcation.

Using the active part the equilibrium paths can be given as

$$\lambda = \pm 2\left(\frac{\varphi^3 + 3\varepsilon}{\varphi}\right)^{1/2} \tag{5.33}$$

and are illustrated in Figure 21a for a positive, the zero and a negative ε values.

Eliminating φ from equations $V_\varphi = 0$, $V_{\varphi\varphi} = 0$ we get the imperfection-sensitivity curve (Figure 21b):

$$\lambda^{cr} = \pm(12\varepsilon)^{1/3} . \tag{5.34}$$

Generalization

The unstable-X point of bifurcation is not typical, so we will deal with the unstable-symmetric point of bifurcation with more types of imperfections. The most general case of the active part we deal with is:

$$V' = c_1 u^4 + \varepsilon_1' u^3 + \varepsilon_2' u^2 + \varepsilon_3' u - \lambda\left(c_2 u^2 + \varepsilon_4' u\right), \tag{5.35}$$

where $c_i \neq 0$ ($c_1 > 0$ for standard and $c_1 < 0$ for dual cusp) , and for perfect systems $\varepsilon_i' = 0$.

Applying the linear transformation

$$u = \frac{v}{\left|4c_1\right|^{1/4}} - \frac{\varepsilon_1'}{4c_1} \tag{5.36}$$

we eliminate the third order terms from Eq. (5.35):

$$V\left(v, \lambda, \varepsilon_1, \varepsilon_2, \varepsilon_3\right) = \alpha\left[\frac{1}{4}v^4 + \varepsilon_1 v^2 + \varepsilon_2 v - \lambda\left(cv^2 + \varepsilon_3 v\right)\right] \tag{5.37}$$

where

$$\alpha = \operatorname{sign} c_1 , \quad c = \frac{\alpha c_2}{\left|2c_1\right|^{1/2}} , \quad \varepsilon_1 = \frac{\alpha \varepsilon_2'}{\left|2c_1\right|^{1/2}} - \frac{3\left(\varepsilon_1'\right)^2}{\left|16c_1\right|^{3/2}} ,$$

$$\varepsilon_2 = \frac{\alpha \varepsilon_3'}{\left|4c_1\right|^{1/4}} - \frac{2\varepsilon_1'\varepsilon_2'}{\left|4c_1\right|^{3/4}} + \frac{2\left(\varepsilon_1'\right)^3}{\left|4c_1\right|^{9/4}} , \quad \varepsilon_3 = \frac{\alpha \varepsilon_4'}{\left|4c_1\right|^{1/4}} - \frac{2c_2\varepsilon_1'}{\left|4c_1\right|^{3/4}} .$$

Function (5.37) can be induced from the canonical form (4.5), because the transformations

$$x = v, \quad a = 2\left(\varepsilon_1 - \lambda c\right), \quad b = \varepsilon_2 - \lambda\varepsilon_3 \tag{5.38}$$

yield

$$f(x, a, b) = V\left(v, \lambda, \varepsilon_1, \varepsilon_2, \varepsilon_3\right) . \tag{5.39}$$

Substituting the transformations (5.38) into Eqs (4.8) of the bifurcation set, one obtains the equations of the imperfection-sensitivity surface:

$$\varepsilon_1 = c\lambda - 3p^2/2, \qquad \varepsilon_2 = \lambda\varepsilon_3 + 2p^3 . \tag{5.40}$$

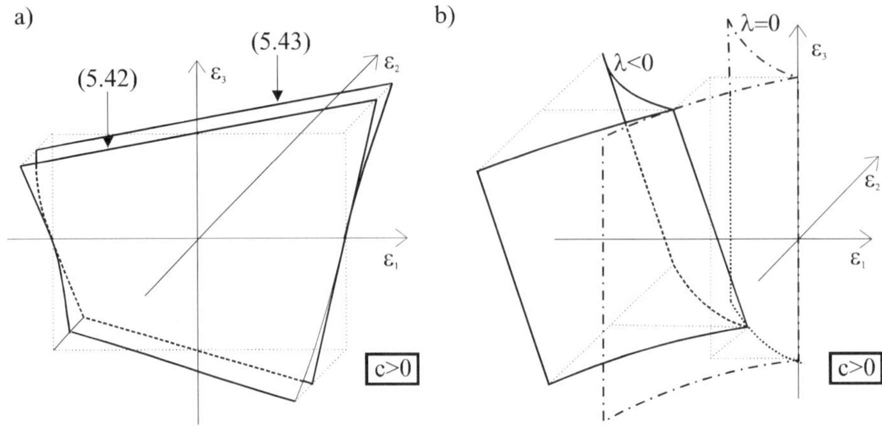

Figure 22. The envelope and the surface of the singular points (a) and the level surfaces (b).

The envelope of the level lines $\left(\lambda = \text{const.}, \varepsilon_3 = \text{const.}\right)$ of this surface also has to satisfy the equation:

$$c \cdot 6 p^2 + 3 p \cdot \varepsilon_3 = 0 \tag{5.41}$$

which is true for two values of p: $p_1 = 0$, $p_2 = -\varepsilon_3 /(2c)$. Eqs (5.40) and (5.41) determine two surfaces:

$$\varepsilon_2 = \varepsilon_1 \varepsilon_3 / c , \tag{5.42}$$

$$\varepsilon_2 = \varepsilon_1 \varepsilon_3 / c + \varepsilon_3^3 /\left(8c^3\right). \tag{5.43}$$

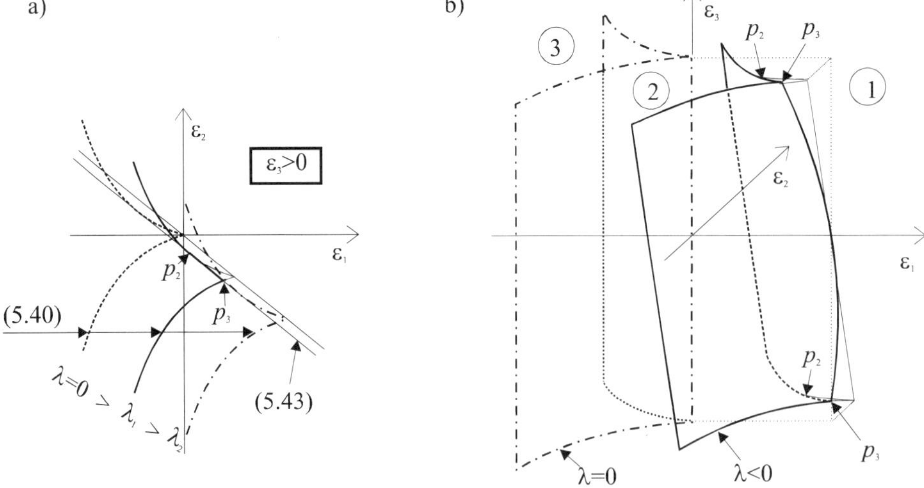

Figure 23. Dual cusp, the critical imperfection territory.

The first surface belongs to the singular points of the level curves, while (5.43) defines the envelope. Both surfaces are shown in Figure 22a, and two level surfaces in Figure 22b.

In the case of a fixed $\varepsilon_3 > 0$ Figure 23a shows the variation of the sections of the level surfaces and the envelope. The point of intersection belongs to parameter $p_3 = \varepsilon_3/(4c)$. Figure 23b shows the three territories of the imperfection space.

Eqs (5.38) show that we have a straight line in the parameter space if we fix the imperfections. Some cases are shown in Figure 24a. The corresponding equilibrium paths are shown in Figures 24b-e. The well-known equilibrium paths (Figure 19a) do not show the possibility of type shown is Figure 24e.

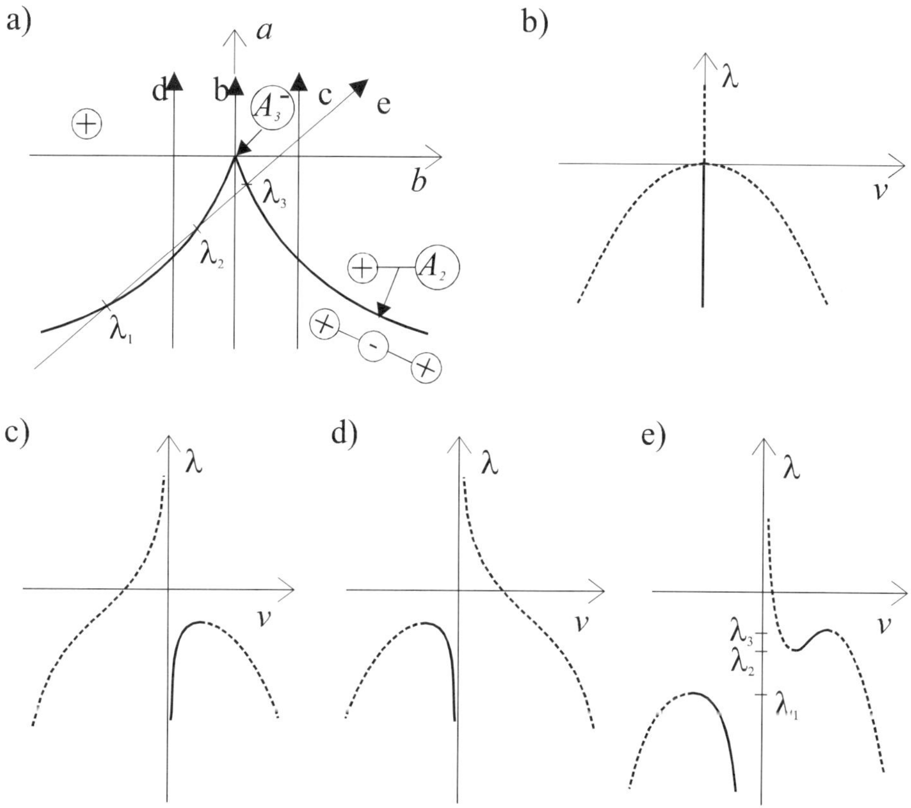

Figure 24. Dual cusp: the parameter space (a) and some equilibrium paths (b-e).

5.3 Standard Cusp Catastrophe

According to Figure 4b the parameter space of the standard cusp catastrophe has two dimensions and the bifurcation set is a curve. Below this curve there are points belong to functions having one maximum and two minima, while the other part gives functions with one minimum. So we can start from any part of the parameter space. Theoretically there are five possibilities for the perfect structure to arrive the cusp point (Figure 25):

- starting from above the curve the λ route cross the bifurcation set (stable-symmetric point of bifurcation),
- the smooth λ route remains always above the bifurcation set (cut-off point),
- the λ route turns back (point-like instability), or
- starting from below the curve the λ route cross the bifurcation set (upside down case),
- it turns back (stable-X point of bifurcation).

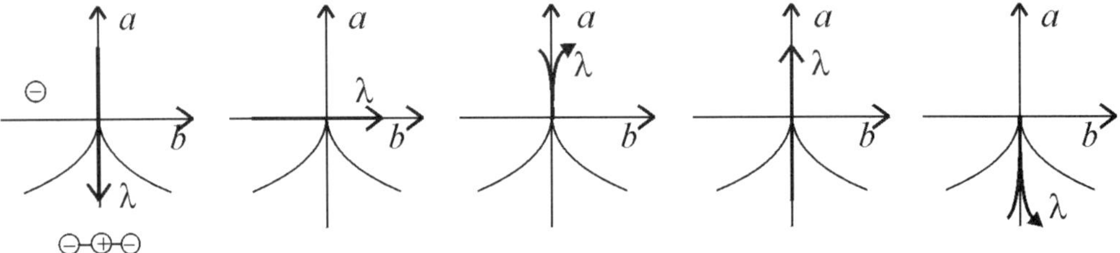

Figure 25. Routes in control space through standard cusp catastrophe.

Now we show examples for these cases.

Example 10. Let as consider again the structure of Example 3a, but we introduce an imperfection for the length of spring 1. So the parameters are:

$$L = 1,\ a = 0,\ c_1 = 2,\ c_2 = 2,\ l_1 = \sqrt{2} + \varepsilon,\ l_2 = \sqrt{2}\ . \tag{5.44}$$

We want to determine the equilibrium paths and the imperfection sensitivity curve in the small vicinity of the critical point of the perfect structure (this point was determined in Example 5).

Similarly to Example 6 we slip the origin into the critical point and determine the truncated Taylor series:

$$V(u,\lambda,\varepsilon) = \frac{1}{16} u^4 - \frac{1}{2} \lambda u^2 - \sqrt{2}\varepsilon u\ . \tag{5.45}$$

The equilibrium paths can be given as

$$\lambda = \frac{0.25 u^3 - \sqrt{2}\varepsilon}{u} \tag{5.46}$$

and are illustrated in Figure 26a for a positive, the zero and a negative ε values.

Eliminating u from equations $V_u = 0$, $V_{uu} = 0$ we get the imperfection-sensitivity curve (Figure 26b):

$$\lambda^{cr} = \left(\frac{3}{2}\varepsilon\right)^{2/3}, \tag{5.47}$$

but the whole imperfection-sensitivity curve shows values where some new equilibrium states arise; the imperfect structure does not lose its stability.

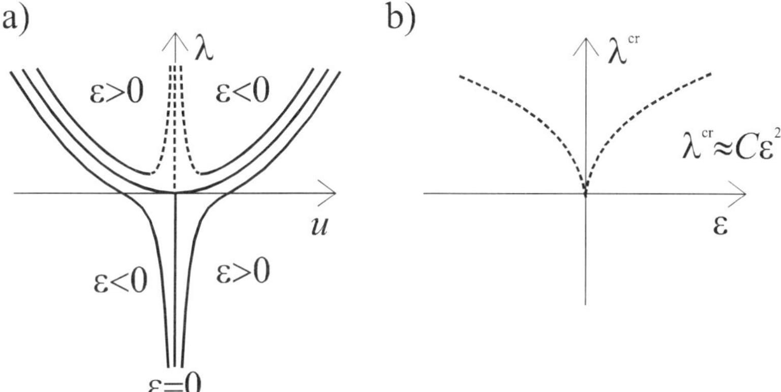

Figure 26. Equilibrium paths for different imperfections (a) and "imperfection-sensitivity curve" (b) in the case of stable-symmetric point of bifurcation.

Example 11. The structure (Figure 27a) is a spring between a hinge and a vertical wall. c denotes the normal stiffness in both tension and compression, the stress free length of the spring is $l + \varepsilon$. A vertical load acts on the structure. Choosing the $c{=}2$, $l{=}1$ values we want to determine the equilibrium paths and the imperfection sensitivity curve in the small vicinity of the critical point of the perfect structure.

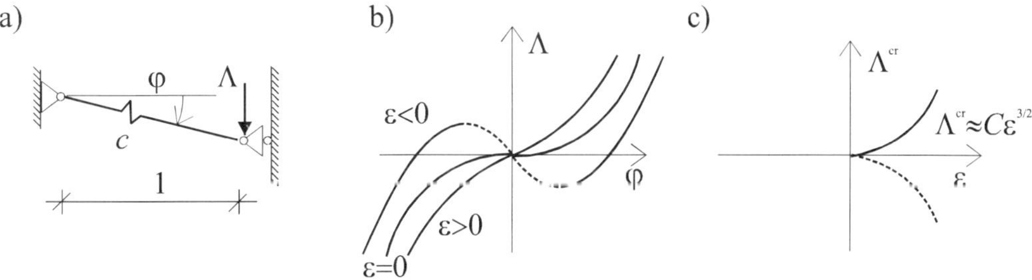

Figure 27. The model (a), the equilibrium paths for different imperfections (b) and imperfection-sensitivity curve (c) in the case of a cut-off point.

The potential energy function is

$$V = \frac{1}{2} c \left(\frac{1}{\cos \varphi} - l - \varepsilon \right)^2 - \Lambda \tan \varphi . \tag{5.48}$$

The critical point of the perfect structure is: $\Lambda^{cr} = 0$, $\varphi^{cr} = 0$. The truncated Taylor series is:

$$V(\varphi, \Lambda, \varepsilon) = 0.25\varphi^4 - \varepsilon\varphi^2 - \Lambda\varphi . \tag{5.49}$$

So the equilibrium paths (Figure 27b) are given by

$$\Lambda = \varphi^3 - 2\varepsilon\varphi , \tag{5.50}$$

and the imperfection-sensitivity curve (Figure 27c) has the form:

$$\Lambda^{cr} = \pm 2\left(\frac{2}{3}\varepsilon\right)^{3/2} , \tag{5.51}$$

but only its upper part is important.

Example 12. The structure (suggested by P. Varkonyi) shown in Figure 28a consists of two telescopic members (without friction), two linear and two rotational springs. The unloaded perfect structure is free of stress, and the linear springs are in vertical, the telescopic members are in horizontal position. The structure is loaded by a vertical dead load of magnitude Λ . We want to determine the equilibrium paths and the imperfection-sensitivity surface in a small vicinity of the critical point of the perfect structure when

$$c_1 = 1, \quad c_2 = 3, \quad l = 1 + \varepsilon_1$$

and we introduce also a horizontal load of magnitude ε_2 .

Position of the structure can be uniquely described by the x and y coordinates of the middle hinge. The total potential function of the imperfect structure is

$$V(x, y, \Lambda, \varepsilon_1, \varepsilon_2) = \frac{1}{2}\left[c_1\left(\sqrt{x^2 + (1+y)^2} - (1+\varepsilon_1)\right)^2 + c_1\left(\sqrt{x^2 + (1-y)^2} - (1+\varepsilon_1)\right)^2 + \right.$$

$$\left. + c_2 \arctan^2 \frac{y}{1+x} + c_2 \arctan^2 \frac{y}{1-x}\right] - \Lambda y - \varepsilon_2 x . \tag{5.52}$$

The position $x = y = 0$ is critical for the unloaded perfect structure hence the Hessian is singular:

$$\mathbf{H} = \begin{bmatrix} 0 & 0 \\ 0 & 8 \end{bmatrix} . \tag{5.53}$$

The truncated Taylor series of the potential function of the imperfect structure is

$$V = \left(\frac{1}{4} + \frac{\varepsilon_1}{4}\right)x^4 + (8 - \varepsilon_1)x^2 y^2 - 2y^4 - \varepsilon_1 x^2 + 4y^2 - \Lambda y - \varepsilon_2 x \tag{5.54}$$

this has a mixed term, so we use the diffeomorphism

$$x = x, \quad y = v + \frac{1}{8}\Lambda\left(-1 + \frac{1}{8}\varepsilon_1\right)x^2 + \left(-1 + \frac{1}{8}\varepsilon_1\right)x^2 v \tag{5.55}$$

to split the passive and active parts of the energy function:

$$V^p(v) = -2v^4 + 4v^2 - \Lambda v , \tag{5.56}$$

$$V^a(x, \Lambda, \varepsilon_1, \varepsilon_2) = \frac{1}{4}x^4 + \left(\frac{1}{8}\Lambda^2 - \varepsilon_1\right)x^2 - \varepsilon_2 x . \tag{5.57}$$

The passive part can be transformed into a Morse saddle. The active part can be induced from the canonical form (4.5) of the cusp catastrophe by the transformations

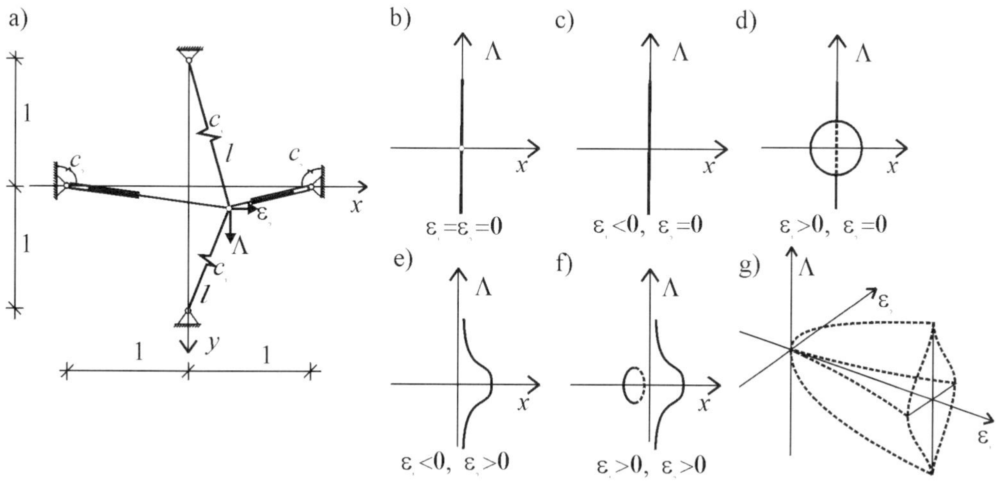

Figure 28. The structure (a), the equilibrium paths (b-f) and the imperfection-sensitivity surface (g).

$$a = \frac{1}{4}\Lambda^2 - 2\varepsilon_1, \quad b = -\varepsilon_2, \tag{5.58}$$

so the Λ route of the perfect structure fits to the positive part of axis a of the parameter space as shown in Figure 25c.

If $\varepsilon_2 = 0$ then $V_x^a = 0$ if $x = 0$, so the primary equilibrium path is vertical (Figure 28b-d). All the points (but the critical one) of the path are stable for the perfect structure (Figure 28b), this is why this case is called as point-like instability. If $\varepsilon_1 < 0$ then the critical point disappears (Figure 28c), and there will be an interval of the load where three equilibrium position exist if $\varepsilon_1 > 0$ (Figure 28d).

If $\varepsilon_2 \neq 0$ then the paths can be given by the following formula:

$$\Lambda = +? \sqrt{\frac{\varepsilon_2 + 2\varepsilon_1 x - x^3}{x}} \tag{5.59}$$

So there is always a continuous stable equilibrium path (Figure 28e), but if ε_1 is large enough then a separate closed path also appears (Figure 28f).

Substituting the transformations (5.58) into the equations of the bifurcation set (see Eqs (4.8)) we get the equation of the imperfection-sensitivity surface in parametric form:

$$\varepsilon_1 = \frac{1}{8}\left(\Lambda^{cr}\right)^2 + 3p^2, \quad \varepsilon_2 = -2p^3. \tag{5.60}$$

The first bifurcation point decreases when $\varepsilon_2 = 0$ and $\varepsilon_1 > 0$

$$\Lambda^{cr} = -\sqrt{8\varepsilon_1}, \tag{5.61}$$

but for other cases the structure will not lose its stability for continuous change of load, so the imperfection-sensitivity surface shows unimportant points (Figure 28g).

Example 13. Let us consider the structure shown in Figure 1 with the following parameters:
$$L = 1,\ a = -1,\ c_1 = 2,\ c_2 = 2,\ l_1 = 1.5 + \varepsilon,\ l_2 = 1.5 . \tag{5.62}$$
The positive vertical load acts upwards. We want to determine the equilibrium paths and the imperfection sensitivity curve in the small vicinity of the critical point of the perfect structure.

The unloaded perfect structure has three equilibrium positions:
$$\varphi_1 = 0,\quad \varphi_{2,3} = \pm 1.338687091 .$$
The first position is unstable, the other two are stable. We have to know in which stable state the structure is when we start to load it. Let us suppose, that the structure is in the state of negative φ. Increasing the load the perfect structure arrives a critical state when $\Lambda^{cr} = 0.3900310562$, $\varphi^{cr} = 0$.

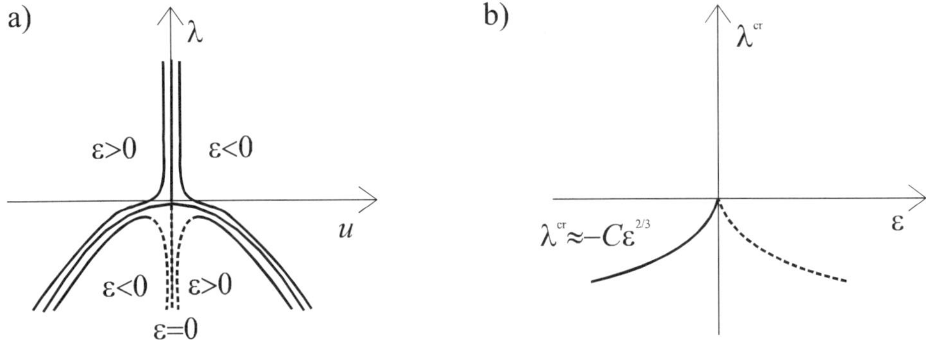

Figure 29. Equilibrium paths for different imperfections (a) and imperfection-sensitivity curve (b) in the up side down case.

We slip the origin into the critical point and determine the truncated Taylor series:
$$V(u,\lambda,\varepsilon) = 0.046957427u^4 + 0.5\lambda u^2 - 0.4472135956\varepsilon u . \tag{5.63}$$
The equilibrium paths (Figure 29a) can be given as
$$\lambda = \frac{-0.1878297104u^3 + 0.4472135956\varepsilon}{u} . \tag{5.64}$$
Now, only the left part of the imperfection-sensitivity curve (Figure 29b) is interesting:
$$\lambda^{cr} = -0.6329450668\varepsilon^{2/3} . \tag{5.65}$$

Example 14. The structure shown in Figure 30a consists of two linear springs. c_i denotes the stiffness in both tension and compression, and l_i is the stress free length of the ith spring (i=1, 2). The structure is loaded by a vertical dead load of magnitude Λ. We want to determine the equilibrium paths and the imperfection sensitivity curve in the small vicinity of the critical point of the perfect structure in the case of the following values:
$$c_1 = 1.985177808 + \varepsilon,\ c_2 = 1,\ l_1 = 0.4,\ l_2 = 0.7 . \tag{5.66}$$

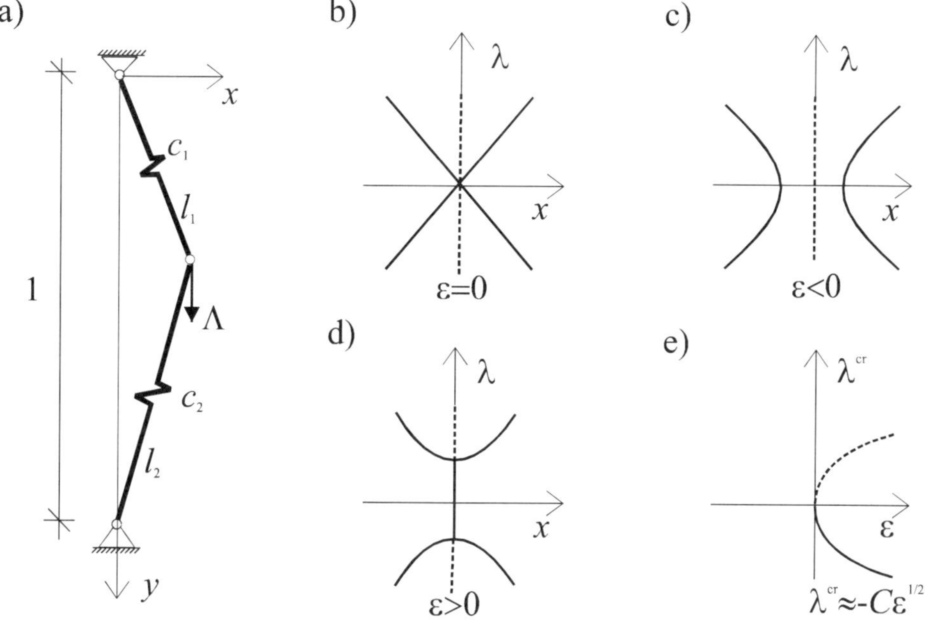

Figure 30. The structure (a), equilibrium paths for perfect (b) and imperfect (c,d) structures and the imperfection-sensitivity curve (e) in the case of a stable-X point of bifurcation.

Position of the structure can be uniquely described by the x and y coordinates of the middle hinge. The unloaded perfect structure has three equilibrium positions:

$$x_1 = 0, \; y_1 = 0.3665011578, \; x_{2,3} = \pm 0.2185749299, \; y_{2,3} = 0.335. \tag{5.67}$$

The first position is unstable, the other two are stable. We have to know in which stable state the structure is when we start to load it. Let us suppose, that the structure is in the state of positive x. Increasing the load the perfect structure arrives a critical state when $\Lambda^{cr} = 0.4455533422$, $x^{cr} = 0, \; y^{cr} = 0.5157563683$.

We slip the origin into the critical point $\left(\Lambda = \Lambda^{cr} + \lambda, \; x = x, \; y = y^{cr} + u\right)$ and determine the truncated Taylor series:

$$V(x,u,\lambda,\varepsilon) = 1.494072598x^4 - 5.976290393x^2u^2 + 1.492588904u^2 + 0.1122200088\varepsilon x^2 +$$
$$+ \left(0.1157563683\varepsilon - \lambda\right)u. \tag{5.68}$$

Using diffeomorphism

$$x = x, \; u = 0.3349884209\lambda - 0.03877704303\varepsilon + w + \left(1.341285652\lambda - 0.1552623560\varepsilon\right)x^2 +$$
$$+ 2.001988082x^2 w \tag{5.69}$$

we can split the function into an active and a passive part:

$$V(x,w,\lambda,\varepsilon) = 1.494072598x^4 + \left(0.1122200088\varepsilon - 0.6706428261\lambda^2\right)x^2 + 1.492588904w^2 \tag{5.70}$$

The equilibrium paths (Figure 30b-d) can be given as

$$\lambda = \pm\left(0.16733200051\varepsilon + 4.455643272x^2\right)^{1/2} \qquad (5.71)$$

Now, only the lower part of the imperfection-sensitivity curve (Figure 30e) is interesting:

$$\lambda^{cr} = \pm 0.4090623487\varepsilon^{1/2} . \qquad (5.72)$$

We remark, that very special imperfection was chosen, which has not appeared in the linear term of the active variable (w), this is why the equilibrium paths of the imperfect structure may have bifurcations. If the imperfection disturbs the symmetry (e. g. the load has a little horizontal component) the equilibrium paths and the imperfection-sensitivity curves will similar to those shown in Figure 21, but stable and unstable positions are changed, and only the right lower part of the imperfection-sensitivity curve is interesting.

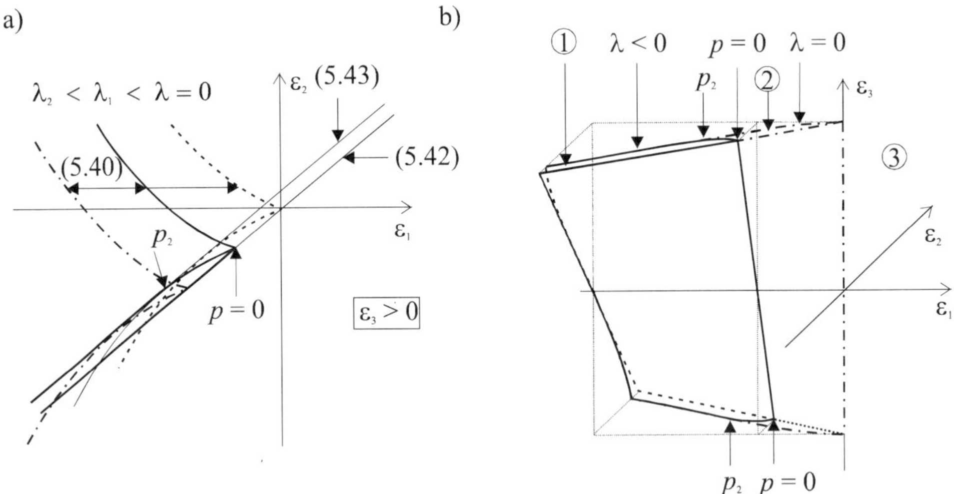

Figure 31. A section (a) and the critical imperfection territory (b) for the stable-symmetric point of bifurcation.

Generalization

Hence cusp catastrophes typically occur when the perfect structure is symmetrical, and the non-zero terms have even degree in the active variable, we only will deal with function (5.35). Of course, we can do the same transformations, as we have done for the dual cusp catastrophe, so the imperfection-sensitivity surface and the envelope have the same forms (see Eqs (5.40), (5.42), (5.43) and Figure 22). But the role of the maximum and minimum points changes, so the critical imperfection territories will differ from the earlier ones. There is a stable-symmetric point of bifurcation if $c > 0$ in (5.37), and Figure 31 shows the critical imperfection territory.

Figure 32a shows different routes the parameter space, and Figure 32b-g give the corresponding equilibrium paths. The first three cases were seen also in Figure 26a, but Figure 32f-g shows new

new cases. According to Figure 32b only very small part of the parameter space belongs to stability problems, but Figure 26b suggested that imperfect structures couldn't lose their stabilities.

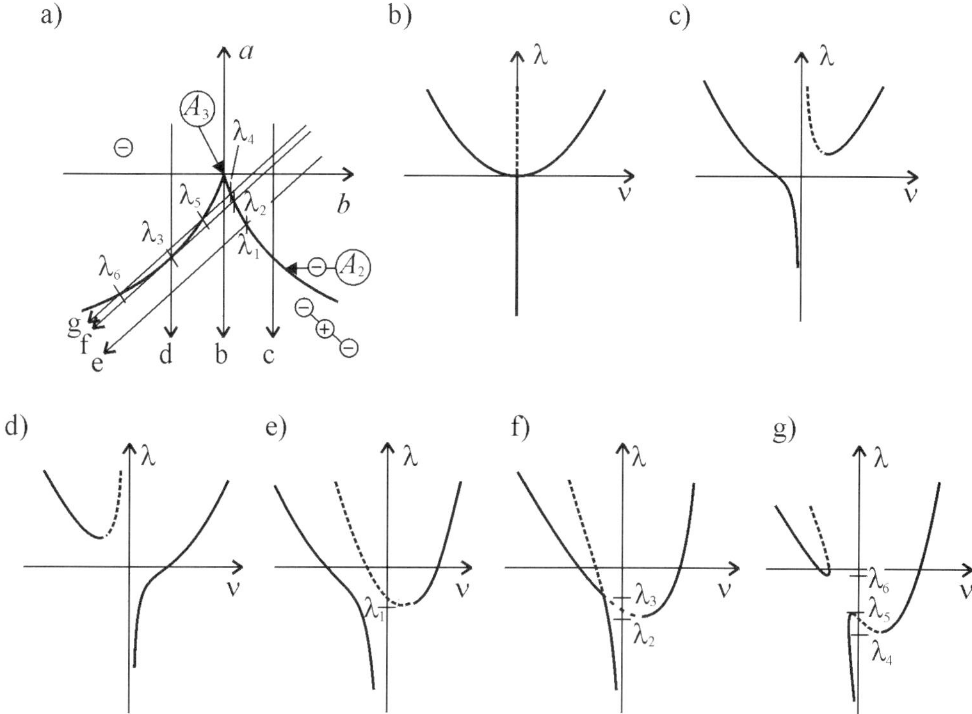

Figure 32. Standard cusp, $c > 0$; routes through the parameter space (a) and the equilibrium paths (b-g).

There is the upside down case if $c < 0$ in (5.37), and supposing that the minimum point characterizing the original state of the structure coincides with the maximum point on the branch $p < 0$ Figure 33 shows the critical imperfection territory. The routes in the parameter space and the equilibrium paths are similar to those shown in Figure 32; only the direction of the λ -axis is reserved.

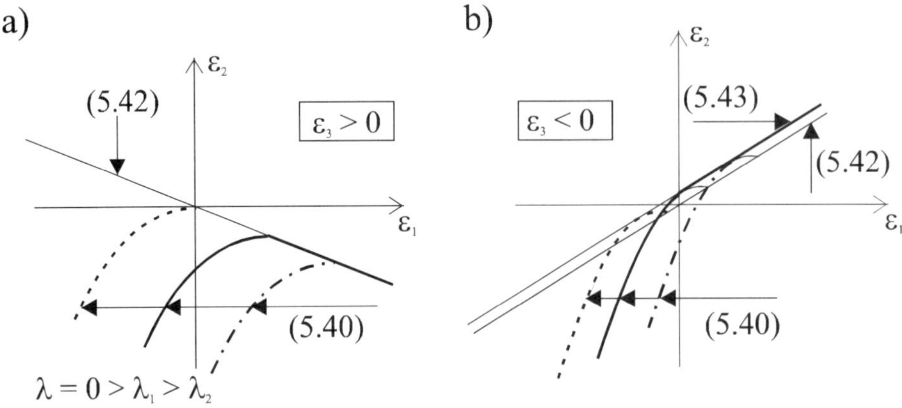

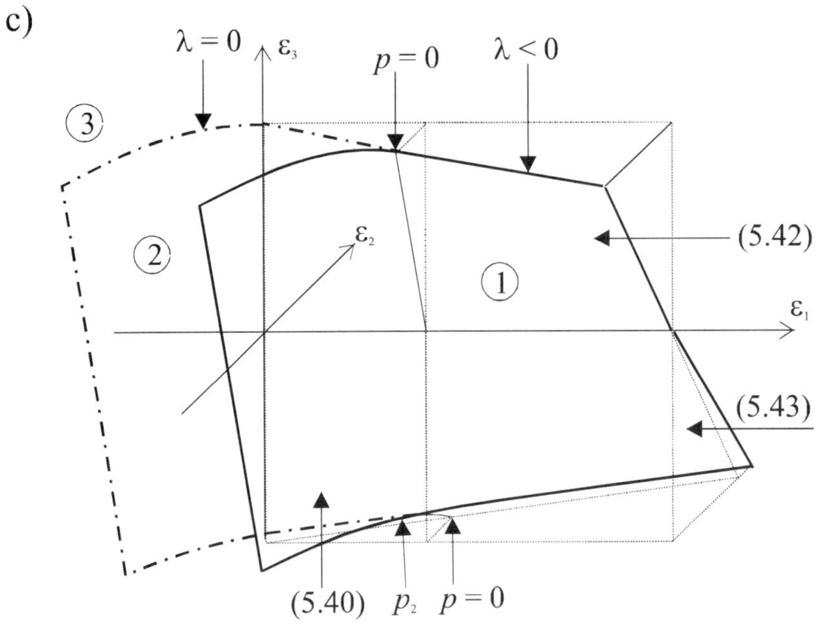

Figure 33. A section (a) and the critical imperfection territory (b) for the upside down case.

5.4 Higher Order Cuspoid Catastrophes

The structure shown in Figure 1 was analysed in the case of different parameter values. Although only the value of parameter *a* was different in Examples 8 and 10, they gave unstable- and stable-symmetric point of bifurcation, respectively. Hence the coefficient of the fourth degree term is a smooth function of parameter a, there must be a special values of a, when the coefficient will be zero. Next we will analyse this special case.

Example 15. Let us consider again the structure shown in Figure 1 with the parameter values

$$c_1 = c_2 = 2, \quad l_1 = l_2 = \sqrt{1 + (1-a)^2}$$

and a remains parameter in the future too. Because of the symmetry all the odd number terms are missing in the potential energy function, so the structure is in equilibrium for any Λ when $\varphi = 0$. The second derivative of the potential energy function is zero at the primary equilibrium path – so we have a critical point – when

$$\Lambda^{cr} = \frac{4}{2 - 2a + a^2} \, . \tag{5.73}$$

Coefficient of the fourth degree term is zero if the parameter has the special value $a_b = 0.5698402910$. Hence the coefficient of the sixth degree term is negative in the critical point we have a dual butterfly catastrophe.

To illustrate that the lower types of catastrophes appear in the neighbourhood of the higher types of catastrophe we introduce three imperfections:

$$a = a_b + \varepsilon_1, \quad l_2 = 1.088594220 + \varepsilon_2, \quad c_2 = 2 + \varepsilon_3 \, . \tag{5.74}$$

At the critical point $\left(\Lambda^{cr} = 3.375420965, \; \varphi^{cr} = 0\right)$ the truncated Taylor series (to show the effects of imperfections, for every imperfections we considered the first two terms which are linear in the imperfections) is:

$$V(\varphi, \lambda, \varepsilon) = -0.02988448601\varphi^4 - 0.5\lambda\varphi^2 + \left(1.424183337\varphi^2 - 0.4766312742\varphi^4\right)\varepsilon_1 +$$

$$+ \left(1.837231877\varphi + 0.2517145008\varphi^2\right)\varepsilon_2 + \left(0.4219276212\varphi^2 + 0.1156144762\varphi^3\right)\varepsilon_3 \, . \tag{5.75}$$

We choose always one imperfection to be non-zero. If $\varepsilon_1 > 0$, the structure has an unstable-symmetric, if $\varepsilon_1 < 0$ then a stable symmetric, if $\varepsilon_3 \neq 0$, then an asymmetric point of bifurcation, and if $\varepsilon_2 \neq 0$, then the structure loses its stability in a limit point.

Example 16. Let us consider the structure shown in Figure 34. The stiffness of the springs is: $c_1 = 3$ and $c_2 = 6$. The unloaded perfect structure is stress free in the position $\varphi = 0$. We introduce an imperfection parameter ε_1, which shows that the horizontal spring is longer by ε_1 than it would be in the vertical position of the rigid link.

The total potential function of the imperfect structure is:

$$V = 0.5\left(c_1\left(\sin\varphi - \varepsilon_1\right)^2 + c_2\varphi^2\right) + \Lambda\cos\varphi \, . \tag{5.76}$$

The primary equilibrium path and the critical load parameter of the perfect structure are $\varphi = 0$ and $\Lambda^{cr} = c_1 + c_2$. The coefficient of the fourth degree term at the critical point is equal to -0.125, so this is an unstable-symmetric point of bifurcation. Neglecting the cubic terms, the imperfection-sensitivity curve (Figure 35a) can be written in parametric form:

$$\lambda^{cr} = -2^{1/2}3p^2, \quad \varepsilon_1 = 2^{3/4}p^3/3 \, . \tag{5.77}$$

Let us notice that if $c_1 = 2$ and $c_2 = 6$ then standard butterfly catastrophe occurs. We introduce another imperfection: ε_2 is the rigidity error of the first spring. To analyse the original structure we choose $\varepsilon_2 = 1$, so the first few terms of the Taylor series of the energy function are:

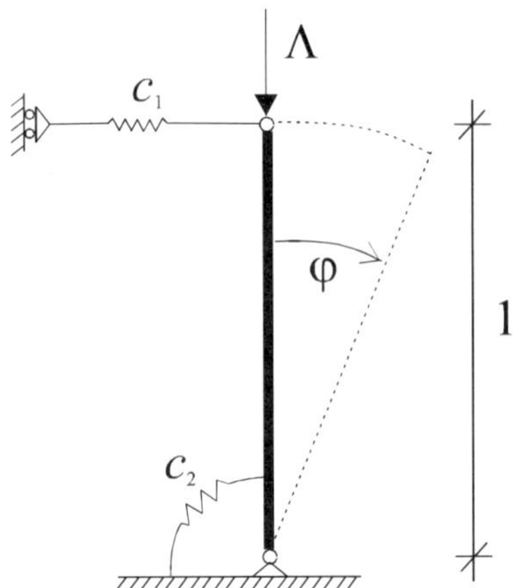

Figure 34. The model for the butterfly catastrophe.

$$V = \frac{1}{30}\varphi^6 - \frac{\varepsilon_1}{40}\varphi^5 + \frac{\lambda-4}{24}\varphi^4 + \frac{\varepsilon_1}{2}\varphi^3 + \frac{1-\lambda}{2}\varphi^2 - 3\varepsilon_1\varphi, \qquad (5.78)$$

which (omitting the fifth degree term) can be induced from Eq. (4.11) by the transformations:

$$x = 5^{-1/6}\varphi, \quad a = 5^{2/3}(\lambda-4)/6, \quad b = 1.5 \cdot 5^{1/2}\varepsilon_1,$$
$$c = 5^{1/3}(1-\lambda), \quad d = -3 \cdot 5^{1/6}\varepsilon_1. \qquad (5.79)$$

Substituting these transformations into Eq. (4.12) we obtain the equations of the imperfection-sensitivity curve (Figure 35b):

$$\lambda^{cr} = \frac{-3.6p^6 + 16p^4 - 18p^2 + 12}{p^4 + 12}, \quad \varepsilon_1 = \frac{0.2p^7 - 9.6p^5 + 12p^3}{3p^4 + 36}. \qquad (5.80)$$

Although the value of ε_2 is large, these curves yield much better approximations of the exact curves (Figure 35c). Enlarged parts are shown in Figure 35d from all the three cases.

Example 17. Let us consider the highly degenerated structure analysed in Gaspar (1999). The perfect structure is shown in Figure 36a. We introduce r imperfection parameters. The eccentricity of the load is ε_1 and the force (S) in the horizontal spring is not a linear function of its elongation (Δ), but

$$S = c\Delta + \sum_{i=2}^{r} \varepsilon_i \Delta^i. \qquad (5.81)$$

The total potential energy function of the imperfect structure is:

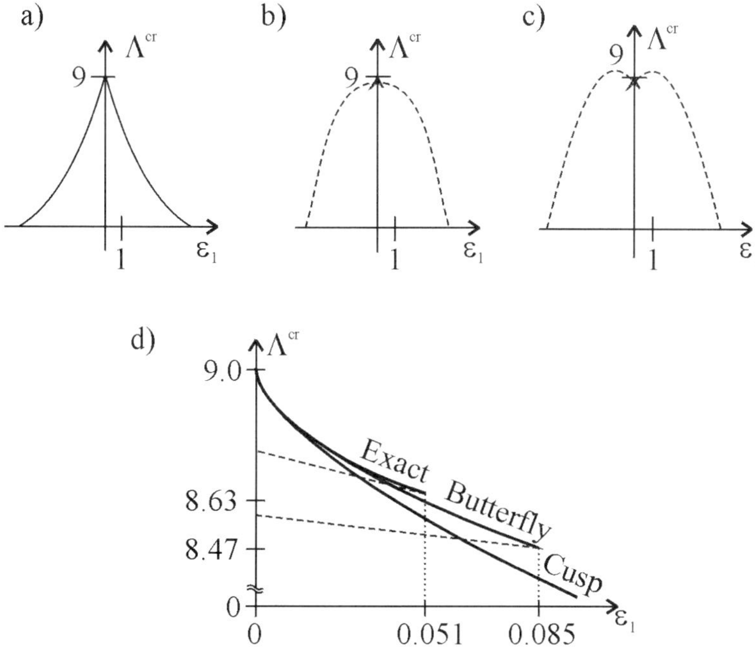

Figure 35. Imperfection-sensitivity curves: cusp (a), butterfly (b), exact (c) and the enlarged parts (d).

$$V = c(1 - \cos\varphi) + \sum_{i=2}^{r} \frac{1}{i+1} \varepsilon_i (\sin\varphi)^{i+1} - \Lambda(1 - \cos\varphi + \varepsilon_1 \sin\varphi). \tag{5.82}$$

In equilibrium position the first, in critical position the first two derivatives are zero:

$$V_\varphi = (c - \Lambda)\sin\varphi - \Lambda\varepsilon_1 \cos\varphi + \sum_{i=2}^{r} \varepsilon_i (\sin\varphi)^i \cos\varphi = 0, \tag{5.83}$$

$$V_{\varphi\varphi} = (c - \Lambda)\cos\varphi + \Lambda\varepsilon_1 \sin\varphi + \sum_{i=2}^{r} \varepsilon_i (\sin\varphi)^{i-1}(i\cos^2\varphi - \sin^2\varphi) = 0. \tag{5.84}$$

The primary equilibrium path is given by $\varphi = 0$, and the critical load of the perfect structure is $\Lambda^{cr} = c$. There is a horizontal secondary equilibrium path for the perfect structure (Figure 36b). The structure does not loose its stability if only ε_1 is different from zero (Figure 36c). If only one $\varepsilon_j (j \geq 2)$ is different from zero, then a cuspoid catastrophe (A_j) occurs at the original critical point (Figure 36d-g). If $j \geq 7$, then there will be catastrophes which do not appear in Thom's theorem. Very waved secondary paths can be produced by appropriate imperfections (Figure 36h).

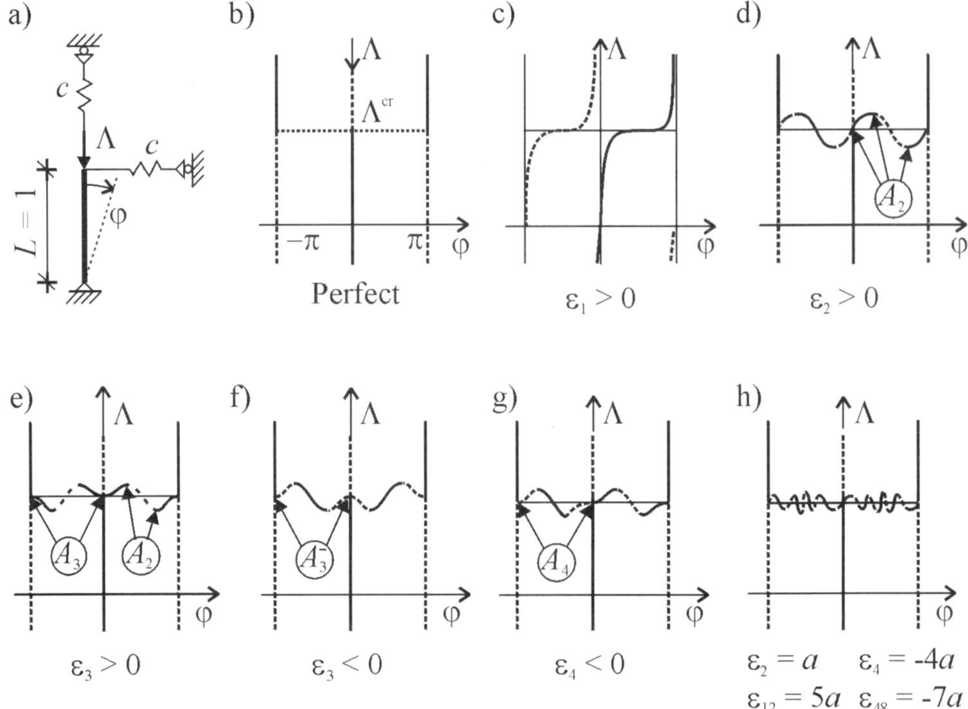

Figure 36. The structure (a) and its equilibrium paths (b-h).

5.5 The Umbilic Catastrophes

According to Figure 9b the parameter space of the elliptic umbilic catastrophe has three dimensions and the bifurcation set is a surface. This surface divides the space into three parts, but only one of them contains points belongs to functions having minima. So if $\lambda < 0$, we must be in this part. Typically there is one possibility for the perfect structure to loose its stability (Figure 37a):

- the λ route cross the umbilic catastrophe point and arrives to the other part, which belongs functions having four stationary points (anticlinal point of bifurcation).

(An atypical case can be when the λ route turns back to the same part.)

The bifurcation set of the hyperbolic umbilic catastrophe divides the space into four part (Figure 10b) and we can start from two of them. So typically there are three possibilities for the perfect structure to loose its stability (Figure 37b):

- the λ route cross both part with functions having two stationary points (monoclinal point of bifurcation),

- •the λ route remains always in the part with functions having four stationary points (homeoclinal point of bifurcation),
- •for negative λ there are four stationary points but for positive λ there is no stationary point (hilltop branching).

a)   b)

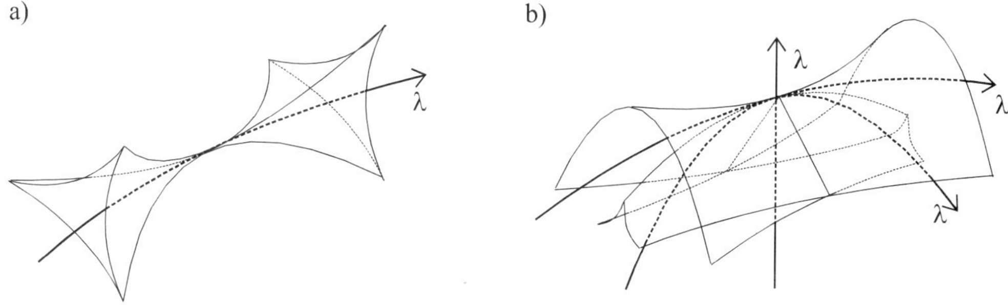

Figure 37. Typical routes in control space of elliptic (a) and hyperbolic (b) umbilic catastrophe.

The following simple model was first analysed by Thompson and Gaspar (1977), then it was developed by others (e. g. Hunt, Reay and Yoshimura (1979), Thompson and Hunt (1984), Gioncu and Ivan (1984), Hackl (1990), Pajunen and Gaspar (1996)).

Example 18. Let us consider the structure (Figure 38) consisting a rigid strut and three springs. The first spring lies in the plane yz, while the others are placed symmetrically to this plane, the horizontal angle is denoted by β. The stiffness of the equal second and third springs is denoted by c_2, and the stiffness of the first spring is $c_1 = 1 - 2c_2$.

The state of the structure is specified by the co-ordinates (x, y) of the top of the strut. The perfect system rests vertically (and the springs are free of stress) under zero load. The total potential energy function is

$$V = c_1\left(\sqrt{1-y}-1\right)^2 + c_2\left(\sqrt{1-x\sin\beta-y\cos\beta}\right)^2 + c_2\left(\sqrt{1+x\sin\beta-y\cos\beta}\right)^2 + \Lambda\sqrt{1-x^2-y^2}\,.$$

The trivial equilibrium path is given by $x = y = 0$ for all load value. There are two critical loads:

$$\Lambda_1^{cr} = c_2\sin^2\beta, \quad \Lambda_2^{cr} = 0.5c_1 + c_2\cos^2\beta\,. \tag{5.85}$$

If $c_2 = 0.25\sin^2\beta$ then the two critical loads will be equal, and we have an umbilic catastrophe. To avoid a negative stiffness, β is restricted to the interval $(45°, 135°)$. The cubic part of the potential energy function at the critical point is

$$\frac{1}{16}\frac{\sin^2\beta+\cos\beta}{1+\cos\beta}y^3 + \frac{3}{16}x^2y\cos\beta\,. \tag{5.86}$$

Applying the transformations shown in Section 4.5 we have

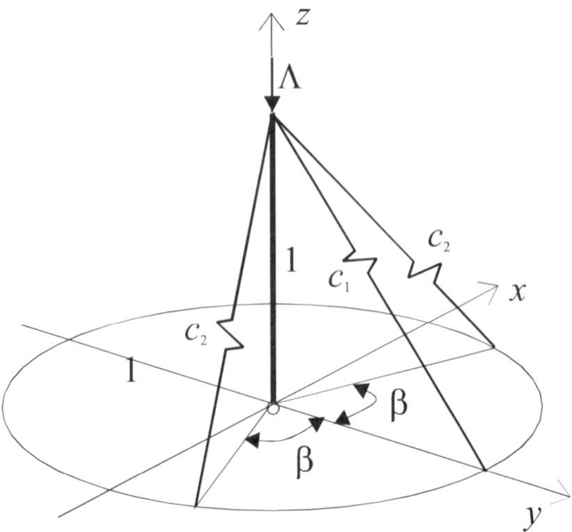

Figure 38. The model for the umbilic catastrophes.

$$\alpha_1 = 1, \quad \alpha_2 = 0, \quad \beta_1 = \frac{3 + 6\cos\beta}{\sin^2\beta - 3\cos^2\beta - 2\cos\beta}, \quad \beta_2 = 0, \tag{5.87}$$

so changing the angle β in the given interval we have all the points of axis β_1 shown in Figure 8b, i. e. the model is suitable to illustrate both the elliptic, hyperbolic, parabolic and symbolic umbilic catastrophe.

Determining the linear approximation of the secondary equilibrium paths in the vicinity of the critical point we have the following cases:

- $45° \leq \beta < 64.26187657°$ or $128.172776° < \beta \leq 135°$, there are three distinct secondary paths and all the three paths rise in the same direction according to y (homeoclinal point of bifurcation, Figure 39a),
- $\beta = 64.26187657°$, the three secondary paths coincide (triple root, Figure 39b),
- $64.26187657° < \beta < 90°$, there is one secondary path (monoclinal point of bifurcation, Figure 39c),
- $\beta = 90°$, there are three secondary paths, two of them coincide, they are parallel with axis x (symbolic umbilic catastrophe, Figure 39d),
- $90° < \beta < 128.1727076°$, there are three distinct secondary paths and two of them rise in opposite direction according to y (anticlinal point of bifurcation, Figure 39e),
- $\beta = 128.1727076°$, there are three distinct secondary paths and one of them is horizontal (parabolic umbilic catastrophe, Figure 39f).

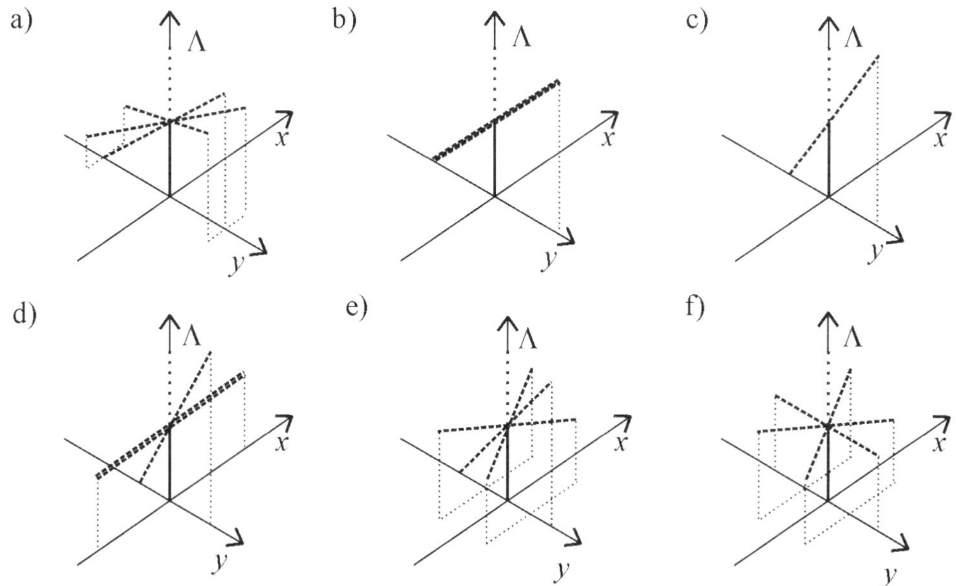

Figure 39. Different types of the equilibrium paths of perfect structures.

(Here dotted lines show positions where the Hessian has two negative eigenvalues.) We remark that in the case of the triple root, the symbolic and the parabolic umbilic catastrophes the linear approximation of the paths is not enough. Pajunen and Gaspar (1996) have determined the global equilibrium paths, their projection to the plane xy are shown in Figure 40 in the vicinity of the triple root and the symbolic umbilic.

Example 19. Let us consider the structure (Figure 41) consisting of three springs. Stiffness of the first two springs is equal to 1, but the third one will be determined to have a hilltop bifurcation.

A state can be given by the co-ordinates (x, y) of the common point of the springs. The total potential energy function is:

$$V(x, y, \Lambda, c_3) = \frac{1}{2}\left(\left(\sqrt{(1-x)^2 + y^2} - \sqrt{37} \right)^2 + \left(\sqrt{(1+x)^2 + y^2} - \sqrt{37} \right)^2 + c_3 x^2 \right) + \Lambda y . \qquad (5.88)$$

Because of the symmetry $V_x = V_{xy} = 0$ if $x = 0$. So the conditions to have a coincident branching point are:

$$V_y(0, y, \Lambda, c_3) = V_{xx}(0, y, \Lambda, c_3) = V_{yy}(0, y, \Lambda, c_3) = 0 . \qquad (5.89)$$

One solution of this system is:

$$y^{cr} = 1.52716137, \quad \Lambda^{cr} = 7.1233582, \quad c_3^{cr} = 2.6644437 . \qquad (5.90)$$

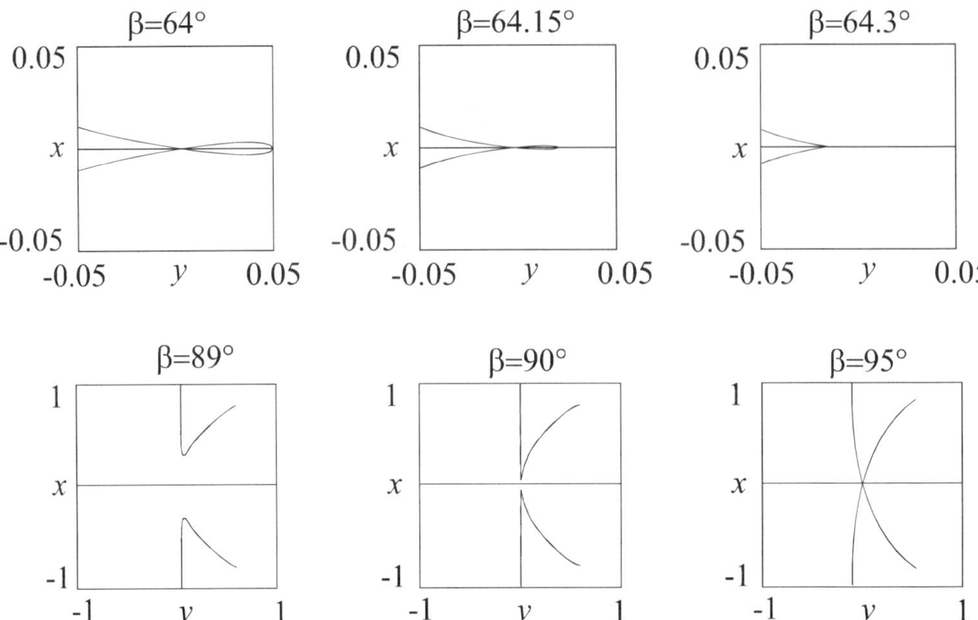

Figure 40. Projections of the equilibrium paths for different values of β .

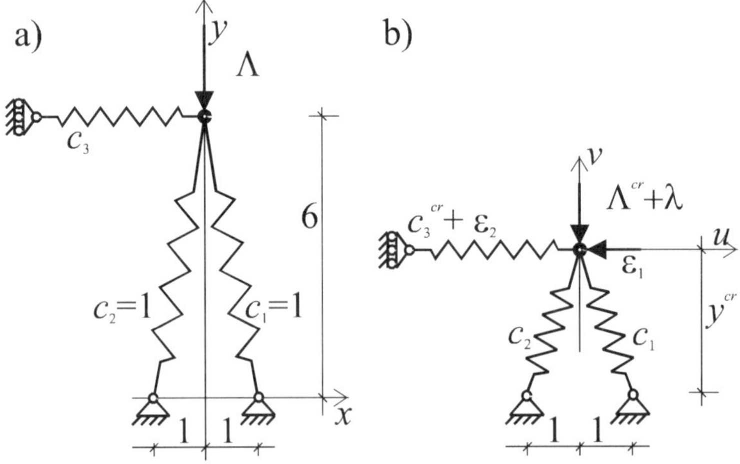

Figure 41. Model for hilltop branching in the unloaded (a) and the critical state (b).

We shift the origin into the critical point and consider the imperfections shown in Figure 41b, and then the truncated Taylor series of the energy function is:

$$V(u,v,\lambda,\varepsilon_1,\varepsilon_2) = 0.45830123v^3 + 0.15225768u^2v + 0.5\varepsilon_2u^2 + \lambda v + \varepsilon_1u.$$
(5.91)

This function can be induced from the canonical form (4.21) of hyperbolic umbilic. This linear transformation is substituted into the equations (4.22) of the bifurcation set, and the equations of the imperfection-sensitivity surface can be expressed in parametric form:

$$\varepsilon_1 = 0.61091632p^2\left(-q^{-2} + 2q^{-1} - 2q + q^2\right), \qquad \varepsilon_2 = -0.99525001p,$$

$$\lambda = -1.8358123p^2\left(2 + q^{-2} + 2q^{-1} + 2q + q^2\right).$$
(5.92)

According to Figure 10c the interesting part of this surface is when $pq > 0$.

The equilibrium paths are shown in Figure 42 both for perfect and an imperfect structure.

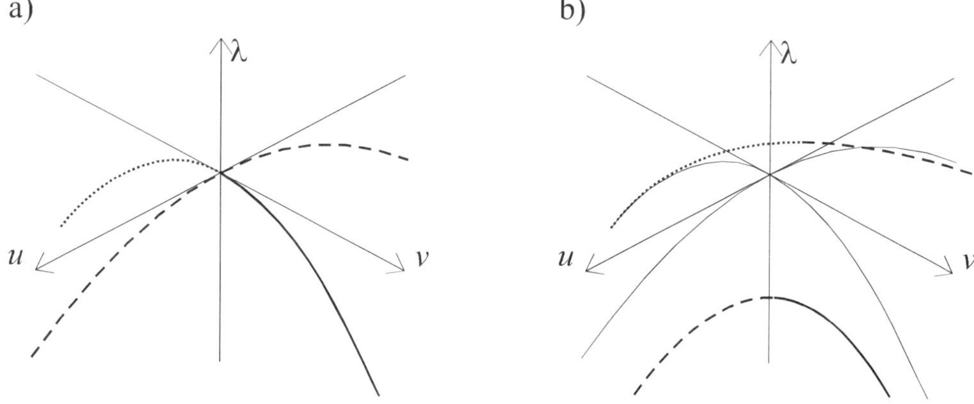

Figure 42. Equilibrium paths of the perfect (a) and an imperfect (b) structure.

Generalization

As in Hunt (1978), the following form will be examined:

$$\hat{V} = c_1\hat{v}^3 + c_2\hat{u}^2\hat{v} - \hat{\lambda}\left(c_3\hat{u}^2 + c_4\hat{v}^2\right) + \hat{\sigma}\hat{u}^2 + \hat{\varepsilon}_1\hat{u} + \hat{\varepsilon}_2\hat{v},$$
(5.93)

where $\hat{u},\hat{v}$ stand for the active generalized state co-ordinates; $c_1c_2 \neq 0, c_3 > 0, c_4 > 0$ are constants; $\hat{\lambda}$ is deviation of the load parameter from the critical value; $\hat{\sigma},\hat{\varepsilon}_1$ and $\hat{\varepsilon}_2$ are functions of imperfections. The parameter $\hat{\sigma}$ shows the difference of both critical loads (Gaspar, 1985), so it is called a splitting parameter.

For simplicity, function (5.93) will be normalized. Using the transformations

$$\hat{u} = c_1^{1/3}c_3^{-1/2}c_4^{1/2}u, \qquad \hat{v} = c_1^{-1/3}v, \qquad \hat{\lambda} = c_1^{2/3}c_4^{-1}\lambda,$$

$$\hat{\sigma} = c_1^{2/3}c_3c_4^{-1}\sigma, \qquad \hat{\varepsilon}_1 = c_1^{1/3}c_3^{1/2}c_4^{-1/2}\varepsilon_1, \qquad \hat{\varepsilon}_2 = c_1^{1/3}\varepsilon_2$$
(5.94)

function (5.93) assumes the form

$$V(u,v,\lambda,\sigma,\varepsilon_1,\varepsilon_2) = v^3 + \alpha Au^2v - \lambda\left(u^2 + v^2\right) + \sigma u^2 + \varepsilon_1u + \varepsilon_2v$$
(5.95)

where

$$\alpha = \mathrm{sign}(c_1 c_2), \quad A = \left| c_1^{-1} c_2 c_3^{-1} c_4 \right|. \tag{5.96}$$

If $\alpha = -1$, then function (4.83) has an elliptic umbilic point at $\hat{\lambda} = \hat{\sigma} = \hat{\varepsilon}_1 = \hat{\varepsilon}_2 = 0$, and if $\alpha = 1$, then there is a hyperbolic umbilic point.

The elliptic umbilic

Function (5.95) can be induced from (4.16) by the transformations

$$x = \sqrt{\frac{A}{3}}\, u, \quad y = v + \frac{(3-A)\lambda + A\sigma}{6A},$$

$$a = \frac{A\sigma - (A+3)\lambda}{2A}, \quad b = \sqrt{\frac{A}{3}}\,\varepsilon_1, \quad c = \varepsilon_2 + \frac{((3-A)\lambda + A\sigma)((A+1)\lambda - A\sigma)}{4A^2} \tag{5.97}$$

if $\alpha = -1$. Let us substitute Eqs (5.97) into Eqs (4.20), which determine the bifurcation set. Ordering the equations we obtain

$$\varepsilon_1 = B\sqrt{A/3}\,(\sin 2\Theta - 2\cos\Theta), \quad \varepsilon_2 = C + B(\cos 2\Theta - 2\sin\Theta) \tag{5.98}$$

where

$$B = \frac{((A+3)\lambda - A\sigma)^2}{12A^2}, \quad C = \frac{((A-3)\lambda - A\sigma)((A+1)\lambda - A\sigma)}{4A^2}. \tag{5.99}$$

Using parameter Θ, Eqs (5.98) define the imperfection-sensitivity surface in a four-dimension-space $(\varepsilon_1, \varepsilon_2, \sigma, \lambda)$. This cannot be drawn unless one parameter is fixed at different values. Fixing σ we obtain the imperfection-sensitivity surface (Figure 43) of a structure at which the critical points of the perfect structure are at a distance σ (near-coincidence of critical loads).

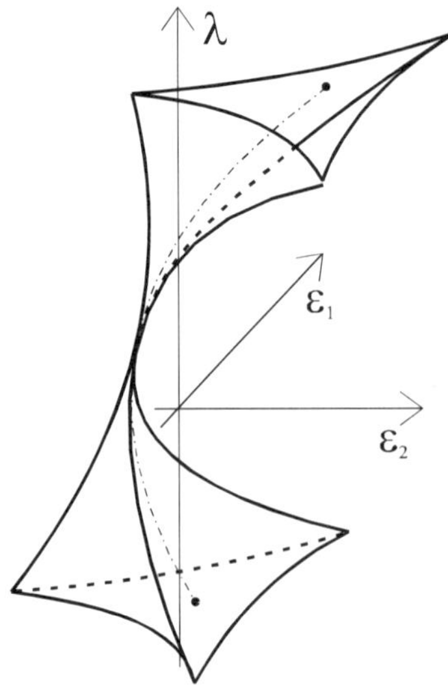

Figure 43. Level surface (σ =const.) of the imperfection-sensitivity surface.

Now let us fix λ rather than σ in Eqs (5.98). According to (5.97a), the condition $a \le 0$ can be written in the form:

$$\sigma \ge \sigma^{(1)} = \frac{A+3}{A}\lambda .$$

$$(5.100)$$

The parts satisfying this condition are shown in Figure 44 for a positive and a negative λ.

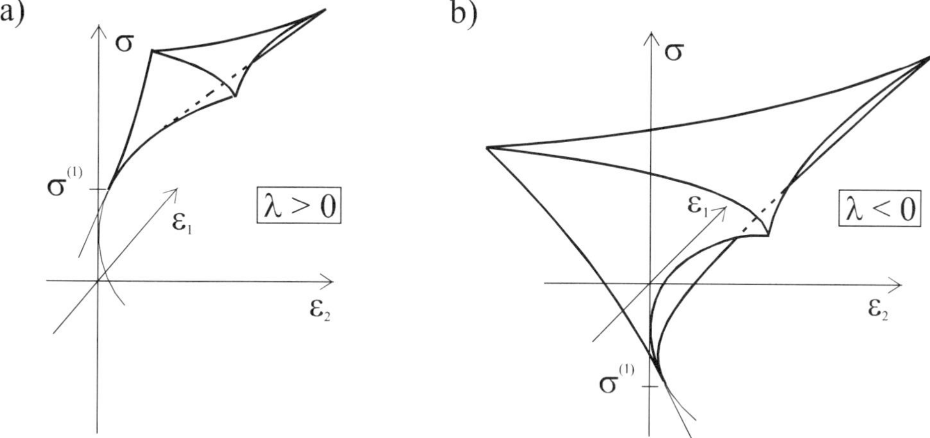

Figure 44. Two level surfaces (λ =const.) of the imperfection-sensitivity surface.

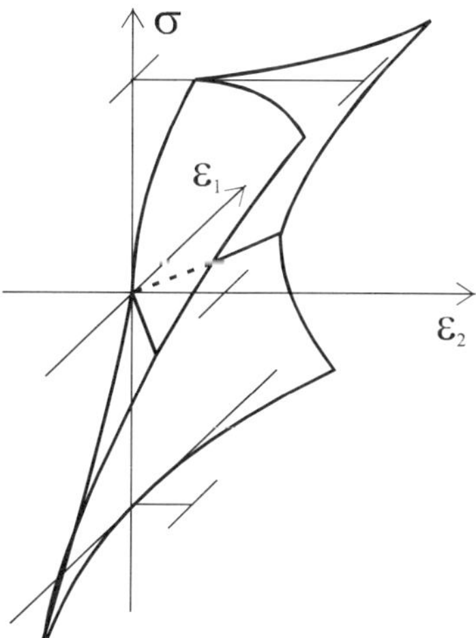

Figure 45. The envelope surface.

The envelope surface of these level surfaces can be given as a function of two parameters (λ, σ). First the value of Θ is computed:

$$\Theta = \arcsin\left(\frac{2A^2(\lambda - \sigma)}{(A+3)((A+3)\lambda - A\sigma)} - 1\right), \tag{5.101}$$

then it is substituted into Eqs (5.98). The envelope is shown in Figure 45.

The critical imperfection territory belonging to a λ_0 is bounded partly by the level surface and partly by the envelope. The different types of the sections $(\sigma = const.)$ of the critical imperfection territories are shown in Figure 46. Here the boundary of the critical imperfection territory is drawn by full line, the level line by dotted line and the envelope by dashed line. The critical imperfection territory lies outside the boundary.

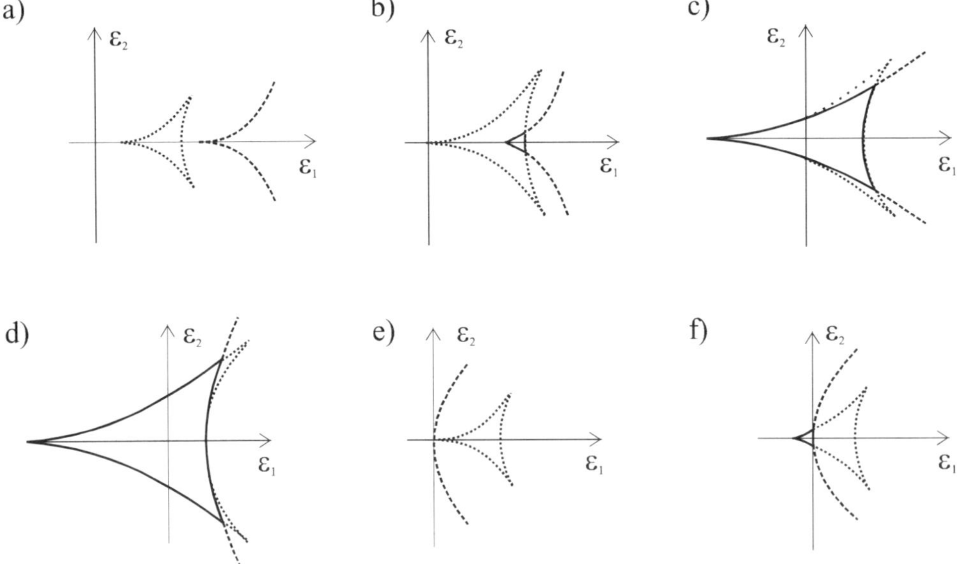

Figure 46. Types of sections $(\sigma = const.)$ of the critical imperfection territories.

The hyperbolic umbilic
Function (5.95) can be induced from (4.21) by the transformations

$$x = 2^{-1/3}\left(v + \sqrt{\frac{A}{3}}u + \frac{3\sigma - (3+A)\lambda}{6A}\right), \quad y = 2^{-1/3}\left(v - \sqrt{\frac{A}{3}}u + \frac{3\sigma - (3+A)\lambda}{6A}\right), \tag{5.102}$$

$$a = 2^{-1/3}\frac{(3-A)\lambda - 3\sigma}{A}, \tag{5.103a}$$

$$b = 2^{-2/3}\left(\varepsilon_2 + \sqrt{\frac{3}{A}}\varepsilon_1 + \frac{((3+A)\lambda - 3\sigma)((1-A)\lambda - \sigma)}{4A^2} \right),$$ (5.103b)

$$c = 2^{-2/3}\left(\varepsilon_2 - \sqrt{\frac{3}{A}}\varepsilon_1 + \frac{((3+A)\lambda - 3\sigma)((1-A)\lambda - \sigma)}{4A^2} \right)$$ (5.103c)

if $\alpha = 1$. Let us substitute Eqs (5.103) into Eqs (4.22), which determine the bifurcation set. Ordering the equations we obtain

$$\varepsilon_1 = B\sqrt{A/3}\left(-q^2 + 2q - 2q^{-1} + q^{-2} \right), \quad \varepsilon_2 = C - B\left(q^2 + 2q + 2q^{-1} + q^{-2} \right)$$ (5.104)

where

$$B = \frac{((3-A)\lambda - 3\sigma)^2}{24A^2}, \quad C = \frac{((3+A)\lambda - 3\sigma)((A-1)\lambda + \sigma)}{4A^2}.$$ (5.105)

Fixing σ we obtain the imperfection-sensitivity surface (see e. g. Hunt, 1978, Figures 8 and 9) of a structure at which the critical points of the perfect structure are at distance σ. Now let us fix λ rather than σ in Eqs (5.104). According to (5.103a) and the condition $pq \geq 0$, it can be seen that if

$$\sigma < \sigma^{(1)} = \frac{3-A}{3}\lambda$$ (5.106)

then only the part corresponding to $q > 0$, in other cases the part corresponding to $q < 0$ has to be considered. This surface is shown in Figure 47.

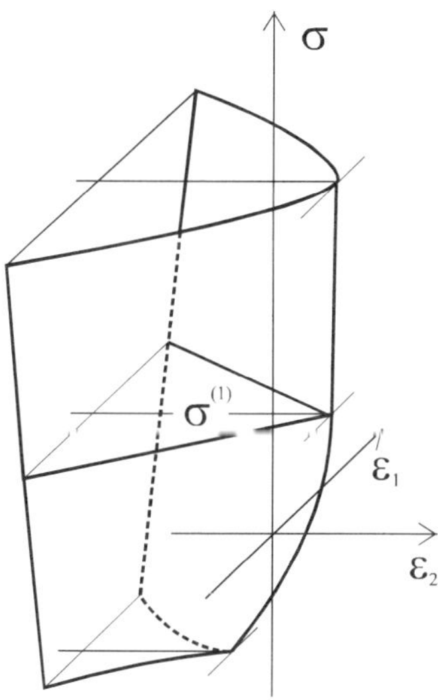

Figure 47. A level surface (λ =const.) of the imperfection-sensitivity surface.

The envelope surface of these level surfaces can be given as a function of two parameters (λ, σ). First the values of q_1 and q_2 are computed:

$$q_{2,3} = \frac{A \pm \sqrt{6A - 9 + (9 - 3A)\sigma / \lambda}}{A \mp \sqrt{6A - 9 + (9 - 3A)\sigma / \lambda}} \tag{5.107}$$

then they are substituted into Eqs (5.104). The envelopes for different cases are shown in Figure 48.

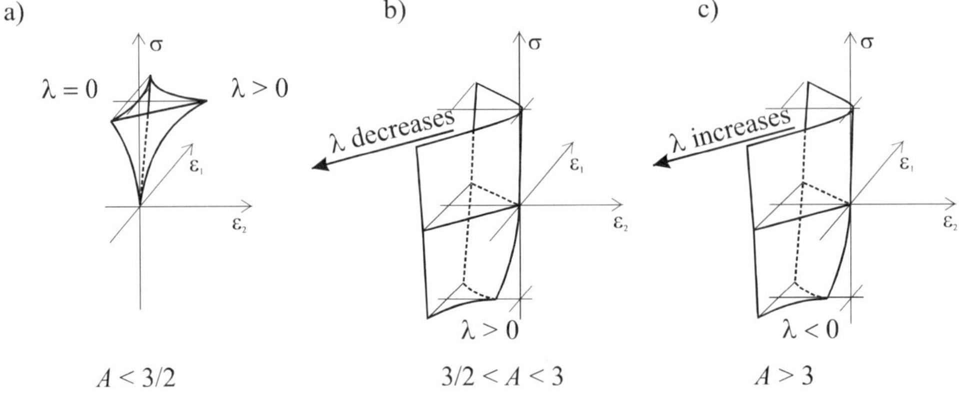

Figure 48. The envelope surfaces $(\sigma > 0)$.

The different types of the sections $(\sigma = const.)$ of the critical imperfection territories are shown in Figure 49. The critical imperfection territory lies on the right side of the boundary.

5.6 Double Cusp Catastrophes

Although double cusp catastrophes do not occur in Thom's theorem some models are shown to illustrate three types of them.

Example 20. Gaspar (1977) investigated a model (Figure 50) consisting a rigid strut and n linear springs with stiffness $1/n$. The structure has cyclic symmetry. A state is specified by the polar coordinates (r, φ) of the top of the strut. The truncated Taylor expansion of the energy function of the perfect structure is

$$V(r, \varphi, \Lambda) = \frac{2}{n} \sum_{i=1}^{n} \left(\frac{1}{2} r c_i + \frac{1}{8} r^2 c_i^2 + \frac{1}{16} r^3 c_i^3 + \frac{1}{128} r^4 c_i^4 + \frac{1}{256} r^5 c_i^5 \right) - \Lambda \left(\frac{r^2}{2} + \frac{r^4}{8} \right), \tag{5.108}$$

where $c_i = \cos(\alpha_i - \varphi)$.

Executing the summation the energy can be written as

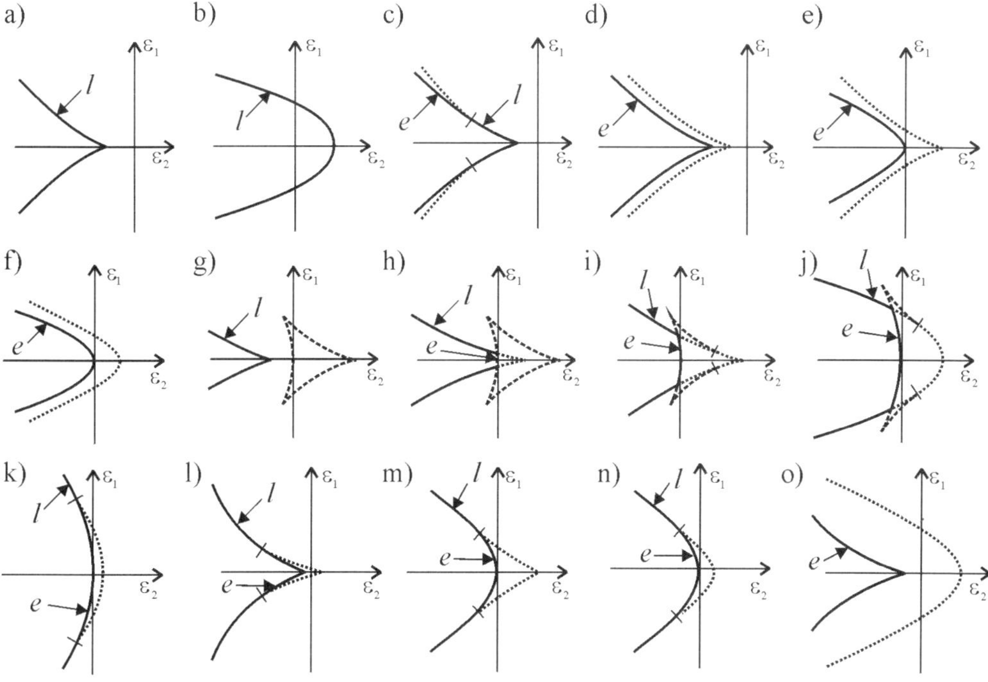

Figure 49. Types of sections $(\sigma = const.)$ of the critical imperfection territories.

$$V(r,\varphi,\Lambda) = \frac{1}{8}r^2 + \frac{1}{8}C_3 r^3 + \frac{15}{512}r^4 + \frac{5}{64}C_4 r^4 + \frac{7}{128}C_5 r^5 - \Lambda\left(\frac{r^2}{2} + \frac{r^4}{8}\right),$$
(5.109)

where

$$C_3 = \begin{cases} \dfrac{1}{4}\cos 3\varphi & \text{if } n = 3 \\ 0 & \text{otherwise} \end{cases}, \qquad C_4 = \begin{cases} \dfrac{1}{8}\cos 4\varphi & \text{if } n = 4 \\ 0 & \text{otherwise} \end{cases},$$

$$C_5 = \begin{cases} \dfrac{5}{16}\cos 3\varphi & \text{if } n = 3 \\ \dfrac{1}{16}\cos 5\varphi & \text{if } n = 5 \\ 0 & \text{otherwise} \end{cases}.$$

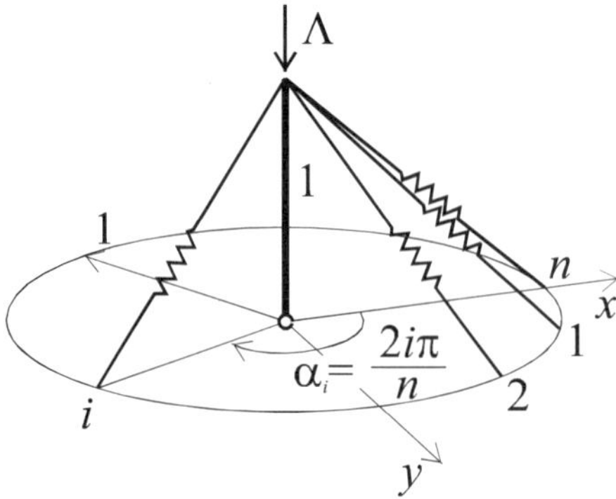

Figure 50. The model with n springs.

There is no linear term in r, so $r = 0$ is an equilibrium position for every Λ values. The quadratic terms do not depend on n, so the critical load parameter is always $\Lambda^{cr} = 0.25$. If $n = 3$, we have the model of Example 18 with $\beta = 120°$, so it has an elliptic umbilic catastrophe.

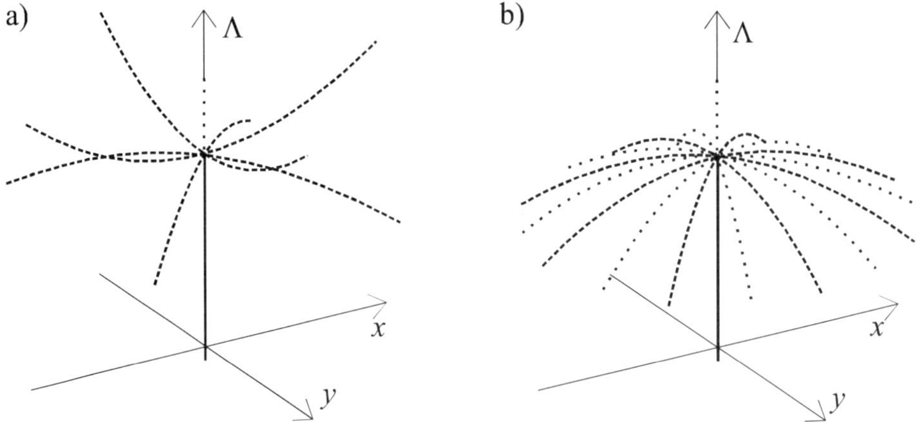

Figure 51. Equilibrium paths of the model with four (a) and more than four (b) springs.

If $n = 4$, there are no cubic terms, so it cannot have umbilic catastrophe. The quartic part is

$$r^4(5\cos 4\varphi - 1)/512 \tag{5.110}$$

which has four real distinct roots in the interval $[0, \pi)$, so it belongs to class 1 of double cusp catastrophes (the characterizing point is inside the tetrahedron shown in Figure 11). There are four unstable secondary paths (Figure 51a). Two of them lie in the vertical planes of the springs ($\Lambda = 0.25 + r^2 / 32$), and the others in the bisector planes ($\Lambda = 0.25 - 3r^2 / 64$). The same type of branching occurs at the Augusti model (Augusti, 1964, Thompson and Hunt, 1973).

If $n \geq 5$ then the quartic part is $-r^4/512$, i.e. it is negative definite and has no real roots, so it belongs to class 13 of double cusp catastrophes (the characterizing point is on the lower whiskers shown in Figure 11). There are n unstable secondary paths (Figure 51b), all of them in the vertical planes of the springs if n is odd, and both in the vertical planes of the springs and in the bisector planes if n is even number. The equation of the path is $\Lambda = 0.25 - r^2/128$. We emphasize that arbitrary large number of secondary equilibrium paths may occur in this special case of double cusp catastrophe, we have to choose the wanted number of springs.

Example 21. The structure (Figure 52a) consists of a rigid strut, a rotational and a linear springs, the later must be always parallel with axis x. The state of the structure is specified by the co-ordinates (x, y) of the top of the strut. The perfect system rests vertically and the springs are free of stress under zero load. The total potential energy function is

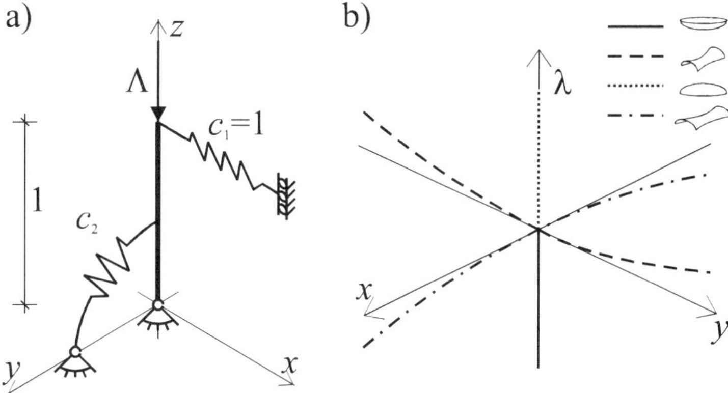

Figure 52. Model with two springs (a) and the equilibrium paths (b).

$$V(x, y, \Lambda, c_2) = 0.5x^2 + 0.5c_2 \arcsin^2 y + \Lambda\sqrt{1 - x^2 - y^2} . \tag{5.111}$$

The trivial equilibrium path is given by $x = y = 0$ for all load value. Both critical loads are equal $(\Lambda^{cr} = 1)$ if $c_2 = 1$. The truncated Taylor expansion of the energy function of the perfect structure is

$$V(x, y, \lambda) = -\frac{1}{8}\left(x^4 + 2x^2y^2 - \frac{1}{3}y^4\right) - \lambda\left(\frac{x^2 + y^2}{2} + \frac{x^4 + 2x^2y^2 + y^4}{8}\right). \tag{5.112}$$

The quartic part has two distinct real and two complex roots so it belongs to class 2 of double cusp catastrophes (the characterizing point is outside the tetrahedron and the bowls shown in Figure 11). There are two secondary paths (Figure 52b) lying in the co-ordinate planes. The Hessian matrix belonging to path in plane (y, λ) has one positive and one negative eigenvalues, while in the case of the other path it has one negative and a zero eigenvalues, which means that every point of it is critical, so this a triple root (similar to Example 18). Gaspar and Nemeth (2002) have shown how many different types of the equilibrium paths can occur when applying different imperfections.

5.7 Special Cases

The perfect structure shown in Example 17 had a horizontal secondary equilibrium path, which is an infinitely degenerated case. Similar structures were shown by Hegedus (1986) and Tarnai (2003).

As we have seen in Sections 5.1-5.2 different exponents may occur in the equations of imperfection curves. Smaller exponents mean larger sensitivity. The question arises: which type is the most dangerous? Kollar (1990) analysed a structure having infinitely great critical load, but a small imperfection results in finite critical load. The next example shows a similar case.

Example 22. The structure (Figure 53a) consists of four rigid struts with unit lengths and a linear spring. Unloaded structure is in equilibrium if two struts are horizontal, the other two are vertical and the stress free spring is vertical. The angle φ determines the position of the structure in the small vicinity of the original state.

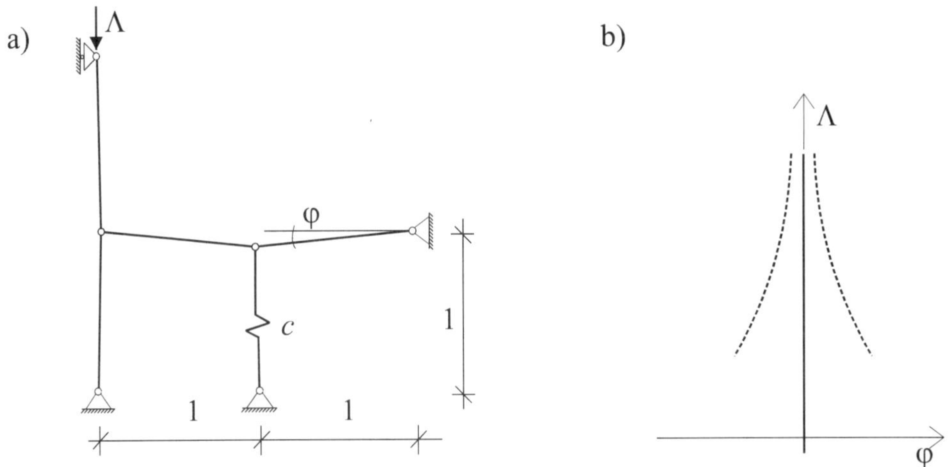

Figure 53. The model (a) and its equilibrium paths (b).

The truncated Taylor series of the potential energy function is:

$$V(\varphi,\Lambda) = \frac{1}{2}c\varphi^2 - \frac{1}{6}c\varphi^4 - \frac{1}{2}\Lambda\varphi^4 .$$

(5.113)

There is no linear term in φ, so $\varphi = 0$ is an equilibrium position for every Λ values. The quadratic term does not depend on Λ, and its coefficient is positive, so the critical load is infinitely great, and the whole primary path is stable. But there are unstable paths infinitely near to it (Figure 53b).

A compressed bar with circular cross-section may buckle in any direction if the load is greater than the critical value, so it has secondary equilibrium surface instead of paths. Next we show a structure having secondary equilibrium surface although it has no rotational symmetry.

Example 23. Gaspar and Mladenov (1996) examined the post-critical behaviour of a bar clamped at the bottom, supported at its top and against rotation (Figure 54a) and loaded by a central (polar) force whose direction passes through the clamped end (Figure 54b).

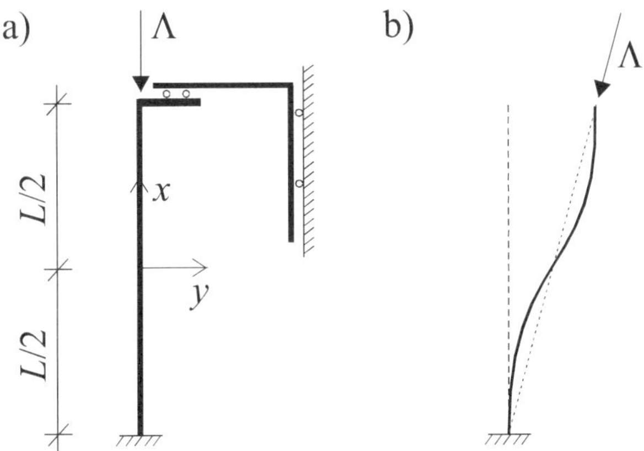

Figure 54. Mladenov's model (a) and the illustration of the central load.

This column has infinite degrees of freedom, but Domokos (1994) has demonstrated that the equilibrium paths of an in-plane column can be always shown in topologically correct way in a suitable 3-D space. For this purpose we choose the space of α, M, Λ, where α is the angle between the force and the vertical axis, M is the bending moment at the bottom of the bar and Λ is the load parameter. It was shown that the two smallest critical load parameters coincide:

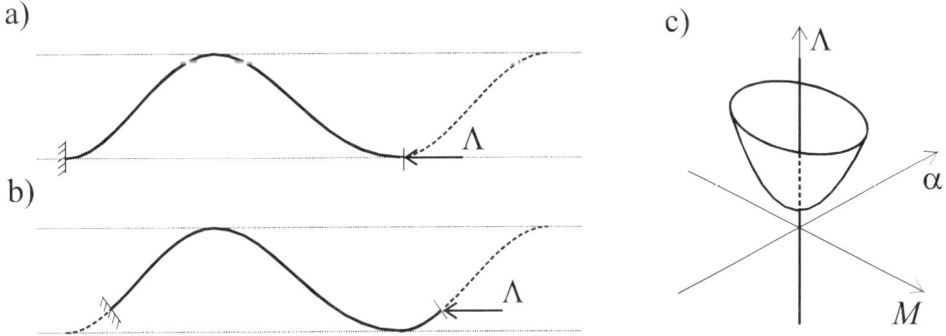

Figure 55. Buckled shapes of the column (a,b) and the equilibrium surface (c).

$$\Lambda_{1,2}^{cr} = \left(2\pi / L\right)^2 EJ .$$ (5.114)

Without any calculation it is easy to show that for every $\Lambda > \Lambda_1^{cr}$ there are infinitely many equilibrium positions. Figure 55a shows a buckling shape of the bar where its top does not translate horizontally. Because of its boundary conditions this shape can be considered as an integer period of a periodic function (illustrated by broken line). But any point of the function can be chosen as the starting point of the period (see Figure 55b) and all the boundary conditions will be satisfied and the total potential energy will not change. So there is a secondary equilibrium surface (Figure 55c) and its points belong to indifferent equilibrium states.

Until now we dealt with smooth energy functions, but the constitutive equation will not be smooth in the next example.

Example 24. Hegedus (1988) investigated a symmetrical version $\left(a = 0\right)$ of the structure shown in Figure 1, but he used two weightless cords instead of the two springs, so they have stiffness when they are in tension, but no stiffness when they are in compression.

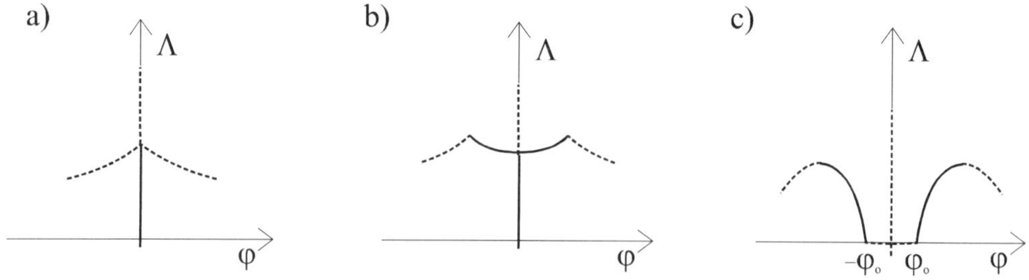

Figure 56. Equilibrium paths when $l = \sqrt{2}$ (a), $l < \sqrt{2}$ (b) and $l > \sqrt{2}$ (c).

It is easy to show that if the stress free lengths of the cords are $l = \sqrt{2}$, than after buckling the structure works as the asymmetric model of (Thompson and Hunt, 1973, p. 15) works in the right direction. So our model has two unstable secondary paths, which stop at the primary path (Figure 56a).

If the cords are shorter then both cords work in the vicinity of the critical point, so it has a stable-symmetric point of bifurcation. This stable secondary path is valid until one of the cords will be stress free, and the structure will follow the equilibrium path of the imperfect structure with one spring (Figure 56b).

If the cords are longer than $\sqrt{2}$ then there is an interval where no cord is working, so there is a horizontal indifferent equilibrium path at zero load, and outside this interval the structure will follow the equilibrium path of the imperfect structure with one spring (Figure 56c).

Finally we remark that Domokos (1991) constructed a rather sophisticated though finite-degree-of-freedom elastic model with continuous spectrum, i. e. every Λ is critical in a finite interval of the load parameter. Similar phenomenon occurs at the Shanley model (Shanley, 1947) where the two springs have bilinear responses.

6   Conclusions

1. Real elastic structures typically loose their stability at a limit point of the equilibrium paths.
2. Because of the symmetry and optimization or at special values of design parameters other types of loss of stability might occur.
3. Elementary catastrophe theory (Thom's theorem) gives a (local) classification of the typical degenerate critical points of the r-parameter family of smooth functions for $r<6$.
4. The local load parameter has a special role among parameters, so we need a subclassification.
5. Imperfection-sensitivity surfaces and critical imperfection territories were shown for the most important subclasses.
6. Simple elastic models were shown for many subclasses:

   - Fold (limit point, asymmetric point of bifurcation),
   - Dual cusp (unstable-symmetric and unstable-X point of bifurcation),
   - Standard cusp (stable-symmetric and stable-X point of bifurcation, cut-off point, point-like instability, upside down case),
   - Butterfly catastrophe,
   - All the other higher order cuspoid catastrophes,
   - Umbilics (monoclinal, homeoclinal and anticlinal point of bifurcation, hilltop branching, triple root, parabolic and symbolic umbilic),
   - 3 types of double cusp catastrophes,
   - 3 infinitely degenerate case (neutral secondary equilibrium path, infinitely large critical value, equilibrium surface),
   - non-smooth constitutive equation.

7. On the analogy between the equilibrium and compatibility paths the singular points of the compatibility paths can be classified (Tarnai, 2001, Gaspar and Lengyel, 2002) for the one degree-of-freedom mechanisms.

References

Arnol'd, V. U. (1972). Normal forms for functions near degenerate critical points, the Weyl groups of A_k, D_k and E_k, and Lagrangian singularities. *Functional Anal. Appl.*, **6**, 254-272.

Augusti, G. (1964). Stabilita' di strutture elastiche elementari in prezenza di grandi spostamenti. *Atti Accad. Sci. fis. mat., Napoli*, Serie 3ᵃ, **4**, No. 5.

Domokos, G. (1991). An elastic model with continuous spectrum. *Int. Series of Numerical Mathematics*, **97**, 99-103.

Domokos, G. (1994). Global description of elastic bars. *ZAMM*, **74**(4), T289-T291.

Gaspar, Z. (1977). Buckling models for higher catastrophes. *J. Struct. Mech.*, **5**, 375-368.

Gaspar, Z. (1982). Critical imperfection territory. *J. Struct. Mech.*, **11**, 297-325.

Gaspar, Z. (1985). Imperfection-sensitivity at near-coincidence of two critical points. *J. Struct. Mech.*, **13**, 45-65.

Gaspar, Z. (1999). Stability of elastic structures with the aid of catastrophe theory. In Kollar, L., ed., *Structural Stability in Engineering Practice*. London: Spon. 88-128.

Gaspar, Z., Domokos, G. (1991) Global description of the equilibrium paths of a simple mechanical model. In Ivanyi, M., ed., *Stability of Steel Structures*. Budapest: Akadémiai Kiadó. 79-86.

Gaspar, Z., Lengyel, A. (2002). Critical points of compatibility paths. In Ivanyi, M., ed., *Stability and Ductility of Steel Structures, SDSS 2002*, Budapest: Akadémiai Kiadó, 779-785.

Gaspar, Z., Mladenov, K. (1996). Post-critical behavior of a column loaded by a polar force. In Ivanyi, M., ed., *Stability of Steel Structures, 1995 Budapest, Further Directions in Stability Research and Design*. Budapest: Akadémiai Kiadó, Vol. II., 897-903.

Gaspar, Z., Nemeth, R. (2002). Models to illustrate special bifurcations. (in Hungarian) In Tassi, G., Hegedus, I., Kovacs, T., eds, Scientific Publ. of Department of Struct. Eng., Faculty of Civ. Eng., Budapest Univ. of Technology and Economics. Budapest: Muegyetemi Kiado. 81-92.

Gioncu, V., Ivan, M. (1984). *Theory of Critical and Post-critical Behaviour of Elastic Structures*. (in Rumanian) Bucuresti: Editura Acad. Resp. Soc. Rom..

Hackl, K. (1990). Ausbreitung von Instabilitäten in einem Knickmodell von Thompson und Gaspar. *ZAMM*, **70**, T189-T192.

Hegedus, I. (1986). Contribution to Gaspar's paper: Buckling model for a degenerated case. *News Letter, Techn. Univ. of Budapest*, **4**(1), 8-9.

Hegedus, I. (1988). The stability of a hinged bar fixed by weightless chords. *News Letter, Techn. Univ. of Budapest*, **6**(3), 15-22.

Hunt, G. W. (1978). Imperfections and near-coincidence for semi-symmetric bifurcations. *Annals of the New York Academy of Sciences*, **316**, 572-589.

Hunt, G. W., Reay, N. A., Yoshimura, T. (1979). Local diffeomorphism in the bifurcational manifestations of the umbilic catastrophes. *Proc. Roy. Soc. Lond. A*, **369**, 47-65.

Koiter, W. T. (1945). *On the Stability of Elastic Equilibrium*. (in Holland) Dissertation, Delft, Holland, (English translation: NASA, Techn. Trans., **F10**, 833, 1967)

Kollar, L. P. (1990). Postbuckling behavior of structures having infinitely great critical loads. *Mech. Struct. Machines*, **18**, 17-31.

Pajunen, S., Gaspar, Z. (1996). Study of an interactive buckling model – From local to global approach. In Rondal, J., Dubina, D., Gioncu, V., eds, *Proc. 2nd Int. Conf. on Coupled Instabilities in Metal Struct. CIMS'96*. London: Imperial College Press. 35-42.

Poston, T., Stewart, I. N. (1976). *Taylor Expansions and Catastrophes*. Research Notes in Mathematics, **7**, London: Pitman.

Poston, T., Stewart, I. N. (1978). *Catastrophe Theory and its Applications*. London: Pitman.

Shanley, F. R. (1947). Inelastic column theory. *J. Aero Sci.*, 14, 261-268.

Tarnai, T. (2001). Kinematic bifurcation. In Pellegrino, S., ed., Deployable structures, *CISM Courses and Lectures No. 412*, Wien: Springer, 143-169.

Tarnai, T. (2003). Zero stiffness elastic structures. *Int. J. of Mechanical Sciences*, **45**, 425-431.

Thom, R. (1972). *Stabilité Structurelle et Morphogenése*. New York: Benjamin.

Thompson, J. M. T., Gaspar, Z. (1977). A buckling model for the set of umbilic catastrophes. *Math. Proc. Camb. Phil. Soc.*, **82**, 497-507.

Thompson, J. M. T., Hunt, G. W. (1973). *A General Theory of Elastic Stability*. London: Wiley.

Thompson, J. M. T., Hunt, G. W. (1984). *Elastic Instability Phenomena*. Chichester: Wiley.

Zeeman, E. C. (1977). *Catastrophe Theory: Selected Papers (1972-1977)*. Reading: Addison-Wesley.